Antibiotic Drug Discovery

New Targets and Molecular Entities

Drug Discovery Series

Editor-in-chief
David Thurston, *King's College*, *UK*

Series editors:
David Fox, *Vulpine Science and Learning*, *UK*
Ana Martinez, *Centro de Investigaciones Biologicas-CSIC, Spain*
David Rotella, *Montclair State University*, *USA*
Sarah Skerratt, *Convergence, UK*

Editorial advisor:
Hong Shen, *Roche Innovation Center Shanghai*, *China*

Titles in the Series:
1: Metabolism, Pharmacokinetics and Toxicity of Functional Groups
2: Emerging Drugs and Targets for Alzheimer's Disease; Volume 1
3: Emerging Drugs and Targets for Alzheimer's Disease; Volume 2
4: Accounts in Drug Discovery
5: New Frontiers in Chemical Biology
6: Animal Models for Neurodegenerative Disease
7: Neurodegeneration
8: G Protein-Coupled Receptors
9: Pharmaceutical Process Development
10: Extracellular and Intracellular Signaling
11: New Synthetic Technologies in Medicinal Chemistry
12: New Horizons in Predictive Toxicology
13: Drug Design Strategies: Quantitative Approaches
14: Neglected Diseases and Drug Discovery
15: Biomedical Imaging
16: Pharmaceutical Salts and Cocrystals
17: Polyamine Drug Discovery
18: Proteinases as Drug Targets
19: Kinase Drug Discovery
20: Drug Design Strategies: Computational Techniques and Applications
21: Designing Multi-Target Drugs
22: Nanostructured Biomaterials for Overcoming Biological Barriers
23: Physico-Chemical and Computational Approaches to Drug Discovery
24: Biomarkers for Traumatic Brain Injury
25: Drug Discovery from Natural Products
26: Anti-Inflammatory Drug Discovery
27: New Therapeutic Strategies for Type 2 Diabetes: Small Molecules
28: Drug Discovery for Psychiatric Disorders
29: Organic Chemistry of Drug Degradation
30: Computational Approaches to Nuclear Receptors

31: Traditional Chinese Medicine
32: Successful Strategies for the Discovery of Antiviral Drugs
33: Comprehensive Biomarker Discovery and Validation for Clinical Application
34: Emerging Drugs and Targets for Parkinson's Disease
35: Pain Therapeutics; Current and Future Treatment Paradigms
36: Biotherapeutics: Recent Developments using Chemical and Molecular Biology
37: Inhibitors of Molecular Chaperones as Therapeutic Agents
38: Orphan Drugs and Rare Diseases
39: Ion Channel Drug Discovery
40: Macrocycles in Drug Discovery
41: Human-based Systems for Translational Research
42: Venoms to Drugs: Venom as a Source for the Development of Human Therapeutics
43: Carbohydrates in Drug Design and Discovery
44: Drug Discovery for Schizophrenia
45: Cardiovascular and Metabolic Disease: Scientific Discoveries and New Therapies
46: Green Chemistry Strategies for Drug Discovery
47: Fragment-Based Drug Discovery
48: Epigenetics for Drug Discovery
49: New Horizons in Predictive Drug Metabolism and Pharmacokinetics
50: Privileged Scaffolds in Medicinal Chemistry: Design, Synthesis, Evaluation
51: Nanomedicines: Design, Delivery and Detection
52: Synthetic Methods in Drug Discovery: Volume 1
53: Synthetic Methods in Drug Discovery: Volume 2
54: Drug Transporters: Role and Importance in ADME and Drug Development
55: Drug Transporters: Recent Advances and Emerging Technologies
56: Allosterism in Drug Discovery
57: Anti-aging Drugs: From Basic Research to Clinical Practice
58: Antibiotic Drug Discovery: New Targets and Molecular Entities

How to obtain future titles on publication:
A standing order plan is available for this series. A standing order will bring delivery of each new volume immediately on publication.

For further information please contact:
Book Sales Department, Royal Society of Chemistry, Thomas Graham House, Science Park, Milton Road, Cambridge, CB4 0WF, UK
Telephone: +44 (0)1223 420066, Fax: +44 (0)1223 420247,
Email: booksales@rsc.org
Visit our website at www.rsc.org/books

Antibiotic Drug Discovery
New Targets and Molecular Entities

Edited by

Steven M. Firestine
Wayne State University, Detriot, USA
Email: sfirestine@wayne.edu

and

Troy Lister
Spero Therapeutics, Cambridge, USA
Email: troy@sperotherapeutics.com

ROYAL SOCIETY OF CHEMISTRY

THE QUEEN'S AWARDS FOR ENTERPRISE: INTERNATIONAL TRADE 2013

Drug Discovery Series No. 58

Print ISBN: 978-1-78262-424-0
PDF eISBN: 978-1-78262-987-0
EPUB eISBN: 978-1-78801-149-5
ISSN: 2041-3203

A catalogue record for this book is available from the British Library

The Royal Society of Chemistry is a charity, registered in England and Wales, Number 207890, and a company incorporated in England by Royal Charter (Registered No. RC000524), registered office: Burlington House, Piccadilly, London W1J 0BA, UK, Telephone: +44 (0) 207 4378 6556.

For further information see our web site at www.rsc.org

Printed in the United Kingdom by CPI Group (UK) Ltd, Croydon, CR0 4YY, UK

Preface

The introduction of penicillin 75 years ago ushered in, arguably, the greatest advancement in the history of medicine. Its discovery by Fleming and subsequent demonstration of efficacy and mass production by Florey and others contributed significantly to the Allied war effort saving many thousands of lives. The broad application of penicillin during and after World War II demonstrated the great benefit antibiotics provided the medical community in treating surgical-, wound-, skin- and systemic-based infections. The continued discovery and development of antibiotics throughout the 1950's to the 1970's bolstered the number of new therapeutics available to doctors to treat a wide variety of infections. This golden age of antibiotic discovery resulted in generations of humans that never feared acquiring a life-threatening infection. Indeed, antibiotics performed so well through the later part of the 20th century that many people in the industrialized world have forgotten the significant threat that bacterial infections pose to humans. Infectious disease was the second leading cause of death in the United States prior to World War II and remains the leading cause of death worldwide today (primarily driven by neglected tropical infections in the developing world). Thus, despite the tremendous discovery and innovation in the field of infectious disease medicine over the past seven decades, bacterial infections remain a significant worldwide scourge.

Even with the successes achieved from 1950–2000, the turn of the century has seen a significant increase in antibiotic-resistance and a heightened awareness of the emergent threat. Antibiotic resistance has always existed. Indeed, resistance to penicillin was detected only a few years after its introduction into clinical practice. Yet, this problem is now more severe because of the rise of resistance in pathogenic bacteria and because we are beginning to identify bacteria that are resistant to all currently available classes of

Drug Discovery Series No. 58
Antibiotic Drug Discovery: New Targets and Molecular Entities
Edited by Steven M. Firestine and Troy Lister

Published by the Royal Society of Chemistry, www.rsc.org

antibiotics. This is particularly troubling because historically, resistance was dealt with by simply switching to another antibiotic. The rise of multi-drug-resistant organisms renders this method of treatment challenging and risky for clinicians.

How did we get here? There are a multitude of factors that played a role. The previously mentioned success in identifying new antibiotics cultivated hubris in thinking that the infectious microbial world had or could be conquered. Economics also played a major role. Many large pharmaceutical companies exited infectious disease drug discovery as research and development became increasingly expensive, protracted clinical trials slowed development and the expectation by payers and the public that antibiotics should be low cost hindered a return on investment. The scientific community too grew weary of a growing list of failed drug discovery and development campaigns, the lack of novel targets and the feeling that nothing innovative remained to be uncovered in antibiotic drug discovery. Lastly, governments worldwide were slow in appreciating the threat posed by antimicrobial resistance and historically have done little to facilitate the advancement of therapeutics. These contributing factors have stymied investment in and discovery of new therapeutics to meet the emerging unmet need of treating antibiotic-resistant infections.

As a result, in 2017, we have arrived at a point where there are few treatment options for many Gram-negative infections, where methicillin-resistant *Staphylococcus aureus* is more common than susceptible strains and where infections due to *Clostridium difficile* are increasingly drug resistant and mortal. Subsequently, clinicians are being forced to resort to older, less well-tolerated drugs to treat resistant microbes. The global community is rapidly approaching the prospect of drug-resistant bacterial epidemics that will have the potential to claim millions of lives and produce an enormous financial burden on health-care systems worldwide. A second golden age of antibiotic drug discovery is desperately needed.

There have been many recommendations for addressing antibiotic resistance including new stewardship control measures, better hygiene, and new methods to identify at-risk patients. Indeed, several valuable, recent initiatives have been started by the U.S. and European governments and regulators to dramatically speed up and reduce the cost antibiotic drug development and many incentives are now available for these pursuits. However, almost everyone that has sought to address this problem agrees that the long-term solution is discovery and introduction of new antibiotic targets and new chemical matter that leads to novel therapeutics. This book presents the latest advancements to this end. The eight chapters in this book provide an exciting discussion of novel targets in purine and isoprenoid biosynthesis, lipopolysaccharide transport, and biofilm production. Also included are reviews of unique targets in tuberculosis and *Clostridium difficile* infections, narrow spectrum antibiotics, as well as a discussion of a new antibiotic isolated from soil bacteria. The chapters focus on targets and inhibitors that are not clinically used, but represent new approaches towards the treatment of resistant infections. It is our hope that the reader will draw inspiration from this collection of scientific advancements and join the critical effort to discover and develop the next era of antibiotics. We are certain that without these new innovations, the next 75 years will not be as healthy as the last.

Steven M. Firestine and Troy Lister
Editors

Contents

Drug Discovery Series No. 58
Antibiotic Drug Discovery: New Targets and Molecular Entities
Edited by Steven M. Firestine and Troy Lister

Published by the Royal Society of Chemistry, www.rsc.org

CHAPTER 1

Treatment of Clostridium difficile *Infections*

CHRISTOPHER YIP[a], JACQUELINE PHAN[a] AND ERNESTO ABEL-SANTOS*[a]

[a]Department of Chemistry and Biochemistry, University of Nevada, Las Vegas, 4505 Maryland Parkway, Campus Box 4003, Las Vegas, NV 89154, USA
*E-mail: ernesto.abelsantos@unlv.edu

1.1 Background

Clostridium difficile infection (CDI) is a nosocomial disease mainly correlated with antibiotic-associated diarrhea. These infections are caused by *Clostridium difficile*, an anaerobic, rod shaped, gram-positive bacterium that is normally found in the gastrointestinal tract (Figure 1.1).[1]

Approximately 5% of healthy adults, and 50% of newborns are asymptomatic carriers of *C. difficile*.[2] *C. difficile* was originally thought to be a commensal bacterium, but due to the recent boom of antibiotic therapies and advancements, it was quickly recognized that *C. difficile* is the leading cause of hospital-acquired diarrhea worldwide.[3] In the United States alone, there are roughly 500 000 cases of CDI annually, with associated costs estimated to be approximately $4.8 billion.[4]

C. difficile has a unique lifecycle such that it can form metabolically dormant, non-reproductive spores when stressed (Figure 1.2).[5]

Drug Discovery Series No. 58
Antibiotic Drug Discovery: New Targets and Molecular Entities
Edited by Steven M. Firestine and Troy Lister

Published by the Royal Society of Chemistry, www.rsc.org

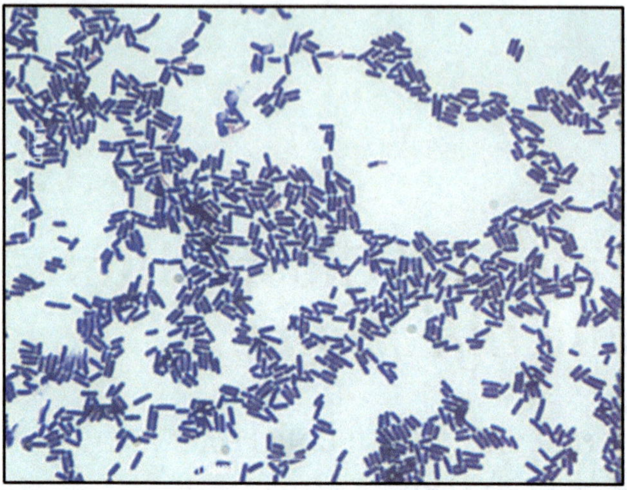

Figure 1.1 **Gram stain of *C. difficile*.** Vegetative *C. difficile* cells are rod shaped and stain purple in the presence of crystal violet, displaying a gram-positive phenotype. Scale = 10 μm.

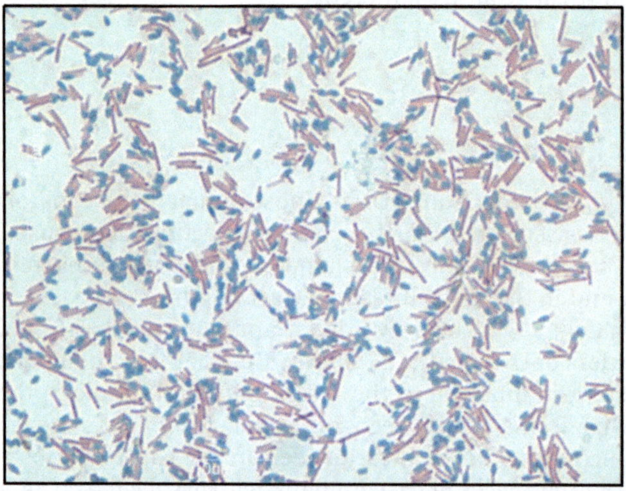

Figure 1.2 **Schaeffer-Fulton stain of *C. difficile*.** Upon environmental stress and starvation, vegetative *C. difficile* cells (pink) begin to produce highly resistant, non-reproductive spores (blue). Scale = 10 μm.

These stressors include, but not limited to, nutrient limitation and desiccation. The resulting spores are highly resistant to harsh environmental factors such as stomach acid, extreme temperatures, and pharmaceutically relevant antibiotics.[6] Spores can persist over prolonged periods of time, while constantly monitoring the environment for favorable conditions. Upon reintroduction to nutrient rich environments, the spores are able to revert back into actively growing cells through a process known as germination.[6] When *C. difficile* has completed its lifecycle, transitioning from a spore to an actively growing cell, the newly germinated cells can now colonize the local environment.

CDI begins with the ingestion of the highly resistant *C. difficile* spores. As these spores travel through the gastrointestinal tract, various endogenous bile salts stimulate the spores to germinate into actively growing, vegetative cells.[7] While the spores are metabolically dormant and act solely as a vehicle for infection, the vegetative cells are metabolically active, can produce toxins, and elicit disease.

The diversity of the endogenous bile salts depends heavily on the intestinal gut flora. Under normal circumstances, the bacteria found naturally in the gastrointestinal tract provide a barrier against *C. difficile* colonization by occupying nutrient-rich niches and by metabolizing specific bile salts required for *C. difficile* germination.[8] With the current pharmaceutical advancements, several new broad-spectrum antibiotics have been developed. Exposure to antibiotics, such as clindamycin, 2nd and 3rd generation cephalosporins, and fluoroquinolones, can disrupt the natural gut flora.[2,9,10] Disruption of the gut flora effectively removes the protective barrier, as the change in the bile salt diversity becomes more favorable for *C. difficile* (Figure 1.3).

As *C. difficile* spores germinate and outgrow, the resulting vegetative cells begin to produce and release two major toxins, TcdA and TcdB. The C-terminal region of both TcdA and TcdB contain a binding domain, which is able to interact with different carbohydrate and protein structures found on the surface of host cell membranes.[11] TcdB binds to chondroitin sulfate proteoglycan 4 (CSPG4) and the poliovirus receptor-like 3 (PVRL3) found on the surface of intestinal epithelial cells.[12] In contrast, TcdA can bind to glycoprotein 96 and sucrase isomaltase, both of which are found on the surface of human colonocytes. Once these toxins interact with the host cell receptors, the toxins are internalized by endocytosis. Acidification of the endosome causes the toxins to undergo conformational changes resulting in translocation into the host cell cytoplasm. Upon entry into the cytoplasm, the toxins undergo autocatalytic cleavage.[11] Cleavage of the toxins allow for the release and activation of their glycosyltransferase domain (GTD) into the host cell.

The GTD can transfer glucose from UDP-glucose to several crucial Rho proteins.[11,13] Glucosylation of Rho proteins result in their inactivation.[11] Since Rho proteins play an essential role in regulating the cell cytoskeleton, inactivation of these proteins can have several cytopathic effects including the disruption of cell-to-cell contacts and tight junctions, as well as increased epithelial permeability.[14] Glucosylation of RhoA also activates the inflammasome and upregulates a pro-apoptotic protein, RhoB.

Expression of *tcdA* and *tcdB* is heavily regulated and dependent on resource availability. When carbon sources and other nutrients are readily available, toxin expression is inhibited.[11] Conversely, toxin production is upregulated during stationary phase when resources are low. This type of regulation suggests that *C. difficile* virulence is a killing strategy used to improve

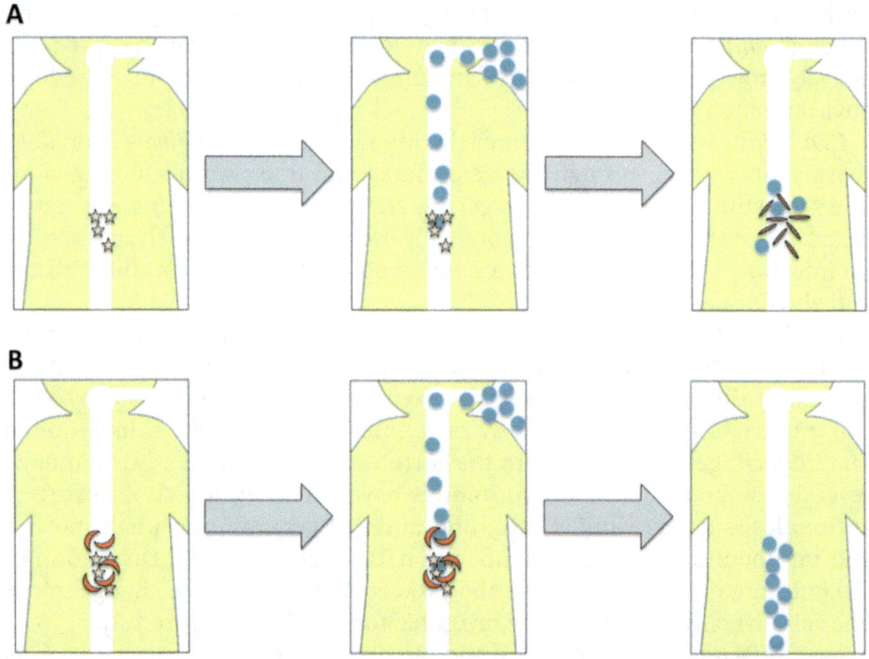

Figure 1.3 (a) **Germination of *C. difficile* spores and the use of anti-germinants.** Upon antibiotic therapy the diversity of bile salts (yellow stars) shift to a population more favorable for *C. difficile* spore germination. As *C. difficile* spores (blue circles) are ingested, the spores travel through the gastrointestinal tract, recognize the specific bile salts, and are able to germinate into toxin producing cells (red rods). (b) In the presence of various bile salt analogs (red moons), as *C. difficile* spores travel through the gastrointestinal tract, the bile salt analogs compete with bile salts that would normally trigger germination, effectively blocking *C. difficile* germination and the progression of CDI.

resource availability by scavenging the host cell for resources. A combination of several factors contributes to *C. difficile's* virulence. TcdA and TcdB are undoubtedly major contributors.[11] Due to genetic variability between *C. difficile* strains, the extensive genotypic variances in the pathogenicity locus (PaLoc), which houses *tcdA* and *tcdB*, currently give rise to at least 31 different toxinotypes.[11,15] These different toxinotypes result from mutations in *tcdA* and *tcdB*, as well as the regulatory factors that ultimately lead to the overexpression or repression of the toxin genes.[15] Several other factors, such as rates of sporulation and toxin release, motility and host cell adherence, can contribute to *C. difficile* virulence.

Interestingly, so called "hypervirulent" strains of *C. difficile* have begun to reveal themselves in the healthcare setting worldwide. Hypervirulent strains are highly variable—some strains may have higher rates of sporulation and toxin production.[13] TcdC, an anti-sigma factor that acts as a negative regulator of toxin production, is upregulated during exponential growth.[15] Several hypervirulent strains contain mutations in *tcdC* that result in the constant, unregulated production of *C. difficile* toxins.[15] Toxins produced by hypervirulent strains can also undergo necessary conformational changes at higher pH ranges, allowing the toxins to enter the host cell cytoplasm earlier and at a faster rate, relative to toxins produced by non-hypervirulent strains.[11] *Clostridium difficile* transferase (CDT), a ribosyltransferase, is a binary toxin produced only by several strains of *C. difficile*. Similar to TcdA and TcdB, CDT binds to a host-cell by interacting with a surface molecule, specifically lipolysis-stimulated lipoprotein receptor (LSR).[16,17] CDT eventually localizes into the host cell cytoplasm where it begins to ribosylate G actin.[15] At low concentrations of the toxin, CDT induces microtubules to form protrusions from the host-cell membrane, facilitating *C. difficile* adherence to the surface of the intestinal epithelial cells.[15,18] At high concentrations of CDT, actin polymerization is inhibited and actin depolymerization is induced ultimately causing the collapse of the host cell cytoskeleton.[17]

1.2 CDI Symptoms and Progression

Symptom severity in CDI patients can range from mild diarrhea, to life-threatening pseudomembranous colitis, a condition which causes exudative plaques on the intestinal mucosa. Mild CDI is defined solely by the presence of diarrhea. Other symptoms that can indicate moderate disease include abdominal pain, loss of appetite, fever, nausea, vomiting, gastrointestinal bleeding, bloody stools, and weight loss.

Severe disease can include some or all of the symptoms associated with mild-to-moderate disease plus additional indicative symptoms. Severe CDI is indicated by a serum albumin < 3 g dl^{-1} with an elevated white blood cell (WBC) count of ≥15 000 cells/mm^3 and/or abdominal tenderness.[19]

In more severe cases of CDI, patients can develop several complications. If symptoms progress, they can lead to hypotension, fever above 38.5 °C, ileus (a condition in which peristaltic activity to move stool through the gastrointestinal tract is diminished) or abdominal distention, altered mental state, WBC count ≥35 000 cells/mm^3, serum lactate levels >2.2 mmol L^{-1}, and ultimately organ failure.[19]

Another complication called pseudomembranous colitis is unique to CDI. Pseudomembranous colitis occurs when toxins produced by *C. difficile* cells damage the walls of the colon causing inflammation and thickening of the colonic mucosa producing yellowish exudates called pseudomembranes to form along the colon. This can lead to other complications such as perforated colon and toxic megacolon. Toxic megacolon can render the colon incapable of expelling gas and stool contents, potentially causing the colon to rupture.

1.3 Relapse

CDI relapse is characterized by the return of symptoms within eight weeks of primary diagnosis after initial symptoms have previously been resolved.[20] CDI recurrence after initial treatment can reach up to 25% in treated patients.[21] Chances of subsequent recurrences nearly doubles to 45% after the first recurrence.[22] One explanation for CDI relapse is that resident *C. difficile* spores may have survived in the gut after completion or discontinuation of antibiotic treatment.[21] *C. difficile* spores may also be picked up *via* contamination of the local environment.[21] Therefore, relapse and reinfection may be difficult to distinguish. However, reinfections can be identified by the diagnosis with a different *C. difficile* strain.[23,24] Other possible reasons for relapse include poor immune response leading to inadequate production of antibodies against to *C. difficile* toxins and frequent disruption of normal gut flora.[22] Moreover, epidemiologic factors such as advanced age, use of other antibiotics, and prolonged hospital stay may also contribute to increased risk for recurrent CDI.[22] Emergence of resistant strains and hypervirulent strains over the past decade has made treatment of recurrent CDI increasingly difficult.[25]

1.4 Diagnosis

CDI may mimic flu-like symptoms or flare-ups of other gastrointestinal diseases. Therefore, early and accurate diagnosis of CDI is important to the successful management of the disease. In 2010, the Society for Healthcare Epidemiology of America (SHEA) and the Infectious Diseases Society of America (IDSA) recommended a two-step algorithm for CDI diagnosis.[26] First, an initial immunoassay screen of stool samples for the presence of *C. difficile* is performed. If *C. difficile* is found to be absent in the initial screen, *C. difficile* should be ruled out as the cause of diarrhea. Following a positive result from the initial screen, detection of toxins from the stool is tested. Positive indications for *C. difficile* toxins along with moderate-to-severe symptoms can warrant the need for additional and more invasive testing.

Initial screening of stools for the presence of *C. difficile* can be performed by a common and relatively inexpensive enzyme immunoassay (EIA) test called the common antigen test or glutamate dehydrogenase (GDH) test. The GDH test looks for the GDH enzyme that is produced in relatively large amounts by *C. difficile* and can be readily detected the stools of CDI patients.[27] Although this EIA can give results within 15–45 min, it tends to have lower sensitivities than other tests since it only test for the presence of the *C. difficile* organisms rather than the production of toxins.

Following a positive GDH test, the second step of the two-step algorithm includes a cell cytotoxicity neutralization assay (CCNA) which tests for the presence of toxin B in stool filtrate. CCNA has long been considered to be the traditional "gold standard" for the detection of *C. difficile* toxins.[28] In a CCNA, filtrate of the collected stool sample is added to mammalian cultures (*e.g.* human fibroblast). If *C. difficile* TcdB is present in the filtrate, the mammalian cells will round up and necrotize.[29] To verify that the cytopathic effect is caused by *C. difficile* toxin, the cell cultures are supplemented with an antitoxin (monoclonal antibodies). If the cytopathic effect is reversed, a test is positive for TcdB. While CCNA is highly sensitive and specific, it has a slow turnaround of 24 to 48 h and requires technical expertise.

Alternatively, a three-step algorithm for CDI diagnosis may be used. The three-step algorithm includes the steps of the two-step algorithm with the addition of an intermediate EIA test that detects the presence of free TcdA and TcdB (TOX-A/BII) in stool.[30] If results from the TOX-A/BII EIA test is positive, the stool is said to be positive for *C. difficile* toxins. If the test is negative, CCNA will be performed in the third step.

Alternately, the third step of the three-step algorithm may employ a molecular diagnostic test instead of CCNA.[31] Nucleic acid amplification tests (NAATs) allow for the detection of *C. difficile* toxin gene fragments *via* real-time quantitative polymerase chain reaction (RT-PCR).[19] Through molecular methods, Toxin B (tcdB) and binary toxin (cdtA and cdtB) can be detected.[31] NAATs have better sensitivity than CCNA to test for non-free toxins. Although this method tests for the presence of the toxin, it does not indicate the expression of the toxins genes. NAATs can be costly and must be interpreted with caution as they may detect toxigenic strains in asymptomatic carriers who may have diarrhea caused by other pathologies.[32]

The three-step algorithm provides an effective and reliable method to diagnosing CDI and may eliminate the need to perform a CCNA test if the TOX-A/BII EIA test is positive. However, due to a lack in sensitivity, it has been recently widely accepted that the TOX-A/BII EIA is not well-suited to be a stand-alone test to diagnose CDI.[26] Therefore, the two-step algorithm is usually preferred over the three-step algorithm as it more practical, cost effective, and requires less workload in comparison to the three-step algorithm.[30]

Colonoscopy and Computed Tomography (CT scan) may also be used to diagnose conditions caused by CDI such as pseudomembranous colitis. These imaging methods are used less often than laboratory tests as they can be more costly, unpleasant to the patient, and less sensitive.[33]

1.5 Prevention Measures—General Hospital Practice and Other Prevention Methods

Although CDI can occasionally occur in healthy individuals, CDI is most prevalent among elderly and immunocompromised patients.[19,34] Patients are more likely to get CDI in healthcare-acquired settings (*e.g.* hospitals and long-term care facilities) than in community-acquired settings. Because *C. difficile* spores can survive for long periods of time on hospital surfaces and patient beds, proper precautions must be taken in these settings to prevent CDI from spreading.

The Association for Professionals in Infection Control and Epidemiology (APIC) suggests that hospitals implement infection control programs.[19] Three recommendations include a criteria index for patients who have risk factors for CDI (*e.g.* malignancy, advanced age, and recent hospitalization or stay at a long-term care facility), advocacy for physicians to use proper *C. difficile* diagnostic testing with rapid turn-around times and high sensitivity for toxin detection, and appropriate notification to staff members of positive

C. difficile test results to ensure that proper precautions and treatments be taken.[19] When a *C. difficile* infection control protocol was implemented at the University of Pittsburgh Medical Center-Presbyterian, a decrease from 7.2 cases per 1000 discharges to 4.8 cases per 1000 discharges was obtained in a 5 year-period.[35]

Proper hand hygiene is critical in the prevention of CDI. Although alcohol antiseptics can be used to kill most vegetative bacteria and viruses, they do not affect *C. difficile* spores due to their intrinsic resistance.[19] Therefore, healthcare providers and visitors should be required to wash their hands with antimicrobial hand soap and water. Since a person can easily contaminate their hands with *C. difficile* spores by contacting an infected patient, healthcare personnel and visitors must also use gloves and gowns upon entry into a CDI patient's room. An intervention study incorporated an infection prevention education program with vinyl gloving wearing surveillance for six months. The study showed a significant decline in CDI rates from 7.7 cases per 1000 patients to 1.5 cases per 1000 patients six months after intervention.[36] Other patient contact precautions include the use of single-use disposable equipment and the limit of patient contact until resolution of diarrhea.

Because the environmental surfaces are common sources for nosocomial infective agents such as *C. difficile*, the use of disinfectants is recommended. The Environmental Protection Agency (EPA) advocates for disinfectants that are sporicidal.[19] Other chlorine-containing agents also have the potential to decontaminate *C. difficile* infected surfaces, but are recommended to have a minimum of 5000 ppm of chlorine.

Another *C. difficile* prevention intervention includes placing CDI patients into private rooms. A cohort study showed that patients in double rooms had a higher acquisition of CDI than patients placed in single rooms.[37] Moreover, hospitalized patients who do not have diarrhea are not recommended to be screened for *C. difficile* as some patients may be asymptomatic carriers of *C. difficile*.[38] In a study about asymptomatic carriers, metronidazole was shown to be ineffective at eliminating *C. difficile* carriage, and vancomycin, though initially clearing up *C. difficile* from stools, was unsuccessful at preventing recolonization of *C. difficile*.[38] The high rate of recolonization often resulted in emergence of new strains and an asymptomatic carrier even developed CDI after vancomycin treatment.[38]

Managing and minimizing the type, frequency, and variety of antibiotics taken for other illness can also reduce the susceptibility of patients to CDI. Although most antibiotics can cause patients to become susceptible to CDI, clindamycin, cephalosporins, and fluoroquinolones pose a higher risk for CDI.[39] The duration of antibiotics and use of multiple antibiotics can also increase susceptibility. One study showed that an antimicrobial stewardship program decreased CDI incidence by 60%.[40]

CDI recurrences can be treated with repeated regimens of metronidazole and vancomycin, though this method may not be successful in preventing future recurrences. Furthermore, metronidazole is usually avoided in recurrent CDI treatment as prolonged use may result in neurotoxicity and hepatic toxicity.[25] Tapered doses followed by pulsed doses of vancomycin may also be used to manage subsequent episodes.[22] For more severe recurrences or in the case of multiple recurrences, novel methods or more drastic measures such as rifamixin "chaser" therapy, nitazoxanide, intravenous immunoglobulin (antibodies), fecal microbiota transplant (FMT), and probiotics may be considered.[22,25]

Rifamixin "chaser" therapy is effective in decreasing recurrent diarrhea among CDI patients who have already undergone conventional treatments. Rifamixin is a nonabsorbed antibiotic that targets *C. difficile* without killing enteric gut bacteria, allowing the gut flora to regrow and potential reduce CDI recurrence.[41] Rifaxmixin, however, has been linked with high resistance levels caused by an amino acid substitution in the β-subunit of RNA polymerase.[41,42] Therefore, rifamixin is used in recurrent CDI treatment after vancomycin treatment as a "chaser" and should be used with caution due to the possibility of resistance.[43]

Nitazoxanide is a nitrothiazolide antiparasitic agent that has been used following standard CDI treatments in a few studies. Due to the small sample size of one study, noninferiority or superiority of nitazoxanide to vancomycin could not be made.[44] Another small study showed 54% clinical cure of patients given a 10 day-course of nitazoxanide after treatment failure with 14 days of metronidazole. However, 20% of patients relapsed and 27% failed to resolve symptoms (succumbing to the disease at various points in the trial).[45] Larger studies are necessary to compare the nitazoxanide with standard CDI treatments.

1.6 Current Treatment and Antibiotics

Treatments for CDI require a rigorous regimen, which normally combines decontamination of the local environment, together with a choice of treatment with metronidazole, vancomycin, and/or fidaxomicin. However, due to *C. difficile's* lifecycle and its ability to form spores, complete decontamination is often difficult and the use of antimicrobials frequently leads to relapse.[46] Due to the high rates of relapse, new forms of treatment and prevention have been evaluated such as the use of anti-toxin, vaccines, and fecal transplants.

Metronidazole (Figure 1.4) is a nitroaromatic, broad-spectrum antibiotic that is highly active against anaerobic bacteria, such as *C. difficile*.[47] Due to metronidazole's low molecular weight, it can enter *C. difficile's* cytoplasm *via* passive diffusion. In anaerobic bacteria, flavodoxin and ferredoxin act as electron acceptors as pyruvate becomes oxidatively decarboxylated.[2] In the presence of metronidazole, electrons are instead donated to the 5-nitro group on the imidazole ring.[47] The resulting nitroso free radical is reactive and will target DNA.[2] While the exact mechanism has not yet been fully elucidated, it is believed that DNA experiences oxidative damage—this damage produces numerous single stranded and double stranded breaks, ultimately leading to DNA degradation and cell death.[47]

While metronidazole is a promising treatment due to its ability to specifically damage anaerobic bacterial DNA, some clinically relevant *C. difficile* strains have shown resistance against metronidazole.[47] Metronidazole resistance is currently believed to result from several different mechanisms, including reduced influx, slow activation of metronidazole, increased DNA repair, inactivation of reduced metronidazole, or increased efflux of the drug.[47] Resistance to metronidazole can also arise from the horizontal transfer of genes conferring resistance of nitroimidazoles.[47]

Vancomycin (Figure 1.4) is a second antibiotic of choice that works relatively well for severe cases of CDI.[48] Vancomycin is a large, hydrophilic, broad-spectrum antibiotic that inhibits the maturation of the peptidoglycan in gram-positive bacteria.[2] In addition to preventing peptidoglycan biosynthesis in vegetative cells, vancomycin also prevents the outgrowth of *C. difficile* spores.[46] Due to the large molecular weight, vancomycin can be administered orally as it is minimally absorbed and can be found at high concentrations in the gut and feces.[48] In addition to having favorable pharmacokinetics, vancomycin resistance has not yet been reported in *C. difficile*.[2]

Because metronidazole broadly targets anaerobic bacteria and vancomycin targets gram-positive organisms, they suppress the growth of the gut microbiota and thereby removing the natural protective barrier against *C. difficile*.[21] Therefore, treatment with metronidazole or vancomycin frequently results in CDI relapse.[46] Fidaxomicin (Figure 1.4), a newer anti-CDI treatment, has been shown to be active against *C. difficile*, and not target gram-negative bacteria.[3,49] Originally isolated from *Dactylosporangium aurantiacum*, fidaxomicin is an 18-membered macrocyclic antimicrobial that inhibits bacterial RNA polymerase, ultimately halting transcription.[3,50] While other RNA synthesis inhibitors prevent the initiation and elongation steps, fidaxomicin halts transcription by binding to RNA polymerase prior to DNA separation.[2,48]

Hydrolysis of an isobutyryl ester by an unknown esterase located at the 4′ position of fidaxomicin yields OP-1118, the major byproduct of fidaxomicin metabolism.[48] Similar to fidaxomicin, OP-1118 also displays potent bactericidal activity against *C. difficile*.[50] Fidaxomicin has also been shown to inhibit spore production and, similar to vancomycin, also prevent the outgrowth of *C. difficile* spores.[3,46,50] Fidaxomicin and OP-1118, like vancomycin, demonstrate low systematic absorption when taken orally, and can be found at high concentrations in the gut and feces.[48] Fidaxomicin and OP-1118 also seem to decrease *C. difficile* toxins.[50] Finally while compared to both vancomycin and metronidazole, fidaxomicin was associated with a lower rate of recurrence.[21]

1.6.1 Anti-Toxins

TcdA and TcdB are the main determinants of CDI and the recent overexpression of these toxins by hypervirulent strains has brought about the idea of using anti-toxins to control the course of the disease. The use of anti-toxins could potentially neutralize toxins, while simultaneously allowing the re-colonization of the natural gut flora and therefore prevent relapse.

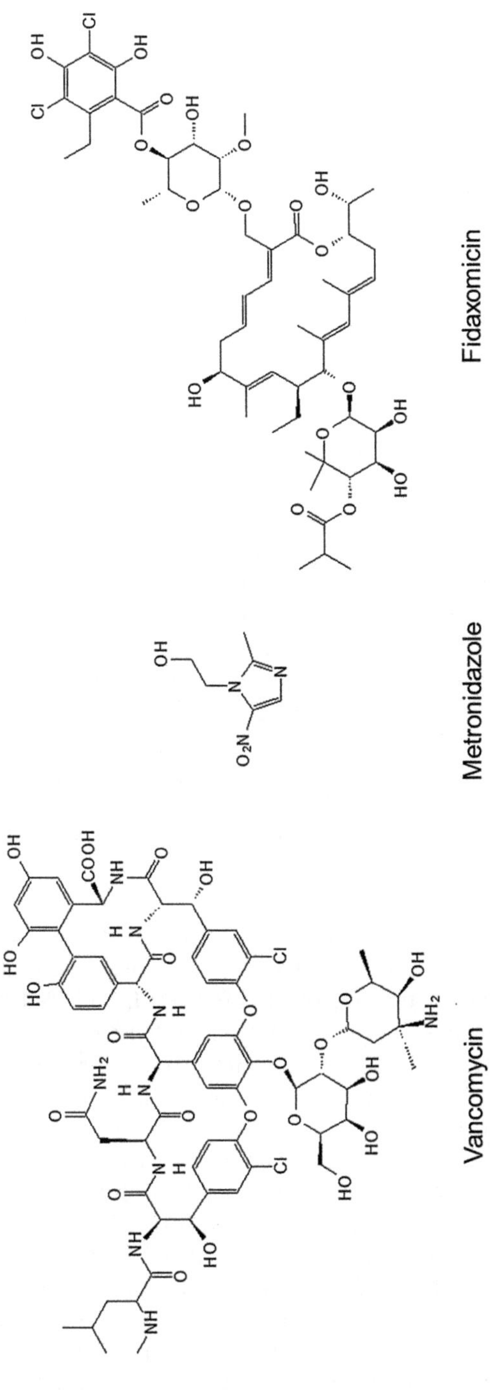

Figure 1.4 Chemical structures of clinically relevant antimicrobials in the treatment of CDI.

The use of two human monoclonal antitoxins, actoxumab and bezlotoxumab, have been shown to neutralize both TcdA and TcdB, respectively.[51] Actoxumab and bezlotoxumab bind to the C-terminal regions of TcdA and TcdB, thereby inhibiting the toxins' ability to bind to their host substrates. Indeed, actoxumab and bezlotoxumab have been shown to prevent epithelial damage and inflammation in mouse models.[51] The use of antitoxins in combination with vancomycin has been reported to decrease the rate of recurrence, compared with the use of vancomycin alone.[2,52] While advances have been made to produce antitoxins against TcdA and TcdB, antitoxins against CDT have not yet been fully addressed. Additionally, while these antitoxins are able to effectively reduce the morbidity and mortality of CDI, the antitoxins play no role in inhibiting or reducing *C. difficile* colonization.[53]

1.6.2 Vaccines

Development and evaluation of *C. difficile* vaccines have been ongoing for the past two decades. Formalin-inactivated whole toxins have shown to be an effective vaccine in protecting hamsters against CDI.[53] Truncated forms of TcdA, TcdB, as well as hybridized versions of the two toxins have been considered to be used as vaccines. Inoculation of Golden Syrian hamsters with TcdB fragments, in combination with TcdA, was shown to generate a strong immunogenic response, ultimately producing antibodies that neutralized the toxins.[52] Interestingly, the C-terminal binding domains are able to produce a greater immunogenic response compared to the N-terminal GTD.[3] As a result, the C-terminal regions of TcdA and TcdB have been hybridized together to form an effective recombinant vaccine without the presence of the glycosyltransferase domains.[3]

While the use of toxoids and recombinant vaccines are able to effectively reduce the symptoms of CDI, asymptomatic colonization by *C. difficile* still occurs. A vaccine that targets surface antigens could potentially reduce *C. difficile* colonization and transmission, rather than directly decreasing the lethal outcome. The surfaces of *C. difficile* vegetative cells contain three highly intricate carbohydrate structures termed PSI, PSII, and PSIII.[5] PSI and PSIII are found only in some *C. difficile* ribotpes and they seem to be expressed stochastically.[54] PSII is an ideal target for vaccine development since it is expressed at higher levels relative to PSI and PSIII, and can be found more abundantly across the *C. difficile* ribotypes.[54] PSII is a polysaccharide composed of repeating hexaglycosyl units, connected by a phosphodiester group. Interestingly, antibodies have been raised against phosphorylated hexaglycosyl units, and these hexaglycosyl units can be found in the biofilms produced by *C. difficile*.[53] In previous studies mice and rabbits were inoculated with conjugated forms of PSII and in both cases, immune responses were raised against the native forms of PSII.[5] Lipoteicholic acid (LTA), another surface antigen that is highly conserved on the surface of *C. difficile* cells, can also stimulate an immune response in mouse and rabbit models.[3] Although antibodies have been raised in response to PSII and LTA, no evidence has yet been reported about the protective efficacy of these antibodies and vaccines.[53]

1.6.3 Anti-Germinants

As the transition from a metabolically dormant spore to a toxin-producing cell is required for disease, the germination process may be a target for CDI treatment and prevention. Upon recognition of taurocholate, a primary bile salt found in the gastrointestinal tract, germination is stimulated in *C. difficile* spores.[1] Gut microbiota, found during normal circumstances, are able to break down and metabolize taurocholate into chenodeoxycholate (CDCA), a secondary bile salt.[8] While taurocholate normally induces germination, CDCA inhibits *C. difficile* spore germination and therefore blocks the onset of disease.[55] All existing treatments of CDI currently focus on combating the established disease. The use of anti-germinants would prevent spore germination, ultimately playing a role in disease prophylaxis (Figure 1.3B).[56]

Indeed, the use of several synthetic bile salts to block *C. difficile's* spore germination has been evaluated. CamSA, a synthetic analog of taurocholate, has been shown to inhibit *C. difficile* spore germination *in vitro* at micromolar concentrations.[1] The use of CamSA has also been shown to protect mice when challenged with *C. difficile* after antibiotic treatments.[56] While the use of anti-germinants shows great potential CDI prophylaxis, different strains of *C. difficile* display different germination phenotypes, and therefore a singular anti-germinant may not be sufficient in completely preventing disease.[57]

1.6.4 Fecal Transplantation

Fecal Microbiota Transplant (FMT) is a non-traditional method to treating multiply-recurrent CDI.[58] FMT is the introduction of stools from a tested healthy donor into the colon of a patient.[59] Fecal transplantation has been used in veterinary medicine for over 100 years and was first performed in humans in 1958. FMT has been studied widely since then and is currently being tested in various other gastrointestinal diseases including Crohn's Disease, Irritable Bowel Syndrome (IBS), and Ulcerative Colitis.

FMT is a method that is used to restore the diverse gut microbiota that have been killed by the use to antibiotics. Patients with recurring CDI have been shown to have abnormally disproportionate gut microbiota. By reintroducing normal gut bacteria back into the patient *via* donor feces, it is suggested that the rich and diverse gut flora and colonization resistance can be restored to correct the imbalance.

Patients who receive FMT usually have had multiple bouts of recurrences and have failed conventional treatment methods. On average, studies have shown a cure rate of 91–93% after 90 days of FMT following a course of vancomycin treatment.[59]

FMT is most commonly done *via* a colonoscopy, endoscopy, or through an enema, but can also be taken as a frozen oral capsule.[60] Administration of fecal samples into the colon *via* colonoscopy or fecal enema show high success rates of 86–100%.[59] Other endoscopic procedures such as fecal sample introduction into the gastric cavity or small intestine yielded slightly lower success rates of 77–94%.[59] A lower success rate of 80% was reported for FMT administration *via* oral capsules.[59]

Although the cost of FMT is less expensive than other CDI treatments, the unappealing aesthetics of the procedure is often a concern of patients.[61,62] Although FMT has been deemed relatively safe, a potential risk is the transmission of infectious agents from the donor feces to the patient. Nevertheless, donor stools undergo rigorous screening for common bacterial and viral enteropathogens.[59] Long-term follow-up studies of FMT are also limited. One study showed that 4 out of 77 patients developed autoimmune diseases such as rheumatoid arthritis after FMT treatment. However, no clear relationship between the autoimmune disease and FMT have been linked. Exclusion of patients from FMT may include contraindications for colonoscopy, need for continuous antibiotic treatment for another diseases, and failure to understand FMT treatment due to conditions such as dementia.[58]

1.6.5 Probiotics

Similarly to FMT, probiotics aim to introduce "good" microorganisms into the GI tract of CDI patient. Probiotics can include combinations of bacteria and yeast. Unlike FMT, probiotics typically include only a limited number of microbial species. Common microorganisms used in probiotic mixtures include *Saccharomyces boulardii*, *Lactobacilli*, *Clostridia*, *Streptococci*, and *Bifidobacteria*. *Lactobacilli* are commonly found in yogurt and other fermented foods. *Bifidobacteria* are found in dairy products and can be used to ease symptoms of Irritable Bowel Syndrome (IBS).[63]

Although the use of probiotic therapy is theoretically useful, there is insufficient data to support the efficacy of probiotic use in the treatment of both primary and secondary CDI. *S. boulardii* has been shown to be well-suited for CDI prevention. In a study of adjunct probiotic use with antibiotics in recurrent CDI, patients showed 35% fewer recurrences than the control group of 65%. However, due to inadequate randomization of antibiotics in the study, a clear link to probiotic effectiveness is questionable. Another criticism of probiotics is the lack of regulation in their manufacturing process. While pharmaceutical drugs must include conditions that the drug is proven to treat along with side effects, contraindications, and adverse drug interactions, probiotics are regulated as dietary supplements. Thus, these products may not contain what is indicated on the labels and are not evaluated for safety.[64] Because dietary supplements are usually self-prescribed, there is no controlled method for reporting adverse reactions and side effects.[65] In general, drugs are considered unsafe until they can be proven safe, whereas probiotics and dietary supplements considered safe until they can be proven unsafe. Another potential risk, as noted in some sporadic case reports, include bloodstream infections due to bacteremia and/or fungemia from the use of live *S. boulardii* and *Lactobacillus* species-based probiotics in immunocompromised patients such as those with HIV and malignancy.[66]

1.7 Conclusion

Incidences of CDI both in healthcare-acquired and community-acquired settings have increased significantly over last several years.[67,68] CDI has surpassed methicillin-resistant *Staphylococcus aureus* (MRSA) as the most common hospital-acquired infection (HAI).[69] Although elderly individuals in healthcare-related settings are traditionally the targets of CDI, younger individuals are now becoming increasingly susceptible to CDI in the community.[69] The rise in CDI is due in part to the emergence of antibiotic-resistant and hypervirulent strains of *C. difficile* contributing to high rates of relapse and virulence.[68,70] CDI is also a multi-faceted problem involving many variables ranging from strains characteristics to patient risk factors to environmental control. Therefore, much is still to be learned about *C. difficile* and new avenues of CDI treatment to be explored.

References

1. A. Howerton, N. Ramirez and E. Abel-Santos, Mapping interactions between germinants and *C. difficile* spores, *J. Bacteriol.*, 2011, **193**(1), 274–282.
2. L. Tsutsumi, Y. B. Owusu, J. G. Hurdle and D. Sun, Progress in the discovery of treatments for *C difficile* infection: a clinical and medicinal chemistry review, *Curr. Top. Med. Chem.*, 2014, **14**(1), 152–175.
3. H. Mathur, M. C. Rea, P. D. Cotter, R. P. Ross and C. Hill, The potential for emerging therapeutic options for *Clostridium difficile* infection, *Gut Microbes*, 2014, **5**(6), 696–710.
4. E. R. Dubberke and M. A. Olsen, Burden of *Clostridium difficile* on the healthcare system, *Clin. Infect. Dis.*, 2012, **55**, S88–S92.
5. M. M. Awad, P. A. Johanesen, G. P. Carter, R. Rose and D. Lyras, *Clostridium difficile* virulence factors: insights into an anaerobic spore-forming pathogen, *Gut Microbes*, 2014, **5**(5), 579–593.
6. C. Ross and E. Abel-Santos, The Ger receptor family from sporulating bacteria, *Curr. Issues Mol. Biol.*, 2010, **12**(3), 147–158.
7. C. Lübbert, E. John and L. von Müller, *Clostridium difficile* infection: guideline-based diagnosis and treatment, *Dtsch. Arztebl. Int.*, 2010, **111**(43), 723–731.
8. J. A. Sorg and A. L. Sonenshein, Bile salts and glycine as cogerminants for *Clostridium difficile* spores, *J. Bacteriol.*, 2008, **190**(7), 2505–2512.
9. D. Surowiec, A. G. Kuyumjian, M. A. Wynd and C. E. Cicogna, Past, present, and future therapies for *Clostridium difficile*-associated disease, *Ann. Pharmacother.*, 2006, **40**(12), 2155–2163.
10. P. Hookman and J. S. Barkin, *Clostridium difficile* associated infection, diarrhea and colitis, *World J. Gastroenterol.*, 2009, **15**(13), 1554–1580.
11. S. Di Bella, P. Ascenzi, S. Siarakas, N. Petrosillo and A. di Masi, *Clostridium difficile* toxins a and b: insights into pathogenic properties and extraintestinal effects, *Toxins (Basel)*, 2016, **8**(5), 134.

12. M. E. LaFrance, M. A. Farrow, R. Chandrasekaran, J. Sheng, D. H. Rubin and D. B. Lacy, Identification of an epithelial cell receptor responsible for *Clostridium difficile* TcdB-induced cytotoxicity, *Proc. Natl. Acad. Sci. U. S. A.*, 2015, **112**(22), 7073–7078.
13. G. P. Carter, J. I. Rood and D. Lyras, The role of toxin A and toxin A in *Clostridium difficile*-associated disease: past and present perspectives, *Gut Microbes*, 2010, **1**(1), 58–64.
14. A. Ofosu, *Clostridium difficile* infection: a review of current and emerging therapies, *Ann. Gastroenterol.*, 2016, **29**(2), 147–154.
15. J. J. Hunt and J. D. Ballard, Variations in virulence and molecular biology among emerging strains of *Clostridium difficile*, *Microbiol. Mol. Biol. Rev.*, 2013, **77**(4), 567–581.
16. D. N. Gerding, S. Johnson, M. Rupnik and K. Aktories, *Clostridium difficile* binary toxin CDT: mechanism, epidemiology, and potential clinical importance, *Gut Microbes*, 2013, **5**(1), 15–27.
17. S. Hemmasi, B. A. Czulkies, B. Schorch, A. Viet, K. Akotires and P. Papatheodorou, Interaction of the *Clostridium difficile* binary toxin CDT and its host cell receptor, lipolysis-stimulated lipoprotein receptor (LSR), *J. Biol. Chem.*, 2015, **290**(22), 14031–14044.
18. A. Rineh, M. J. Kelso, F. Vatansever, G. P. Tegos and M. R. Hamblin, *Clostridium difficile* infection: molecular pathogenesis and novel therapeutics, *Expert Rev. Anti-Infect. Ther.*, 2014, **12**(1), 131–150.
19. C. M. Surawicz, L. J. Brandt, D. G. Binion, A. N. Ananthakrishnan, S. R. Curry and P. H. Galligan, *et al.*, Guidelines for diagnosis, treatment, and prevention of *Clostridium difficile* infections, *Am. J. Gastroenterol.*, 2013, **108**(8), 478–498.
20. T. Larrainzar-Coghen, D. Rodriguez-Pardo, M. Puig-Asensio, V. Rodríguez, C. Ferrer and R. Bartolomé, *et al.*, First recurrent of *Clostridium difficile* infection: clinical relevance, risk factors, and prognosis, *Eur. J. Clin. Microbiol. Infect. Dis.*, 2016, 371–378.
21. O. A. Cornely, M. A. Miller, T. J. Louie, D. W. Crook and S. Gorbach, Treatment of first recurrence of *Clostridium difficile* infection: Fidaxomicin versus vancomycin, *Clin. Infect. Dis.*, 2012, **55**, S154–S161.
22. S. Johnson, Recurrent *C. difficile* infection: a review of risk factors, treatments, and outcomes, *J. Infect.*, 2009, **58**(6), 403–410.
23. F. Barbut, B. Richard, K. Hamadi, V. Chomette, B. Burghoffer and J. C. Petit, Epidemiology of recurrences or reinfections of *Clostridium difficile*-associated diarrhea, *J. Clin. Microbiol.*, 2000, **38**(6), 2386–2388.
24. I. Figueroa, S. Johnson, S. P. Sambol, E. J. Goldstein, D. M. Citron and D. N. Gerding, Relapse versus reinfection: recurrent *Clostridium difficile* infection following treatment with fidaxomicin or vancomycin, *Clin. Infect. Dis.*, 2012, **55**, S104–S109.
25. C. S. Cocanour, Best strategies in recurrent or persistent *Clostridium difficile* infection, *Surg. Infect. (Larchmt.)*, 2011, **12**(3), 235–239.
26. S. H. Cohen, D. N. Gerding, S. Johnson, C. P. Kelly, V. G. Loo and L. C. McDonald, *et al.*, Clinical practice guidelines for *Clostridium difficile* infection in adults: 2010 update by the society for healthcare epidemiology

of America (SHEA) and the infectious diseases of America (IDSA), *Infect. Control Hosp. Epidemiol.*, 2010, **31**(5), 431–455.
27. D. M. Lyler, L. A. Barroso and T. D. Wilkins, Identification of the latex test-reactive protein of *Clostridium difficile* as glutamate dehydrogenase, *J. Clin. Microbiol.*, 1999, **29**(11), 2639–2642.
28. T. Planche and M. Wilcox, Reference assays for *Clostridium difficile* infection: One or two gold standards? *J. Clin. Pathol.*, 2011, **64**(1), 1–5.
29. J. R. Ticehurst, D. Z. Aird, L. M. Dam, A. P. Borek, J. T. Hargove and K. C. Carroll, Effective detection of toxigenic *Clostridium difficile* by a two-step algorithm including tests for antigen and cytotoxin, *J. Clin. Microbiol.*, 2006, **44**(3), 1145–1149.
30. M. O. Qutub, N. AlBaz, P. Hawken and A. Anoos, Comparison between the two-step and the three-step algorithms for the detection of toxigenic *Clostridium difficile*, *Indian J. Med. Microbiol.*, 2011, **29**(3), 293–296.
31. A. Schneider, A. Mól, K. Lisowska, M. Jax-Dambek, D. Lachowicz, P. Obuch-Woszczatński and H. Pituch, Three-step diagnostic algorithm in diagnosing patients suspected of *Clostridium difficile*-associated diarrhea, *Int. J. Dev. Biol.*, 2014, **68**(4), 669–674.
32. C. A. Burnham and K. C. Carroll, Diagnosis of *Clostridium difficile* infection: an ongoing conundrum for clinicians and for clinical laboratories, *Clin. Microbiol. Rev.*, 2013, **26**(3), 604–630.
33. T. G. Fraser and J. F. Swiencicki, *Clostridium difficile*, Cleveland Clinic: Center for Continuing Education, Feb 2013.
34. R. L. Jump, *Clostridium difficile* infection in older adults, *Aging Health*, 2013, **9**(4), 403–414.
35. C. A. Muto, M. K. Blank, J. W. Marsh, E. N. Vergis, M. M. O'Leary and K. A. Shutt, *et al.*, Control of an outbreak of infection with the hypervirulent *Clostridium difficile* BI strain in a university hospital using a comprehensive "bundle" approach, *Clin. Infect. Dis.*, 2007, **45**(10), 1266–1273.
36. S. Johnson, D. N. Gerding, M. M. Olson, M. D. Weiler, R. A. Hughes, C. R. Clabots and L. R. Peterson, Prospective, controlled study of vinyl glove use to interrupt *Clostridium difficile* nosocomial transmission, *Am. J. Med.*, 1990, **88**(2), 137–140.
37. S. Kundrapu, V. C. Sunkesula, L. A. Jury, A. K. Sethi and C. J. Donskey, Utility of perirectal swab specimens for diagnosis of *Clostridium difficile* infection, *Clin. Infect. Dis.*, 2012, **55**(11), 1527–1530.
38. S. Johnson, S. R. Homann, K. M. Bettin, J. N. Quick, C. R. Clabots, L. R. Peterson and D. N. Gerdin, Treatment of asymptomatic *Clostridium difficile* carriers (fecal excretors) with vancyomycin or metronidazole: a randomized, placebo-controlled trial, *Ann. Intern. Med.*, 1992, **117**(4), 297–302.
39. D. N. Gerding, Clindamycin, cephalosporins, fluoroquinolones, and *Clostridium difficile*-associated diarrhea: This is an antimicrobial resistance problem, *Clin. Infect. Dis.*, 2004, **38**(5), 646–648.
40. L. Valiquette, B. Cossette, M. P. Garant, H. Diab and J. Pepin, Impact of a reduction in the use of high-risk antibiotics on the course of an epidemic of *Clostridium difficile*-associated disease caused by the hypervirulent NAP1/027 strain, *Clin. Infect. Dis.*, 2007, **45**, S1112–S1121.

41. S. R. Curry, J. W. Marsh, K. A. Suhtt, C. A. Muto, M. M. O'Leary and M. I. Saul, *et al.*, High frequency of rifampin resistance identified in an epidemic *Clostridium difficile* clone from a large teaching hospital, *Clin. Infect. Dis.*, 2009, **48**(4), 425–429.
42. J. R. O'Connor, M. A. Galang, S. P. Sambol, D. W. Hecht, G. Vedantam, D. N. Gerding and S. Johnson, Rifampin and rifaximin resistance in clinical isolates of *Clostridium difficile*, *Antimicrob. Agents Chemother.*, 2008, **52**(8), 2813–2817.
43. A. A. Venugopal and S. Johnson, Current state of *Clostridium difficile* treatment options, *Clin. Infect. Dis.*, 2012, **55**, S71–S77.
44. D. M. Musher, N. Logan, A. M. Bressler, D. P. Johnson and J. F. Rossignol, Nitazoanide versus vancomycin in *Clostridium difficile* infection: a randomized, double-blind study, *Clin. Infect. Dis.*, 2009, **48**(4), 41–46.
45. D. M. Musher, N. Logan, V. Mehendiratta, N. A. Melgarejo, S. Garud and R. J. Hammil, *Clostridium difficile* colitis that fails conventional metronidazole therapy: Response to nitazoxanide, *J. Antimicrob. Chemother.*, 2007, **59**(4), 705–710.
46. C. A. Allen, F. Babakhani, P. Sears, L. Nguyen and J. A. Sorg, Both fidaxomicin and vancomycin inhibit outgrowth of *Clostridium difficile* spores, *Antimicrob. Agents Chemother.*, 2012, **57**(1), 664–667.
47. S. Löfmark, C. Edlund and C. E. Nord, Metronidazole is still the drug of choice for treatment of anaerobic infections, *Clin. Infect. Dis.*, 2010, **50**, S16–S23.
48. G. G. Zhanel, A. J. Walkty and J. A. Karlowsky, Fidaxomicin: a novel agent for the treatment of *Clostridium difficile* infection, *Can. J. Infect. Dis. Med. Microbiol.*, 2015, **26**(6), 305–312.
49. E. J. Goldstein, F. Babakhani and D. M. Citron, Antimicrobial activities of fidaxomicin, *Clin. Infect. Dis.*, 2012, **55**, S143–S148.
50. F. Babakhani, L. Bouillaut, A. Gomez, P. Sears, L. Nguyen and A. L. Sonenshein, Fidaxomicin inhibits spore production in *Clostridium difficile*, *Clin. Infect. Dis.*, 2012, **55**, S162–S169.
51. Z. Yang, J. Ramsey, T. Hamza, Y. Zhang, S. Li and H. Yfantis, *et al.*, Mechanisms of protection against *Clostridium difficile* infection by the monoclonal antitoxin antibodies actoxumab and bezlotoxumab, *Infect. Immun.*, 2015, **83**(2), 822–831.
52. H. Qiu, R. Cassan, D. Johnstone, X. Han, A. G. Joyee and M. McQuoid, *et al.*, Novel *Clostridium difficile* anti-toxin (TcdA and TcdB) humanized monoclonal antibodies demonstrate *In Vitro* neutralization across a broad spectrum of clinical strains and *In vivo* potency in a hamster spore challenge model, *PLoS One*, 2016, **11**(6), 1–21.
53. R. Leuzzi, R. Adamo and M. Scarselli, Vaccines against *Clostrdium difficile*, *Hum. Vaccines Immunother.*, 2014, **10**(6), 1466–1477.
54. M. A. Monteiro, Z. Ma, L. Bertolo, Y. Jiao, L. Arroyo and D. Hodgins, *et al.*, Carbohydrate-based *Clostridium difficile* vaccines, *Expert Rev. Vaccines*, 2013, **12**(4), 421–431.
55. J. A. Sorg and A. L. Sonenshein, Chenodeoxycholate is an inhibitor of *Clostrdium difficile* spore germination, *J. Bacteriol.*, 2009, **191**(3), 1115–1117.

56. A. Howerton, M. Patra and E. Abel-Santos, A new strategy for the prevention of *Clostridium difficile* infection, *J. Infect. Dis.*, 2013, **207**(1), 1498–1504.
57. D. Heeg, D. A. Burns, S. T. Carman and N. P. Minton, Spores of *Clostridium difficile* clinical isolates display a diverse germination response to bile salts, *PLoS One*, 2012, **7**(2), 1–9.
58. R. Satokari, E. Mattila, V. Kainulainen and P. Arkkila, Simple faecal preparation and efficacy of frozen inoculum in faecal microbiota transplantation for recurrent *Clostridium difficile* infection-an observational cohort study, *Aliment. Pharmacol. Ther.*, 2015, **41**(1), 46–53.
59. J. S. Bakken, T. Borody, L. J. Brandt, J. V. Brill, D. C. Demarco and M. Franzos, *et al.*, Treating *Clostridium difficile* infection with fecal microbiota transplantation, *Clin. Gastroenterol. Hepatol.*, 2011, **9**(12), 1044–1049.
60. I. Youngster, G. H. Russel, C. Pindar, T. Ziv-Baran, J. Sauk and E. L. Hohmann, Oral, capsulized, frozen fecal microbiota transplantation for relapsing *Clostridium difficile* infection, *JAMA*, 2014, **312**(17), 1772–1778.
61. R. U. Varier, E. Biltaji, K. J. Smith, M. S. Roberts, M. K. Jensen, J. LaFleur and R. E. Nelson, Cost-effectiveness analysis of fecal microbiota transplantation for recurrent *Clostridium difficile* infection, *Infect. Control Hosp. Epidemiol.*, 2015, **36**(4), 438–444.
62. K. Sampath, L. C. Levy and T. B. Gardner, Fecal transplantation: beyond the aesthetic, *Gastroenterology*, 2013, **145**(5), 1151–1153.
63. G. Aragon, D. B. Graham, M. Borum and D. B. Doman, Probiotic therapy for irritable bowel syndrome, *Gastroenterol. Hepatol. (N. Y.)*, 2010, **6**(1), 39–44.
64. S. Johnson, P. Maziade, L. V. McFarland, W. Trick, C. Donskey and B. Currie, *et al.*, Is primary prevention of *Clostridium difficile* infection possible with specific probiotics?, *Int. J. Infect. Dis.*, 2013, **16**(11), 786–792.
65. P. B. Fontanarosa, D. Rennie and C. D. DeAngelis, The need for regulation of dietary supplements – lessons from ephedra, *JAMA*, 2003, **289**(12), 1568–1570.
66. P. Muñoz, E. Bouza, M. Cuenca-Estrella, J. M. Eiros, M. Perez and M. Sanchez-Somolinos, *et al.*, *Saccharomyces cerevisiae* fungemia: an emerging infectious disease, *Clin. Infect. Dis.*, 2005, **40**(11), 1625–1634.
67. S. Khanna, D. S. Pardi, S. L. Aronson, P. P. Kammer, R. Orenstein and J. L. St Sauver, *et al.*, The epidemiology of community-acquired *Clostridium difficile* infection: a population-based study, *Am. J. Gastroenterol.*, 2011, **107**(1), 89–95.
68. S. M. Vindigni and C. M. Surawicz, *C. difficile* infection: Changing epidemiology and management paradigms, *Clin. Transl. Gastroenterol.*, 2016, **6**(7), 41–53.
69. A. Gupta and S. Khanna, Community-acquired *Clostridium difficile* infection: an increasing public health treat, *Infect. Drug Resist.*, 2014, **7**, 63–72.
70. C. Ghose, *Clostridium difficile* infection in the twenty-first century, *Emerging Microbes Infect.*, 2013, **2**(9), 198–207.

CHAPTER 2

Targeting Purine Biosynthesis for Antibacterial Drug Design

GYAN MODI[a,b], SHIBIN CHACKO[a] AND LIZBETH HEDSTROM*[a]

[a]Brandeis University, Departments of Biology and Chemistry, MS 009, 415 South St, Waltham, MA 02453, USA; [b]Current address: Indian Institute of Technology, Banaras Hindu University, Department of Pharmaceutics, Room #23, Varanasi-221005, India
*E-mail: hedstrom@brandeis.edu

2.1 Introduction

Intense research on antibiotics was conducted during the 1940s to the 1970s as part of an empirical screening program and most of the currently available antibiotics were discovered during this period. These drugs saved millions of lives from bacterial infections such as tuberculosis, syphilis, pneumonia and rheumatic fever. Unfortunately, many bacterial pathogens have become resistant to these antibiotics and thus present a serious threat to human life. Innovative antibacterial drugs with new modes of action are required to combat the continuing emergence and spread of antibiotic resistant bacteria.

Purine nucleotides play critical roles in many vital cellular processes. They participate in virtually every biochemical pathway, as building blocks for DNA and RNA, energy donors, signaling molecules, precursors for carbohydrate and vitamin synthesis and cofactors. Proliferation is directly dependent on the availability of adenine and guanine nucleotides. Enzymes of the purine

Drug Discovery Series No. 58
Antibiotic Drug Discovery: New Targets and Molecular Entities
Edited by Steven M. Firestine and Troy Lister

Published by the Royal Society of Chemistry, www.rsc.org

nucleotide pathways are clinically validated targets for the development of drugs against cancer and viral infections, but have not yet been exploited for microbial infection. Here, we will comment on the antibacterial potential of purine nucleotide biosynthesis and review the current state of antibiotic discovery targeting these pathways. It is worth noting that promising targets can also be found in the enzymes of the purine pathways not covered in this review: PRPP synthetase, ribonucleotide reductases, nucleotide kinases and purine nucleoside phosphorylases catalyze critically important and potentially vulnerable reactions.

2.1.1 The *De novo* Purine Nucleotide Biosynthesis Pathway

Purine nucleotide biosynthesis proceeds through two pathways: the *de novo* pathway where the synthesis begins with metabolic precursors such as amino acids, ribose 5-phosphate, CO_2 and NH_3, and the salvage pathways that recycle the free purine bases and nucleosides. These pathways contain a smattering of transformations unique to bacteria and a few highly diverged enzymes relative to the host, both of which can be good opportunities for target-based antibiotics. Deoxynucleotide and oligonucleotide biosynthesis are also promising (validated in the case of RNA polymerase) targets for antibiotics, but for the purposes of this review, we will confine our discussion to the synthesis of ribonucleotides and the salvage of purine bases. The reader is referred to an earlier review focusing on IMP dehydrogenase (IMPDH).[1]

In mammals, the *de novo* pathway proceeds through 10 enzymatic conversions, using four ATP molecules, to complete the synthesis of IMP (see Figure 2.1 and Table 2.1 for pathway and abbreviations; the reader is also referred to the excellent reviews by Zhang *et al.*[2] and Kappock *et al.*[3]). These transformations require six enzymes, including the trifunctional GARS-AIRS-GARTfase (steps 2, 3 and 5), the bifunctional CAIRS-SAICARS (steps 6 and 7) and the bifunctional AICARTfase-IMPCH (steps 9 and 10). Bacteria require 11 steps to make IMP and, except for the bifunctional AICARTfase-IMPCH, use monofunctional enzymes. Several enzymes differ substantially between humans and bacteria, providing potential opportunities for drug design. The first enzyme, PPAT, is a Fe_4S_4 protein in humans and Gram positive bacteria, but lacks the Fe_4S_4 cluster in Gram negative bacteria.[3] In virtually all organisms, step 3 is a folate-dependent formylation reaction catalyzed by GARTFase. However, many pathogenic bacteria also produce FGAR *via* an ATP-dependent formylation reaction catalyzed by FGARS. Variations are also observed in the fourth step, which is catalyzed by FGAMS1 in eukaryotes and Gram negative bacteria. Gram positive bacteria utilize FGAMS2, which is a complex of three proteins with the additions of a separate glutaminase and a regulatory protein.[4] The biggest difference between the human and bacterial pathways is in the sixth step, the carboxylation of AIR to CAIR.[5] In mammals, this transformation utilizes CO_2 and is catalyzed by CAIRS. In most, but not all bacteria,[3] the synthesis of CAIR

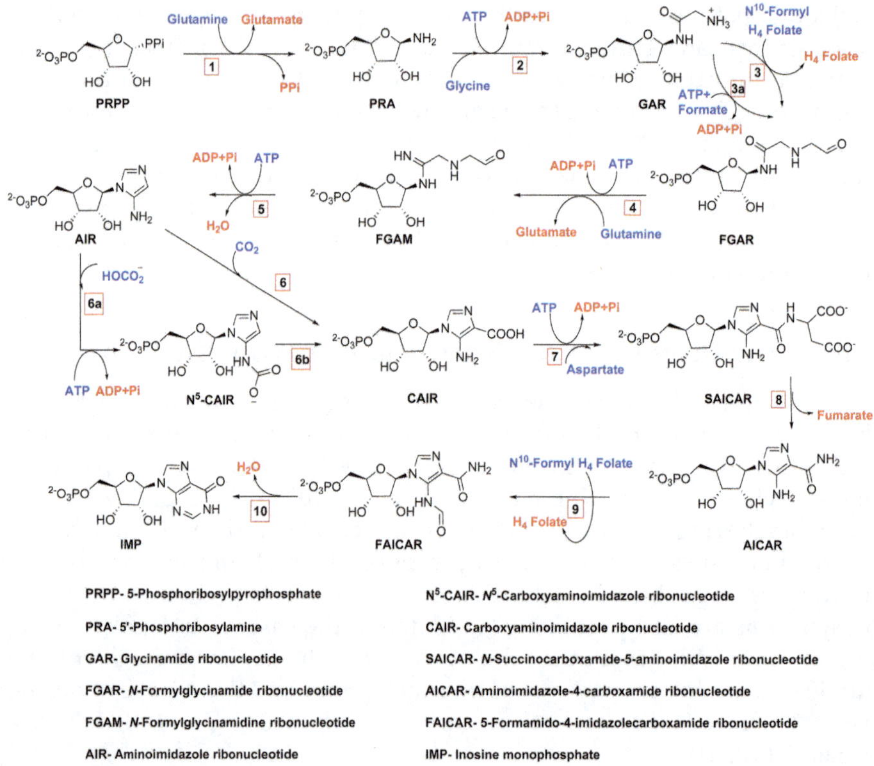

Figure 2.1 *De novo* biosynthesis of IMP.

uses bicarbonate and ATP and is carried out by two enzymes, NCAIRS and NCAIRM. These unique bacterial enzymes are tempting candidates for antibiotic discovery.

2.1.2 The Purine Conversion Pathways

The purine biosynthetic pathways bifurcate after IMP to produce both adenine and guanine nucleotides (Figure 2.2). Interestingly, the synthesis of AMP requires GTP while the synthesis of GMP requires ATP, necessitating that the ATP and GTP pools are balanced. The synthesis of AMP replaces of the 6-carbonyl group of IMP with an amino group in two steps catalyzed by ASS (step 11) and ASL (step 12) (Figure 2.2). Note that ASS also catalyzes step 8 in the *de novo* pathway. Guanine nucleotide biosynthesis also consists of two steps. The rate limiting step in guanine nucleotide biosynthesis is the oxidation of IMP to XMP, which is catalyzed by IMPDH (step 13). XMP is subsequently converted to GMP by the action of GMPS (step 14). Adenine and guanine nucleotides can be recycled to IMP by the deamination of AMP and the reduction of GMP (steps 15 and 16, respectively; Figure 2.3).

Table 2.1 *Purine biosynthesis*. The genes and enzymes of *de novo* purine biosynthesis and selected salvage pathways are listed with abbreviations. The names of compounds can be found in Figure 2.1.

Step	Gene	Enzyme	Abbrev
1	*purF*	Amidophosphoribosyltransferase	PPAT
2	*purD*	GAR synthetase	GARS
3	*purN*	GAR transformylase	GARTfase
3a	*purT*	FGAR synthetase	FGARS
4	*purL*	FGAM synthetase I	FGAMS1
4a	*purL* *purQ* *purS*	FGAM synthetase II	FGAMS2
5	*purM*	AIR synthetase	AIRS
6	*purE2*	AIR carboxylase	CAIRS
6a	*purK*	NCAIR synthetase	NCAIRS
6b	*purE1*	NCAIR mutase	NCAIRM
7	*purC*	SAICAR synthase	SAICARS
8, 11	*purB*	Adenylosuccinate lyase	ASL
9	*purH*	AICAR transformylase	AICARTfase
10	*purJ*	IMP cyclohydrolase	IMPCH
12	*purA*	Adenylosuccinate synthetase	ASS
13	*guaB*	IMP dehydrogenase	IMPDH
14	*guaA*	GMP synthetase	GMPS
15	*apt*	Adenine phosphoribosyltransferase	APRT
16	*hpt*	Hypoxanthine-guanine phosphoribosyltransferase	HGPRT
17	*gpt, xpt*	Xanthine-guanine phosphoribosyltransferase	XGPRT

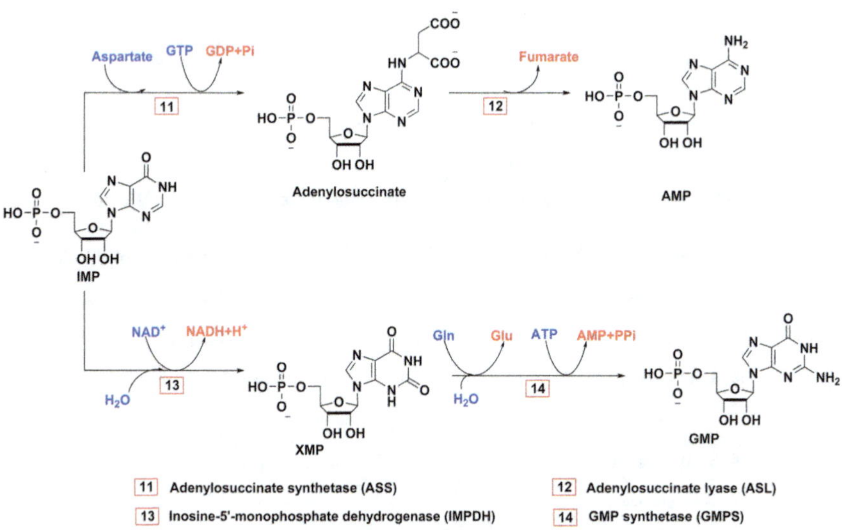

Figure 2.2 Biosynthesis of GMP and AMP from IMP.

2.1.3 The Purine Salvage Pathways

In addition to the *de novo* synthesis of purine nucleotides, most organisms are capable of salvaging purine bases and nucleosides from their surroundings (Figure 2.3). While purine nucleoside kinases are known (steps 17 and 18), the salvage of purine bases *via* phosphoribosyltransferase reactions is the more important reaction for most organisms (steps 19, 20 and 21). In fact, nucleosides are often metabolized to free bases by hydrolases and phosphorylases (steps 22 and 22a), which are then salvaged by phosphoribosyltransferase. Phosphoribosyltransferases are typically adenine-specific (APRTs), hypoxanthine-guanine-specific (HGPRTs) or xanthine-guanine-specific (XGPRTs), though some hypoxanthine-guanine-xanthine phosphoribosyltransferases (HGXPRTs) have also been reported. Limited pathways exist for nucleoside and base interconversion. Both adenosine and adenine can be deaminated to inosine and hypoxanthine, respectively (steps 23 and 24), which can then be converted into IMP and enter both adenine and guanine nucleotide pools. However, while guanine can be deaminated to xanthine (step 25), xanthine cannot be reduced to hypoxanthine. Guanosine deaminases are found in plants, but as yet have not been reliably identified

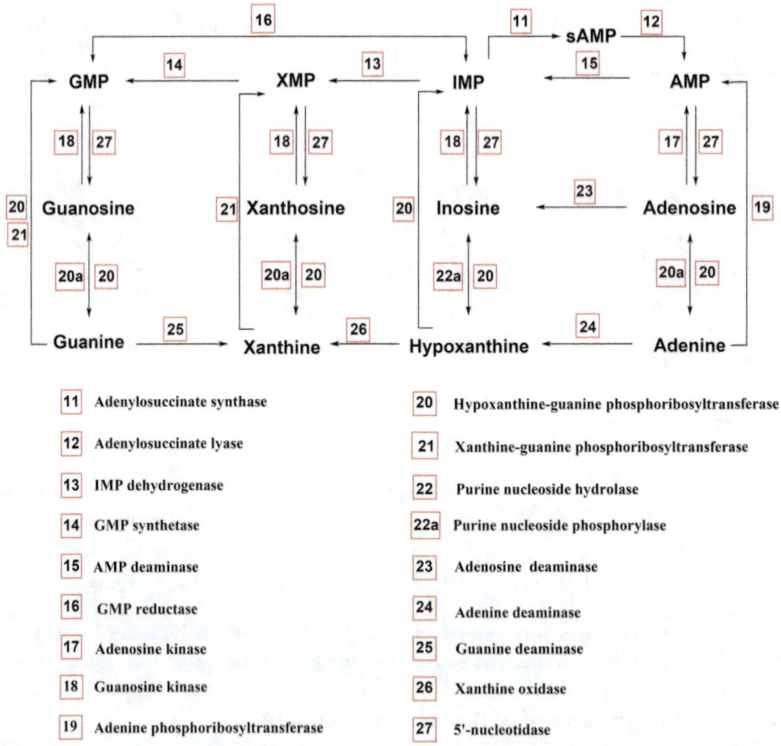

Figure 2.3 Purine salvage pathways.

outside that kingdom.[6] Note that hypoxanthine salvage requires the action of the downstream purine conversion enzymes to produce adenine and guanine nucleotides. Thus, ASS, ASL, IMPDH and GMPS must also be considered when evaluating salvage pathways. Purine base and nucleoside transporters are also critically important for salvage pathways, and these proteins are very difficult to identify by sequence. Whether salvage pathways can support infection is the major unanswered question in the development of antibiotics targeting purine biosynthesis.

2.1.4 Availability of Purine Bases and Nucleosides

The relative importance of *de novo* synthesis *versus* salvage pathways depends on nutrient availability within a site of infection as well as the enzyme repertoire of the pathogen. Except for guanine, purine bases and nucleosides are reasonably abundant in blood (Table 2.2). Guanine is undetectable, in keeping with its low solubility at neutral pH. Hypoxanthine, inosine and xanthine are the most abundant purines in cerebrospinal fluid. Little is known about the purine content in other sites of infection, though they are generally assumed to be like that of plasma. The purine pathways might appear to be poor candidates for antibiotic development given the availability of both adenosine and guanosine. However, the proliferation of mammalian cells requires *de novo* purine biosynthesis, suggesting that these concentrations cannot support the growth of even comparatively slow growing organisms.

2.1.5 The Complex Interplay Between Salvage Pathways and Precursor Availability

Importantly, the availability of purines and presence of salvage enzymes does not necessarily mean that the *de novo* pathways can be bypassed. The difficulty in predicting the vulnerability of an organism to inhibitors of

Table 2.2 Concentrations of purines in human blood and cerebrospinal fluid (CF). Values are the range of determinations reported in the Human Metabolome Database (accessed July 2016).[81] n.d. = not detected; n.a. = not available.

Purine	Blood (μM)	CF (μM)
Adenosine	0.04–5	0.01–0.1
Guanosine	0.8	n.a.
Inosine	0.05–5	0.6
Xanthosine	5	n.a.
Adenine	0.3–0.6	0–0.2
Guanine	n.d.	n.d.
Hypoxanthine	0.4–34	2–6
Xanthine	0.4–5	5–13

purine nucleotide biosynthesis can be illustrated with the protozoan parasite *Tritrichomonas foetus*.[7] This parasite can salvage xanthine, yet is sensitive to IMPDH inhibitors when cultured on rich media containing high concentrations of adenine, hypoxanthine and xanthine. *T. foetus* has an HGXPRT that prefers hypoxanthine, which out-competes xanthine, thus preventing xanthine salvage. Mutations that block hypoxanthine uptake allow xanthine salvage to proceed and render the parasite resistant to IMPDH inhibitors. Unfortunately, specificity of transporters is difficult to predict from sequence, adding another degree of difficulty to the identification of vulnerable biochemical pathways.

2.2 The Essentiality of Enzymes of the Purine Biosynthetic Pathways

The development of antibiotics targeting the purine nucleotide pathways is a risky proposition due to the redundancy of the *de novo* and salvage pathways together with the presence of purine bases and nucleosides. Plenty of evidence, both pro and con, can be found in the literature as microbiologists have labored to construct vaccine strains and identify drug targets. A massive effort is currently underway to catalog essential genes in bacterial pathogens on a genome wide scale. The first investigations used comparatively cumbersome methods to create libraries with individual gene knockouts. The advent of random mutagenesis methods has allowed identification of genes required for fitness on a genome-wide scale (TnSeq or TraDis).[8,9] These methods exploit mobile genetic elements called transposons that can insert randomly into DNA, usually inactivating the gene at the site of insertion. The *mariner* systems are especially efficient, so that most genes contain multiple transposon insertions, decreasing the chance that insertion in a permissive position will result in misidentification of nonessential genes. These transposons insert at TA sequences, and a robust measure of essentiality and fitness can be derived by comparing the frequency of insertion to the frequency of TA sites.

2.2.1 Essentiality of Purine Nucleotide Biosynthesis for Growth in Rich Media

Initial investigations focused on growth in nutrient rich media that usually contain millimolar concentrations of purines. Not surprisingly, the enzymes of the *de novo* purine biosynthesis pathway are seldom essential under these conditions (Table 2.3). Inspection of the Database of Essential Genes reveals that the *de novo* purine biosynthesis genes are found to be essential with a frequency of 25–30% under these growth conditions.[10] Only the genes encoding FGAMS2 (PurQ subunit only), HGPRT, ASL, IMPDH and GMPS are essential with a frequency of ≥50%. These observations highlight the importance of the enzymes that convert IMP to adenine and guanine nucleotides in both *de novo* and salvage pathways.

Table 2.3 Enzymes essential for growth in minimal media. From the Database of Essential genes (accessed June 2016;[10]). Note: *purH* and *purJ* are fused; only *purH* is listed. E, essential; NE, nonessential.

Gene	Enzyme	#E	#NE	%
purF	PPAT	3	16	16
purD	GARS	4	13	24
purN	GARTfase	4	17	19
purT	FGARS	0	17	0
purL	FGAMS	3	12	20
purQ	FGAMS2	3	2	60
purS	FGAMS2	2	4	33
purM	AIRS	3	14	18
purE	NCAIRM	3	12	20
purK	NCAIRS	3	11	21
purC	SAICARS	4	11	27
purB	ASL	10	10	50
purH	AICARTF	10	16	38
purA	ASS	5	13	28
guaB	IMPDH	9	10	47
guaA	GMPS	17	12	59
guaC	GMPR	1	5	17
apt	APRT	8	15	35
hpt	HGPRT	4	4	50
xpt	XGPRT	0	5	0

2.2.2 Essentiality of Purine Nucleotide Biosynthesis During Infection

These experiments are beginning to interrogate essentiality and fitness under more clinically relevant growth conditions and even during *in vivo* infection.[11–17] Frustratingly, genes of the purine pathways are often unscored (Table 2.4), perhaps because they are required in the construction of the mutant libraries, but more likely because their loss causes a growth deficit that leaves them under-represented in the library. More problematic is the varying stringency in the assignment of essentiality among different laboratories. Nonetheless, some generalities are beginning to emerge as the methods are refined and libraries with deeper coverage are interrogated. The *de novo* purine biosynthetic pathway is required for lung infection by three pathogens, *Haemophilus influenza*, *Klebsiella pneumoniae* and *Streptococcus pneumoniae*. Reassuringly, at least seven *de novo* purine nucleotide biosynthesis genes are scored in all three pathogens, and all are required with the exception of *purT*, which encodes the redundant FGAR synthetase. These observations provide strong evidence for the importance of *de novo* purine biosynthesis in these organisms, and possibly also in this niche. Interestingly, *Str. pneumoniae* infections of the nasopharynx do not require *de novo* purine nucleotide biosynthesis, demonstrating the context dependence of essentiality. Although *de novo* purine biosynthesis appears to be required for the growth of *Escherichia coli*, *Bacillus anthracis* and *Salmonella typhimurium* in blood,[18] *Staphylococcus aureus* does not require *de novo* purine nucleotide biosynthesis in either septicemia or abscess infections.[12]

Table 2.4 Essentiality of *de novo* purine nucleotide biosynthesis and purine salvage genes under clinically relevant growth conditions and during *in vivo* infection. Infections are in mice. No information was available for *purQ*, so it was omitted from the table (note that *purJ* is fused to purH, only *purH* is listed). Y, identified as essential or required for fitness by the criteria in each reference; N, identified as nonessential; blank, no information available, possibly indicating that these genes are essential under the conditions used to make the library.

Bacteria	*purF*	*purD*	*purN*	*purT*	*purL*	*purS*	*purM*	*purK*	*purE*	*purC*	*purB*	*purH*	*purA*	*guaB*	*guaA*	*apt*	*hpt*	*xpt*	Ref.	Comments
A. baumannii				N							N		N		N		N	N	14	Lung
B. anthracis								Y	Y										18	Septicemia
	N[a]												Y	N[a]					21	Peritoneal
E. coli	N	N	N	N	N		N	Y	Y	Y		N	N	N	N				18	Human serum
H. influenza	Y	Y			Y		Y		Y	Y[b]	Y	Y	Y	Y					13	Lung; gene
K. pneumoniae	Y	Y	Y	N	Y			Y				Y				N	Y		16	Lung
P. aeruginosa	N		N		N		N			N	Y		Y	Y	Y				17	Cystic fibrosis patient sputum
Sta. aureus	N		N		N	N	N		N	N	N			Y	Y		Y	N	12	Septicemia
	N		N		N	N	N		N	N	Y			Y	Y		Y	N		Abscess
Str. Group A		Y											Y	Y					15	Human blood
Str. pneumoniae	Y		Y		Y		Y	Y	Y		Y	Y	Y	Y			Y		11	Lung
	N		N		Y		N	N	N		Y	Y	N	Y			N			Nasopharynx

[a]unspecified mutation.
[b]annotated as *hemH*, but correct enzyme name and pathway.

Similarly, *de novo* purine nucleotide biosynthesis also does not appear to be important for growth of *Pseudomonas aeruginosa* in sputum from a cystic fibrosis patient. The essentiality of *de novo* purine nucleotide biosynthesis is less certain in other pathogens like *Acinetobacter baumanni* and Group A *Streptococcus*, where only a couple genes are scored (Table 2.4). These observations suggest that drugs targeting the *de novo* purine nucleotide pathways may have a limited spectrum of action both in terms of pathogens and sites of infections.

The pathways converting IMP to adenine and guanine nucleotides are better candidates for antibiotic development than *de novo* purine nucleotide biosynthesis (Table 2.4). The genes encoding the guanine nucleotide biosynthetic enzymes are essential in many clinical contexts. Indeed, *guaB* is required in all seven of the scored examples in Table 2.4. At least three of these also require GMPS, providing further confidence that the synthesis of GMP is the critical function. The enzymes of adenine nucleotide biosynthesis are also required in several pathogens, though their importance is more sensitive to the environment. *Sta. aureus* requires *purB* for abscess infections, but not for infections in blood, while *Str. pneumoniae* does not appear to need *purA* in nasopharynx infections. Surprisingly, the salvage pathways appear to be essential in some contexts. HPRT is required for *K. pneumoniae*, *Sta. aureus* and *Str. pneumoniae* lung infections. XPRT does not appear to be essential in any pathogen, and APRT has not been scored enough times to make a conclusion. The above observations further substantiate the earlier conclusions of McFarland and Stocker, that disruption of the purine conversion pathways, particularly the adenine conversion pathways, are more likely to render a pathogen avirulent than the disruption of the *de novo* biosynthesis pathway.[19]

Single gene knockouts provide a more rigorous test of virulence, and here again the results are mixed. Knockouts in both the *de novo* and conversion pathways in *K. pneumoniae* are avirulent in lung and peritoneal cavity infections,[20] confirming the findings of genomic studies. However, knockouts in the *de novo* and guanine nucleotide conversion pathways have little effect on virulence of *Salmonella dublin* and *Sa. typhimurium* in peritoneal infections, while knockout of the adenine nucleotide pathway results in complete loss of virulence.[19] Perhaps the most puzzlingly data come from *B. anthracis*. In peritoneal infections, only mutations in the adenine conversion pathways decrease virulence.[18] The guanine conversion pathway is required for growth, but not virulence, while mutations in the *de novo* pathway have no effects. In contrast, Δ*purE* decreases virulence in *B. anthracis* septecimia by 10^3-fold, though Δ*purK* has only a modest 20-fold effect.[21] Analysis of the *de novo* pathway defects in *Yersinia pestis* has produced similarly confusing results, but mutations in guanine conversion produced avirulent *Y. pestis* strains.[22–24] Mutations in *purF*, *purL*, *guaB* but not *purA*, attenuate *Francisella tularensis*[25–28] while the losses of *purD* and *purF* attenuate *Brucella abortus*.[29] The guanine conversion pathways have also been validated as potential targets in *Borrelia burgdorferi*,[30,31] *Shigella flexneri*,[32] and Group B *Streptococcus*.[33] The gene for *guaB2* is essential in *M. tuberculosis*,[34] and conditional knockouts are bacteriocidal in macrophages as well as to establish an infection in mice (V. Mizrahi, personal communication). On balance, these observations further suggest inhibitors of the purine conversion pathways may have the best chance to be developed into moderate-to-broad spectrum antibiotics.

2.2.3 Inhibitors May Not Have the Same Phenotypes as Gene Knockouts

Several caveats must be recognized. First, these experiments measure competitive fitness, not virulence *per se*. It is not hard to imagine that an organism that can both salvage hypoxanthine and make IMP *de novo* will have a growth advantage over one that cannot salvage, but this difference may not translate into a defect in virulence. It is also possible that a gene required to establish infection might not be needed for maintenance. Conditional gene knockouts that can ablate a gene after infections are established are needed to more rigorously validate these targets. Last, but not least, small molecule inhibition is not equivalent to a gene knockout. A small molecule may not be able to achieve the 100% inhibition that complete ablation of the target protein provides. In addition, many proteins have multiple functions, and while a small molecule typically affects just enzymatic activity, a gene knockout will disrupt all functions. To take an example from the pyrimidine nucleotide pathways, in addition to its enzymatic activity, CTP synthase forms cytoskeletal filaments that regulate bacterial cell shape independent of its catalytic function.[35] Gene disruption would eliminate both enzymatic activity and CTPS filaments, while a small molecule inhibitor is likely to only affect enzymatic activity. Similar functions could well be present in the enzymes of the purine nucleotide pathways—indeed, IMPDH is known to form filaments in mammalian cells,[36] and bacterial IMPDHs form filaments *in vitro*.[46] Alternatively, a small molecule could also be more effective than a gene knockout by triggering downstream effects that would not occur in the absence of the target protein. These considerations illustrate the importance of validating targets *in vivo* early in a drug discovery program.

2.2.4 The Problem of Resistance

Single target inhibitors are susceptible to the emergence of resistance. This problem is usually addressed with cocktails of several drugs so that a pathogen must acquire multiple mutations to become resistant to treatment. One concern with inhibitors of the nucleotide pathways is their potential to distort the balance of the nucleotide pools and cause mutations, threatening the useful lifetime of partner drugs. Indeed, deletions of ASS and IMPDH increase the incorporation of inosine into DNA.[37] However, all antibiotics, indeed all forms of stress, induce mutations in bacteria.[38,39] Whether this risk is actually greater for inhibitors of the purine nucleotide biosynthesis pathways will need to be carefully assessed.

2.3 Progress Targeting the Enzymes of *De novo* Purine Nucleotide Biosynthesis

The enzymes of the *de novo* purine salvage pathways have been largely ignored as potential antibiotic targets. The likelihood that salvage pathways would bypass inhibition of *de novo* synthesis greatly discouraged research in

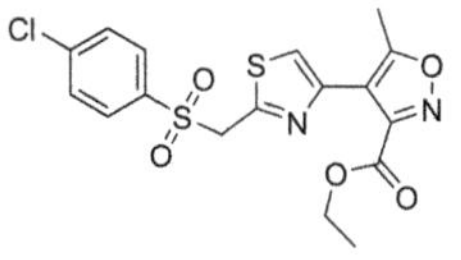

Figure 2.4 Representative NCAIRS inhibitor.[41]

this area. Moreover, the assays are generally cumbersome and not amenable to high throughput screens. This problem is further exacerbated by unstable intermediates such as PRA ($t_{1/2} < 5$ s) and N^5-CAIR ($t_{1/2} \sim 15$ s).[3] Indeed, few of the mammalian enzymes have been characterized in detail despite their critical role in proliferation and cancer. The active sites are conserved, increasing the challenge of developing selective inhibitors for the bacterial enzymes that will not affect the host orthologs. The possibility that recombinant enzymes may not accurately reflect the behavior of *in vivo* targets is another concern, especially for the *de novo* purine nucleotide biosynthesis enzymes that may function in large complexes.[40] The one bright spot is that X-ray crystal structures are available for at least one representative of every step in the *de novo* synthesis pathway,[2] opening the door for structure-based drug design. Nonetheless, there are substantial hurdles to overcome for a successful drug discovery program targeting most of the *de novo* purine nucleotide biosynthetic enzymes.

The two unique enzymes in microbial purine biosynthesis NCAIRS and NCAIRM are the best candidates in the *de novo* pathway, and some progress has been made with these targets. Selective inhibitors of NCAIRS were discovered in a high throughput screen, although no antibacterial activity was reported (Figure 2.4).[41] Inhibitors of NCAIRM have also been disclosed. A thermal shift screen identified several inhibitors with antibacterial activity against *B. anthracis*, although the antibacterial activity appears to be due to the inhibition of another target. More recently, fragments have been reported that bind to NCAIRM with affinities ranging between 14 and 700 μM, and ligand efficiencies greater than 0.3.[42] The recent development of a new assay makes a conventional high throughput screen feasible.[43] These are promising first reports, but clearly more effort is needed to develop inhibitors if these targets are to be validated.

2.4 Progress Targeting Enzymes in the Purine Nucleotide Conversion Pathways

Although the enzymes of the purine conversion pathways are more promising targets for antibiotic discovery, they have also been largely ignored, except for IMPDH. The dual function of ASS makes it a particularly appealing potential target, but few inhibitors, and no specific inhibitors of bacterial enzymes, have been reported. A similar situation exists for ASL and GMPS. As above, the assays are relatively difficult and the similarity of the active sites of these enzymes may be discouraging. Indeed, Fyfe *et al.* concluded that the active site of *Sta. aureus* ASL was too similar for structure-based inhibitor design.[44] However, this enzyme forms a different oligomer than the human ortholog, presenting an unsuspected opportunity for selective inhibitors that could only have been revealed after purification and characterization. Better assays and more characterization are required to assess the feasibility of developing selective inhibitors of these enzymes.

2.4.1 IMPDH Structure and Mechanism

IMPDH is by far the most advanced antibiotic target in the purine pathways by its importance in other human diseases, ease of assay, structural characterization and a quirk of evolution. The human IMPDHs are well-characterized, clinically validated drug targets for immunosuppressive, anticancer, and antiviral therapy.[45] As yet, drugs do not distinguish between the two human isozymes, hIMPDH1 and hIMPDH2, which are 84% identical.[46] hIMPDH2 appears to be the critical isozyme in rapidly proliferating cells, and is the best characterized, with several X-ray crystal structures,[47–49] and detailed kinetic and mechanistic investigations.[45] IMPDH has also received attention as a potential target for parasite infections, and selective inhibitors of the protozoan parasite *Cryptosporidium* IMPDH have been reported (detailed below). Many of these compounds display antiparasitic activity *in vitro*, and at least two are active in a mouse model of acute infection.[50] This work provides a good foundation for drug discovery targeting bacterial IMPDHs.

IMPDH is usually a square planar homotetramer (Figure 2.5), though higher order oligomers have also been reported, including filaments, rods

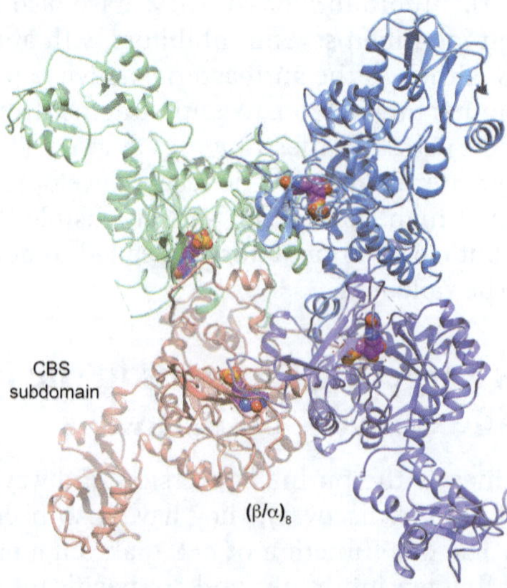

Figure 2.5 Tetramer of IMPDH. Colors denote monomers. IMP is shown in space fill. A tetramer of IMPDH from *Streptococcus pyogenes* (pdb accession number 1zfj).[82] Figure rendered with UCSF Chimera.[83]

and rings.[36,51] The physiological role of these higher order structures is not understood. Most monomers have two domains: the catalytic domain, which is a $(\beta/\alpha)_8$ structure, and the subdomain containing two cystathionine beta synthase domains (CBS).[52] The CBS domains bind nucleotides and probably regulate filament formation. The nucleotide preference appears to vary in IMPDHs from different organisms, so the role of these interactions is not currently understood. Curiously, deletion of the subdomain in *E. coli* IMPDH makes the bacteria sensitive to high concentrations of adenosine.[53,54] The adenine nucleotide pools swell out of proportion to the guanine nucleotide pools, indicating subdomain maintains nucleotide pool balance by an unknown mechanism. IMPDH has several moonlighting functions in eukaryotic cells involving translation, transcription, lipid vesicles and protein kinase B that are also not understood.[45,55–57] How these moonlighting functions will be affected by an inhibitor of enzymatic activity is an open question.

The mechanism of IMPDH further complicates inhibitor design. IMPDH catalyzes conversion of IMP to XMP with concomitant reduction of NAD^+ to NADH.[45] The reaction requires two sequential chemical transformations: (1) a conserved Cys acts as nucleophile to attack the activated C2 position of IMP to form NADH and the covalent intermediate E-XMP* and (2) the hydrolysis of E-XMP* to liberate XMP (Figure 2.6).[58] To carry out these two steps, the enzyme adopts two different conformations. The "open" conformation accommodates both substrate and cofactor (NAD^+) for the redox reaction whereas, in the "closed" conformation, a mobile flap binds in the cofactor site, carrying a conserved Arg that acts as the general base for the hydrolysis of E-XMP*. The kinetic mechanism favors IMP binding first, which means that inhibitors that occupy in the cofactor binding site can bind to the E•IMP, E-XMP* or E•XMP complexes. Thus, these inhibitors will display competitive, uncompetitive or noncompetitive mechanisms depending on their relative affinities for each complex. The hydrolysis step is rate-determining for most IMPDHs, which makes E-XMP* the most abundant, and therefore most vulnerable, enzyme complex. Unfortunately, it is the least easily accessed in X-ray crystallography experiments, though it has been captured in few instances.[59–61]

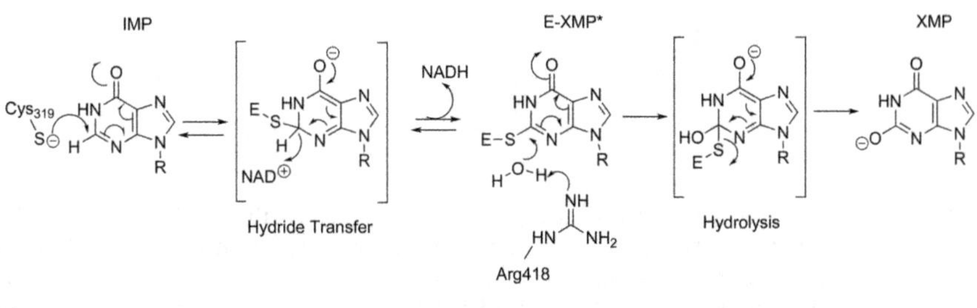

Figure 2.6 Mechanism of the IMPDH reaction. *Tritrichomonas foetus* IMPDH numbering is shown.

2.4.2 Prokaryotic and Eukaryotic IMPDHs: A Comparative Analysis of Rational Drug Design

Prokaryotic and eukaryotic IMPDHs have significant mechanistic and structural differences that indicated it would be possible to design selective agents. In general, the values of K_m and k_{cat} are higher for bacterial IMPDHs. The well-known IMPDH inhibitor mycophenolic acid is a selective inhibitor of eukaryotic IMPDHs, demonstrating that selectivity was possible, and many selective inhibitors of prokaryotic IMPDHs have been reported (see below). Structural studies showed that the IMP site and the nicotinamide subsite (N-site) of the cofactor site are strongly conserved, as expected since those sites are where the chemical transformations occur. In contrast, the adenosine subsite (A-site) of the cofactor site is highly diverged in eukaryotic and prokaryotic IMPDHs).[60,62] Recent structures of IMPDHs from *Vibrio cholera* (*Vc*IMPDH) and *M. tuberculosis* (*Mtb*IMPDH2) show that the adenosine subsite has migrated relative to its position in human IMPDH (Figure 2.7(a)). Moreover, the cofactor is found

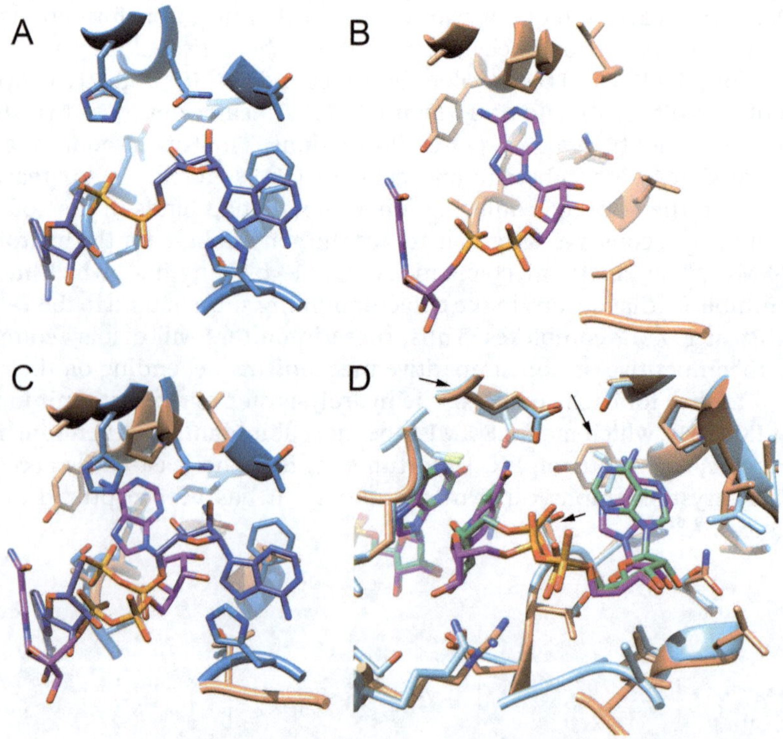

Figure 2.7 The novel cofactor binding sites of bacterial IMPDHs. The residues that interact with the adenine rings are shown. (a) Human IMPDH2 (1nf8) steel blue, NAD slate blue; (b) *M. tuberculosis* IMPDH2 (4zqm) coral, NAD magenta; (c) Overlay of (a) and (b). (d) *M. tuberculosis* IMPDH2 (4zqm) coral, NAD magenta and *V. cholerae* IMPDH (4qne) steel blue, NAD green. The arrows mark the position of the key residues that interact with *Cp*IMPDH inhibitors. Figure rendered with UCSF Chimera.[83]

in an unusually compact conformation that is observed in few other dehydrogenases (Figure 2.7(b)). This unprecedented difference in the cofactor binding sites explains inhibitor selectivity and provides an attractive opportunity for the further design of selective inhibitors. Note however that two classes of bacterial A-sites have been observed, exemplified by the *M. tuberculosis* and *V. cholerae* enzymes (Figure 2.7(c)). This difference will likely limit the potential spectrum of prokaryotic IMPDH inhibitors.

IMPDH-targeted antibiotic discovery developed from anti-parasitic drug discovery. *Cryptosporidium* spp. are obligatory enteric intracellular parasites, and cryptosporidiosis is major cause of severe diarrhea and malnutrition in children.[63] Unlike most bacteria, *Cryptosporidium* has a streamlined purine salvage pathway that relies on *Cp*IMPDH to produce guanine nucleotides, making this enzyme a promising target. Moreover, the gene for *Cp*IMPDH was obtained from bacteria, creating the possibility that *Cp*IMPDH inhibitors might also have antibiotic activity. Selective *Cp*IMPDH inhibitors were discovered by high throughput screening, and further optimized, so that nanomolar inhibitors are available in six different frameworks (Figure 2.8).[64–70] At least two *Cp*IMPDH inhibitors display antiparasitic activity in a mouse model of acute cryptosporidiosis, validating the enzyme as a target.[50]

Crystal structures are available for examples of each *Cp*IMPDH inhibitor series, defining the structural determinants of selective inhibition.[69–73] All of the *Cp*IMPDH inhibitors interact with the purine base of IMP, bend around Ala253, form a hydrogen bond or dipole interaction with Glu416 and have pi interactions with Tyr445' in the adjacent subunit (*Ba*IMPDH numbering; Figure 2.9(a)). Glu416 is conserved among all bacterial IMPDHs, but Ala253 and Tyr445' are variable (Figure 2.7(d)). Importantly, Ala253 and Tyr445' are found in IMPDHs from a wide variety of pathogenic bacteria, including *M. tuberculosis*, *B. anthracis*, *Campylobacter jejuni*, *Clostridium perfringes*, *Sta. aureus*, *Str. pyogenes*, *Francisella tularensis* and *P. aeruginosa*, to name but a few.[74] Encouragingly, the inhibitor binding site is highly conserved within the set of enzymes containing Ala253 and Tyr445 (Figure 2.9(b)). Importantly, X-ray crystal structures are available for *Cp*IMPDH inhibitors bound to IMPDHs from *B. anthracis*, *M. tuberculosis* and *C. jejuni* and *Cl. perfringes*.[60,62]

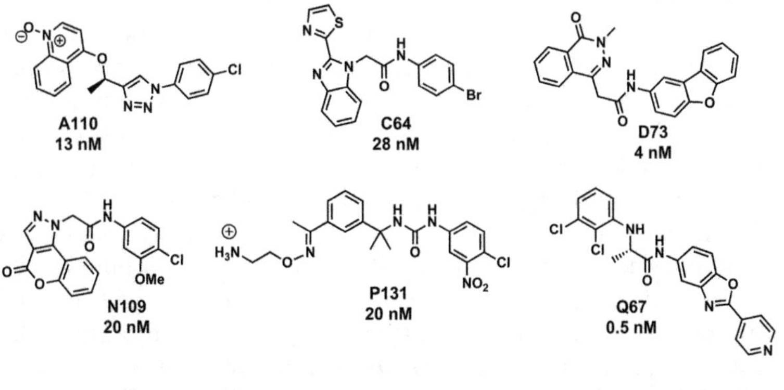

Figure 2.8 Representative *Cp*IMPDH inhibitors from each compound series.

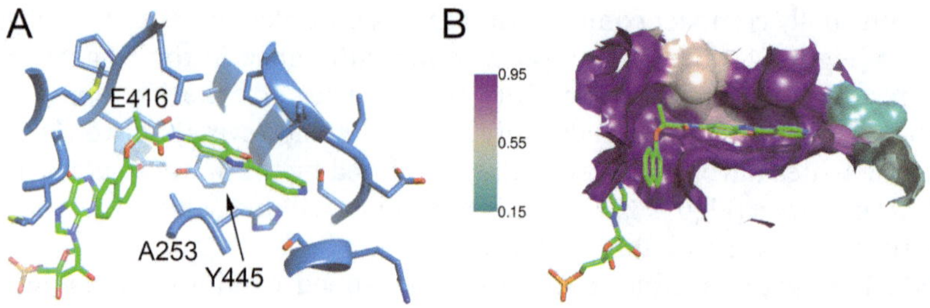

Figure 2.9 The *Cp*IMPDH inhibitor binding site. (a) Interactions of inhibitor Q21 with *B. anthracis* IMPDH (4my8). (b) Conservation of the inhibitor binding site. The surface of the Q21 binding site is shown, colored by conservation among IMPDHs from *C. parvum* and 19 bacterial pathogens: *Acinetobacter baumannii*, *Arcobacter butzleri*, *B. anthracis*, *Borrelia burgdorferi*, *Burkholderia cenocepacia*, *B. pseudomallei*, *Campylobacter jejuni*, *Clostridium perfringes*, *C. parvum*, *Coxiella burnetii*, *Francisella tularensis*, *Helicobacter pylori*, *Listeria monocytogenes*, *Mycobacterium tuberculosis*, *Neisseria gonorrhoeae*, *N. menigitidis*, *Pseudomonas aeruginosa*, *Staphylococcus aureus*, and *Streptococcus pneumoniae* and *Streptococcus pyogenes*. Figure rendered with UCSF Chimera.[83]

2.4.3 Repurposing *Cryptosporidium* IMPDH Inhibitors as Antibiotics

The ability to "piggy-back" bacterial IMPDH inhibitor discovery on the parasite IMPDH program is a great advantage. A structure activity relationship (SAR) study was carried out for repurposing *Cp*IMPDH inhibitors as antibacterial agents.[75] The authors selected 140 compounds from five structurally diverse compound series (A, C, D, P and Q) to determine IC_{50} against *B. anthracis* IMPDH (*Ba*IMPDH) and antibacterial activity (minimal inhibitory concentration, MIC). Enzyme inhibition correlated with that observed for *Cp*IMPDH, although potency was usually weaker. Nonetheless, seven compounds displayed K_{iapp} values less than or equal to 10 nM, and 12 compounds are more potent inhibitors of *Ba*IMPDH than *Cp*IMPDH. However, only 16 compounds displayed antibiotic activity with MIC $\leq$ 12 μM when bacteria were cultured on minimal media. These compounds also displayed activity against *Sta. aureus*. The MICs increased in the presence of guanine, as expected if antibacterial activity resulted from the inhibition of *Ba*IMPDH (Table 2.5, Figure 2.10). These observations suggest that most of the *Cp*IMPDH inhibitors do not enter the bacteria, which is not surprising given that they were designed to cross lipid membranes, not bacterial cell walls. *Cp*IMPDH inhibitors also display activity against *M. tuberculosis*, and again antibacterial activity is decreased by the presence of guanine indicating that the antibacterial activity results from inhibition of IMPDH (Table 2.5).[62] Interestingly, there is very little overlap in the compounds that display antibacterial activity against *M. tuberculosis* with those that are active against *B. anthracis*/*Sta. aureus*, perhaps also highlighting differences in the

Table 2.5 Antibacterial activity of selected *Cp*IMPDH inhibitors. TPSA, topological polar surface area; *Cp, Cryptosporidium parvum*; *Ba*, *Bacillus anthracis*; *Mtb*, *Mycobacterium tuberculosis*.

Cmpd	cLogP	TPSA (Å^2)	$K_{i,app}$ (nM)			MIC (μM)			
						Ba[a]		*Mtb*[b]	
			Cp[a]	*Ba*[a]	*Mtb*[b]	− Gua	+ Gua	− Gua	+ Gua
A98	4.1	49.9	9 ± 1	15 ± 4	n.d.	1	8	>20	n.d.
P67	3.34	94.0	4.2 ± 0.8	40 ± 20	13 ± 5	12	n.d.	2.9 ± 0.4	≥40
P150	4.42	94.0	25 ± 10	40 ± 20	35 ± 3	0.5	>30	14 ± 5	>50
Q33	3.68	81.5	28 ± 2	150 ± 50	37 ± 9	>50	n.d.	5.3 ± 0.9	24 ± 16
Q46	4.01	124.1	2.3 ± 0.9	130 ± 40	12 ± 5	>50	n.d.	12 ± 3	38 ± 12
Q67	5.02	75.1	0.5 ± 0.1	14 ± 3	5 ± 1	12	n.d.	7 ± 3	50
Q77	4.21	72.3	13	130 ± 20	100 ± 20	n.d.	n.d.	6.3 ± 3.0	37

[a]Data from ref. 75.
[b]Data from ref. 62; n.d. = not determined.

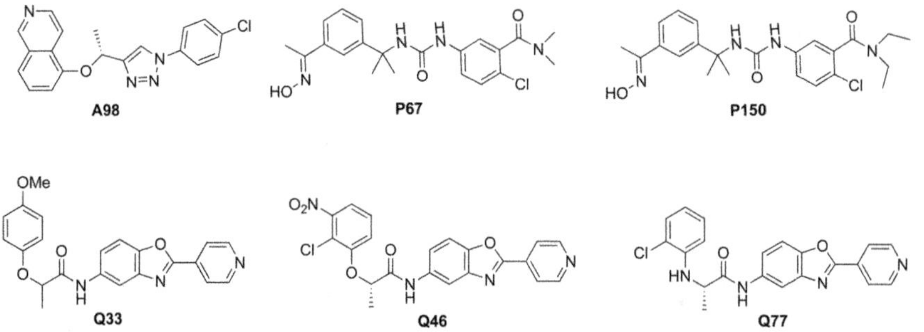

Figure 2.10 Structures of selected *Cp*IMPDH inhibitors with antibacterial activity.

SAR of uptake. Redesign of the compounds to increase polarity and reduce size should improve uptake.[76]

IMPDH-targeted antibiotics appear to be a promising strategy for the treatment of tuberculosis. As noted earlier, *Mtb*IMPDH2 is an essential enzyme.[34] Moreover, *M. tuberculosis* cannot salvage xanthine.[77] Though its genome encodes both a purine nucleoside phosphorylase and a purine nucleoside hydrolase,[77] it also cannot salvage guanosine (V. Mizrahi, personal communication). Selective inhibitors of *Mtb*IMPDH2 have also been discovered in phenotypic screens (H. Boshoff and V. Mizrahi, personal communications), further justifying interest in this target. A focused screen of diphenyl urea (DPU) compounds for inhibitors of *Mtb*IMPDH2 identified three compounds that also inhibited the growth of *M. tuberculosis* with MICs in the range of 0.2 to 0.4 μg mL^{-1} (Figure 2.11).[78,79] The MIC increased by 16-fold when *M. tuberculosis guaB2* was over-expressed in *M. smegmatis*, suggesting that the antibacterial activity resulted from the inhibition of *Mtb*IMPDH2. Most excitingly, one compound reduced bacterial load in the spleen in a mouse infection, but failed to reduce the bacterial load in the lungs.

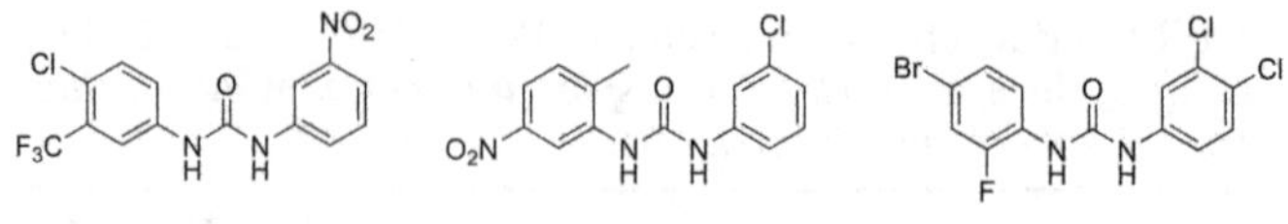

Figure 2.11 Structures of selective MtbIMPDH inhibitors.[78]

The further optimization of inhibitors will be greatly assisted by the availability of structures of E•I complexes for IMPDHs from *B. anthracis*, *M. tuberculosis*, *Ca. jejuni*, *Cl. perfringes* as well as *Cryptosporidium*.[62,73] Structures are also available for the *P. aeruginosa* and *Vibrio cholerae* enzymes, though the latter is resistant to *Cp*IMPDH inhibitors.[51,60,80] Curiously, the potency of inhibitors varies by as much as 50-fold despite the close structural similarity of these complexes.[60] Perhaps regions that are disordered in the structures contribute to affinity. Worryingly, all the structures are E•IMP complexes, which may not represent the highest affinity complex. If the inhibitor does bind preferentially to E-XMP*, the kinetics of E-XMP* formation and decomposition will also influence inhibitor affinity. In addition, the flap folds into the vacant cofactor site for the hydrolysis of E-XMP*, and this interaction will compete with the inhibitor. Lastly, as is widely observed in many systems, residues far from the binding site can influence ligand binding, but such residues are difficult to identify by inspection.

2.5 Conclusion

The enzymes of the purine nucleotide pathways are tempting, but risky, targets for antibiotic design due to the redundancy of the *de novo* synthesis and salvage pathways and the availability of purine bases and nucleosides at the site of infection. Nonetheless, these pathways appear to be essential for certain pathogens at some sites of infection. The enzymes of the purine conversion pathways may be the most promising candidate targets. Except for IMPDH, these targets have not been examined in detail. Further validation in the form of conditional gene knockouts to assess the vulnerability of these targets would help mitigate the risk and justify the effort required to develop small molecule inhibitors suitable for testing in animal models.

Acknowledgements

The support of NIH RO1 AI093459 (LH) is gratefully acknowledged. Molecular graphics and analyses were performed with the UCSF Chimera package. Chimera is developed by the Resource for Biocomputing, Visualization, and Informatics at the University of California, San Francisco (supported by NIGMS P41-GM103311).

References

1. L. Hedstrom, G. Liechti, J. B. Goldberg and D. R. Gollapalli, *Curr. Med. Chem.*, 2011, **18**, 1909–1918.
2. Y. Zhang, M. Morar and S. E. Ealick, *Cell. Mol. Life Sci.*, 2008, **65**, 3699–3724.
3. T. J. Kappock, S. E. Ealick and J. Stubbe, *Curr. Opin. Chem. Biol.*, 2000, **4**, 567–572.
4. M. Morar, A. A. Hoskins, J. Stubbe and S. E. Ealick, *Biochemistry*, 2008, **47**, 7816–7830.
5. S. M. Firestine, S. W. Poon, E. J. Mueller, J. Stubbe and V. J. Davisson, *Biochemistry*, 1994, **33**, 11927–11934.
6. K. Dahncke and C. P. Witte, *Plant Cell*, 2013, **25**, 4101–4109.
7. L. Hedstrom, K. S. Cheung and C. C. Wang, *Biochem. Pharmacol.*, 1990, **39**, 151–160.
8. T. van Opijnen, K. L. Bodi and A. Camilli, *Nat. Methods*, 2009, **6**, 767–772.
9. G. C. Langridge, M. D. Phan, D. J. Turner, T. T. Perkins, L. Parts, J. Haase, I. Charles, D. J. Maskell, S. E. Peters, G. Dougan, J. Wain, J. Parkhill and A. K. Turner, *Genome Res.*, 2009, **19**, 2308–2316.
10. H. Luo, Y. Lin, F. Gao, C. T. Zhang and R. Zhang, *Nucleic Acids Res.*, 2014, **42**, D574–D580.
11. T. van Opijnen and A. Camilli, *Genome Res.*, 2012, **22**, 2541–2551.
12. M. D. Valentino, L. Foulston, A. Sadaka, V. N. Kos, R. A. Villet, J. Santa Maria Jr, D. W. Lazinski, A. Camilli, S. Walker, D. C. Hooper and M. S. Gilmore, *mBio*, 2014, **5**, e01729-14.
13. J. D. Gawronski, S. M. Wong, G. Giannoukos, D. V. Ward and B. J. Akerley, *Proc. Natl. Acad. Sci. U. S. A.*, 2009, **106**, 16422–16427.
14. N. Wang, E. A. Ozer, M. J. Mandel and A. R. Hauser, *mBio*, 2014, **5**, e01163-14.
15. Y. Le Breton, P. Mistry, K. M. Valdes, J. Quigley, N. Kumar, H. Tettelin and K. S. McIver, *Infect. Immun.*, 2013, **81**, 862–875.
16. M. A. Bachman, P. Breen, V. Deornellas, Q. Mu, L. Zhao, W. Wu, J. D. Cavalcoli and H. L. Mobley, *mBio*, 2015, **6**, e00775.
17. S. A. Lee, L. A. Gallagher, M. Thongdee, B. J. Staudinger, S. Lippman, P. K. Singh and C. Manoil, *Proc. Natl. Acad. Sci. U. S. A.*, 2015, **112**, 5189–5194.
18. S. Samant, H. Lee, M. Ghassemi, J. Chen, J. L. Cook, A. S. Mankin and A. A. Neyfakh, *PLoS Pathog.*, 2008, **4**, e37.
19. W. C. McFarland and B. A. Stocker, *Microb. Pathog.*, 1987, **3**, 129–141.
20. E. D. Garber, A. J. Hacket and R. Franklin, *Genetics*, 1952, **38**, 693–697.
21. G. Ivanovics, E. Marjai and A. Dobozy, *J. Gen. Microbiol.*, 1968, **53**, 147–162.
22. T. W. Burrows and G. A. Bacon, *Br. J. Exp. Pathol.*, 1954, **35**, 134–143.
23. R. R. Brubaker, *Infect. Immun.*, 1970, **1**, 446–454.
24. Y. Flashner, E. Mamroud, A. Tidhar, R. Ber, M. Aftalion, D. Gur, S. Lazar, A. Zvi, T. Bino, N. Ariel, B. Velan, A. Shafferman and S. Cohen, *Infect. Immun.*, 2004, **72**, 908–915.

25. J. E. Quarry, K. E. Isherwood, S. L. Michell, H. Diaper, R. W. Titball and P. C. Oyston, *Vaccine*, 2007, **25**, 2011–2018.
26. K. Kadzhaev, C. Zingmark, I. Golovliov, M. Bolanowski, H. Shen, W. Conlan and A. Sjostedt, *PLoS One*, 2009, **4**, e5463.
27. A. E. Santiago, L. E. Cole, A. Franco, S. N. Vogel, M. M. Levine and E. M. Barry, *Vaccine*, 2009, **27**, 2426–2436.
28. A. E. Santiago, B. J. Mann, A. Qin, A. L. Cunningham, L. E. Cole, C. Grassel, S. N. Vogel, M. M. Levine and E. M. Barry, *Pathog. Dis.*, 2015, **73**, ftv036.
29. Q. L. Truong, Y. Cho, A. K. Barate, S. Kim, M. Watarai and T. W. Hahn, *Microb. Pathog.*, 2015, **79**, 1–7.
30. D. J. Botkin, A. N. Abbott, P. E. Stewart, P. A. Rosa, H. Kawabata, H. Watanabe and S. J. Norris, *Infect. Immun.*, 2006, **74**, 6690–6699.
31. M. W. Jewett, K. A. Lawrence, A. Bestor, R. Byram, F. Gherardini and P. A. Rosa, *J. Bacteriol.*, 2009, **191**, 6231–6241.
32. F. R. Noreiga, G. Losonsky, C. Lauderbaugh, F. M. Liao, J. Y. Wang and M. M. Levine, *Infect. Immun.*, 1996, **64**, 3055–3061.
33. L. Rajagopal, A. Vo, A. Silvestroni and C. E. Rubens, *Mol. Microbiol.*, 2005, **56**, 1329–1346.
34. J. E. Griffin, J. D. Gawronski, M. A. Dejesus, T. R. Ioerger, B. J. Akerley and C. M. Sassetti, *PLoS Pathog.*, 2011, **7**, e1002251.
35. M. Ingerson-Mahar, A. Briegel, J. N. Werner, G. J. Jensen and Z. Gitai, *Nat. Cell Biol.*, 2010, **12**, 739–746.
36. W. C. Carcamo, M. Satoh, H. Kasahara, N. Terada, T. Hamazaki, J. Y. Chan, B. Yao, S. Tamayo, G. Covini, C. A. von Muhlen and E. K. Chan, *PLoS One*, 2011, **6**, e29690.
37. B. Pang, J. L. McFaline, N. E. Burgis, M. Dong, K. Taghizadeh, M. R. Sullivan, C. E. Elmquist, R. P. Cunningham and P. C. Dedon, *Proc. Natl. Acad. Sci. U. S. A.*, 2012, **109**, 2319–2324.
38. R. S. Galhardo, P. J. Hastings and S. M. Rosenberg, *Crit. Rev. Biochem. Mol. Biol.*, 2007, **42**, 399–435.
39. P. L. Foster, *Crit. Rev. Biochem. Mol. Biol.*, 2007, **42**, 373–397.
40. H. Zhao, J. B. French, Y. Fang and S. J. Benkovic, *Chem. Commun. (Camb.)*, 2013, **49**, 4444–4452.
41. S. M. Firestine, H. Paritala, J. E. McDonnell, J. B. Thoden and H. M. Holden, *Bioorg. Med. Chem.*, 2009, **17**, 3317–3323.
42. H. Lei, C. Jones, T. Zhu, K. Patel, N. M. Wolf, L. W. Fung, H. Lee and M. E. Johnson, *Bioorg. Med. Chem.*, 2016, **24**, 596–605.
43. K. L. Sullivan, L. C. Huma, E. A. Mullins, M. E. Johnson and T. J. Kappock, *Anal. Biochem.*, 2014, **452**, 43–45.
44. P. K. Fyfe, A. Dawson, M. T. Hutchison, S. Cameron and W. N. Hunter, *Acta Crystallogr., Sect. D: Biol. Crystallogr.*, 2010, **66**, 881–888.
45. L. Hedstrom, *Chem. Rev.*, 2009, **109**, 2903–2928.
46. Y. Natsumeda, S. Ohno, H. Kawasaki, Y. Konno, G. Weber and K. Suzuki, *J. Biol. Chem.*, 1990, **265**, 5292–5295.
47. M. D. Sintchak, M. A. Fleming, O. Futer, S. A. Raybuck, S. P. Chambers, P. R. Caron, M. Murcko and K. P. Wilson, *Cell*, 1996, **85**, 921–930.

48. M. D. Sintchak and E. Nimmesgern, *Immunopharmacology*, 2000, **47**, 163–184.
49. T. D. Colby, K. Vanderveen, M. D. Strickler, G. D. Markham and B. M. Goldstein, *Proc. Natl. Acad. Sci. U. S. A.*, 1999, **96**, 3531–3536.
50. S. K. Gorla, N. N. McNair, G. Yang, S. Gao, M. Hu, V. R. Jala, B. Haribabu, B. Striepen, G. D. Cuny, J. R. Mead and L. Hedstrom, *Antimicrob. Agents Chemother.*, 2014, **58**, 1603–1614.
51. G. Labesse, T. Alexandre, L. Vaupre, I. Salard-Arnaud, J. L. Him, B. Raynal, P. Bron and H. Munier-Lehmann, *Structure*, 2013, **21**, 975–985.
52. E. Nimmesgern, J. Black, O. Futer, J. R. Fulghum, S. P. Chambers, C. L. Brummel, S. A. Raybuck and M. D. Sintchak, *Protein Expression Purif.*, 1999, **17**, 282–289.
53. M. Pimkin and G. D. Markham, *Mol. Microbiol.*, 2008, **69**, 342–359.
54. M. Pimkin, J. Pimkina and G. D. Markham, *J. Biol. Chem.*, 2009, **284**, 7960–7969.
55. S. E. Mortimer, D. Xu, D. McGrew, N. Hamaguchi, H. C. Lim, S. J. Bowne, S. P. Daiger and L. Hedstrom, *J. Biol. Chem.*, 2008, **283**, 36354–36360.
56. J. P. Whitehead, F. Simpson, M. M. Hill, E. C. Thomas, L. M. Connolly, F. Collart, R. J. Simpson and D. E. James, *Traffic*, 2004, **5**, 739–749.
57. E. N. Kozhevnikova, J. A. van der Knaap, A. V. Pindyurin, Z. Ozgur, W. F. van Ijcken, Y. M. Moshkin and C. P. Verrijzer, *Mol. Cell*, 2012, **47**, 133–139.
58. L. Hedstrom, *Chem. Rev.*, 2009, **109**, 2903–2928.
59. M. D. Sintchak, M. C. Badia, O. Futer and E. Nimmesgern, *Antiviral Res.*, 1999, **41**, A56.
60. M. Makowska-Grzyska, Y. Kim, N. Maltseva, J. Osipiuk, M. Gu, M. Zhang, K. Mandapati, D. R. Gollapalli, S. K. Gorla, L. Hedstrom and A. Joachimiak, *J. Biol. Chem.*, 2015, **290**, 5893–5911.
61. R. M. Buey, R. Ledesma-Amaro, M. Balsera, J. M. de Pereda and J. L. Revuelta, *Appl. Microbiol. Biotechnol.*, 2015, **99**, 9577–9589.
62. M. Makowska-Grzyska, Y. Kim, S. K. Gorla, Y. Wei, K. Mandapati, M. Zhang, N. Maltseva, G. Modi, H. I. Boshoff, M. Gu, C. Aldrich, G. D. Cuny, L. Hedstrom and A. Joachimiak, *PLoS One*, 2015, **10**, e0138976.
63. W. Checkley, A. C. White Jr, D. Jaganath, M. J. Arrowood, R. M. Chalmers, X. M. Chen, R. Fayer, J. K. Griffiths, R. L. Guerrant, L. Hedstrom, C. D. Huston, K. L. Kotloff, G. Kang, J. R. Mead, M. Miller, W. A. Petri Jr, J. W. Priest, D. S. Roos, B. Striepen, R. C. Thompson, H. D. Ward, W. A. Van Voorhis, L. Xiao, G. Zhu and E. R. Houpt, *Lancet Infect. Dis.*, 2015, **15**, 85–94.
64. N. N. Umejiego, D. Gollapalli, L. Sharling, A. Volftsun, J. Lu, N. N. Benjamin, A. H. Stroupe, T. V. Riera, B. Striepen and L. Hedstrom, *Chem. Biol.*, 2008, **15**, 70–77.
65. S. K. Maurya, D. R. Gollapalli, S. Kirubakaran, M. Zhang, C. R. Johnson, N. N. Benjamin, L. Hedstrom and G. D. Cuny, *J. Med. Chem.*, 2009, **52**, 4623–4630.
66. S. Kirubakaran, S. K. Gorla, L. Sharling, M. Zhang, X. Liu, S. S. Ray, I. S. Macpherson, B. Striepen, L. Hedstrom and G. D. Cuny, *Bioorg. Med. Chem. Lett.*, 2012, **22**, 1985–1988.

67. S. K. Gorla, M. Kavitha, M. Zhang, X. Liu, L. Sharling, D. R. Gollapalli, B. Striepen, L. Hedstrom and G. D. Cuny, *J. Med. Chem.*, 2012, **55**, 7759–7771.
68. C. R. Johnson, S. K. Gorla, M. Kavitha, M. Zhang, X. Liu, B. Striepen, J. R. Mead, G. D. Cuny and L. Hedstrom, *Bioorg. Med. Chem. Lett.*, 2013, **23**, 1004–1007.
69. S. K. Gorla, M. Kavitha, M. Zhang, J. E. Chin, X. Liu, B. Striepen, M. Makowska-Grzyska, Y. Kim, A. Joachimiak, L. Hedstrom and G. D. Cuny, *J. Med. Chem.*, 2013, **56**, 4028–4043.
70. Z. Sun, J. Khan, M. Makowska-Grzyska, M. Zhang, J. H. Cho, C. Suebsuwong, P. Vo, D. R. Gollapalli, Y. Kim, A. Joachimiak, L. Hedstrom and G. D. Cuny, *J. Med. Chem.*, 2014, **57**, 10544–10550.
71. I. S. MacPherson, S. Kirubakaran, S. K. Gorla, T. V. Riera, J. A. D'Aquino, M. Zhang, G. D. Cuny and L. Hedstrom, *J. Am. Chem. Soc.*, 2010, **132**, 1230–1231.
72. Y. Kim, M. Makowska-Grzyska, S. K. Gorla, D. R. Gollapalli, G. D. Cuny, A. Joachimiak and L. Hedstrom, *Acta Crystallogr., Sect. F: Struct. Biol. Commun.*, 2015, **71**, 531–538.
73. M. Makowska-Grzyska, Y. Kim, R. Wu, R. Wilton, D. R. Gollapalli, X. K. Wang, R. Zhang, R. Jedrzejczak, J. C. Mack, N. Maltseva, R. Mulligan, T. A. Binkowski, P. Gornicki, M. L. Kuhn, W. F. Anderson, L. Hedstrom and A. Joachimiak, *Biochemistry*, 2012, **51**, 6148–6163.
74. D. R. Gollapalli, I. S. Macpherson, G. Liechti, S. K. Gorla, J. B. Goldberg and L. Hedstrom, *Chem. Biol.*, 2010, **17**, 1084–1091.
75. K. Mandapati, S. K. Gorla, A. L. House, E. S. McKenney, M. Zhang, S. N. Rao, D. R. Gollapalli, B. J. Mann, J. B. Goldberg, G. D. Cuny, I. J. Glomski and L. Hedstrom, *ACS Med. Chem. Lett.*, 2014, **5**, 846–850.
76. R. O'Shea and H. E. Moser, *J. Med. Chem.*, 2008, **51**, 2871–2878.
77. R. G. Ducati, A. Breda, L. A. Basso and D. S. Santos, *Curr. Med. Chem.*, 2011, **18**, 1258–1275.
78. V. Usha, J. V. Hobrath, S. S. Gurcha, R. C. Reynolds and G. S. Besra, *PLoS One*, 2012, **7**, e33886.
79. V. Usha, S. S. Gurcha, A. L. Lovering, A. J. Lloyd, A. Papaemmanouil, R. C. Reynolds and G. S. Besra, *Microbiology*, 2011, **157**, 290–299.
80. V. A. Rao, S. M. Shepherd, R. Owen and W. N. Hunter, *Acta Crystallogr., Sect. F: Struct. Biol. Cryst. Commun.*, 2013, **69**, 243–247.
81. D. S. Wishart, T. Jewison, A. C. Guo, M. Wilson, C. Knox, Y. Liu, Y. Djoumbou, R. Mandal, F. Aziat, E. Dong, S. Bouatra, I. Sinelnikov, D. Arndt, J. Xia, P. Liu, F. Yallou, T. Bjorndahl, R. Perez-Pineiro, R. Eisner, F. Allen, V. Neveu, R. Greiner and A. Scalbert, *Nucleic Acids Res.*, 2013, **41**, D801–D807.
82. R.-G. Zhang, G. Evans, F. J. Rotella, E. M. Westbrook, D. Beno, E. Huberman, A. Joachimiak and F. R. Collart, *Biochemistry*, 1999, **38**, 4691–4700.
83. E. F. Pettersen, T. D. Goddard, C. C. Huang, G. S. Couch, D. M. Greenblatt, E. C. Meng and T. E. Ferrin, *J. Comput. Chem.*, 2004, **25**, 1605–1612.

CHAPTER 3

Inhibitors of Biofilm Production

ROBERTA J. MELANDER AND CHRISTIAN MELANDER*

Department of Chemistry, North Carolina State University, Raleigh, NC, 27695, USA
*E-mail: ccmeland@ncsu.edu

3.1 Introduction

The inhibition of biofilm formation is a promising novel target for the discovery of new antibacterial agents. Biofilms are multicellular communities of microorganisms that are adhered to a surface and encased in a self-produced matrix. Bacteria within a biofilm exhibit different phenotypic behaviors when compared to planktonic bacteria, including altered growth rates, altered gene expression, and perhaps most significantly, altered responses to antibiotics and the host-immune system. Bacteria within a biofilm can display up to 1000-fold decreased sensitivity to antibiotics, making biofilm-based infections a considerable challenge to treat. The ability to prevent biofilm formation therefore has considerable potential to aid in the treatment of numerous infections, rendering bacteria susceptible to antibiotics and the host immune response. This chapter will discuss various strategies that have been explored to inhibit biofilm production, and describe several promising biofilm inhibitors that have been identified.

3.2 Biofilms

Biofilms are sessile multicellular communities of microorganisms, often likened to tissues of higher organisms regarding their complexity; biofilms

Drug Discovery Series No. 58
Antibiotic Drug Discovery: New Targets and Molecular Entities
Edited by Steven M. Firestine and Troy Lister

Published by the Royal Society of Chemistry, www.rsc.org

contain pores and channels that allow the circulation of water and nutrients, and elimination of waste, and cells in different regions of the biofilm typically exhibit different gene expression patterns and growth rates. Residing within a biofilm confers a considerable degree of protection to the microorganism, allowing survival in hostile environments.[1] Biofilms are formed in response to various environmental cues, such as nutrient depletion, ecological competition, and the presence of sub-lethal levels of antibiotics.[2,3] Biofilms play a role in a number of infections including: infection of indwelling medical devices, wound infections, bacterial endocarditis, ear infections, tooth decay, lung infections of cystic fibrosis patients, and likely many more.[4] The National Institutes of Health (NIH) estimates that 80% of bacterial infections are biofilm related, comprising 17 million new biofilm infections each year in the U.S. that result in up to 550 000 fatalities.[5] Additionally, biofilm related infections impart a considerable economic burden, with device related biofilm infections alone resulting in an increase in US hospitalization costs of over one billion dollars annually.[6]

The ability of bacteria within a biofilm to tolerate levels of antibiotics that would eradicate planktonic bacteria is a result of several factors:[1] as mentioned earlier, bacteria within a biofilm exhibit changes in gene expression, including upregulation of efflux pumps,[2,7] the biofilm matrix limits penetration of antibiotics, meaning that they cannot reach cells deep within the biofilm,[8,9] and[3] most antibiotics target processes that occur in actively growing and dividing bacterial cells and are ineffective against cells exhibiting reduced growth rates.[10,11] These factors combine to result in most antibiotics being largely ineffective against bacteria within a biofilm, and while there are some examples of conventional antibiotics exhibiting some activity, such as rifampin against staphylococcal biofilm cells,[12] and colistin against *Pseudomonas aeruginosa* biofilm cells,[13] sub-MIC levels of most antibiotics (including the most commonly used antibiotic classes—aminoglycosides, β-lactams, fluoroquiolones, glycopeptides, rifamycins, tetracyclines) can induce biofilm formation. There is therefore a considerable need to look to alternative strategies to the traditional approach of simply killing bacteria to treat infections caused by bacteria that have adopted the biofilm lifestyle.[14]

3.3 Strategies for Combating Biofilms

There have been several diverse approaches investigated for the prevention and eradication of bacterial biofilms, ranging from small molecule inhibitors of bacterial signaling pathways to enzymatic degradation of the biofilm matrix. Several of these strategies will be discussed below.

3.3.1 Quorum Sensing Inhibitors

Perhaps the most widely studied approach for combating biofilms is the disruption of the communication pathways used by bacteria to coordinate the formation and development of the complex biofilm community. Quorum sensing is a population density based bacterial intercellular communication method mediated by the production and detection of diffusible small molecules. Quorum sensing allows the bacteria to coordinate gene expression and, in addition to other phenotypes, form a biofilm.[15] There are both intraspecies and interspecies quorum sensing pathways that utilize different small molecules (Figure 3.1). Quorum sensing pathways in Gram-negative bacteria employ acyl homoserine lactones (AHL) as the signaling molecule such as 3-oxo-C12-AHL **1** from *P. aeruginosa*, while Gram-positive bacteria communicate using autoinducer peptides (AIP), for example the *Staphylococcus aureus* AIP **2**. Interspecies communication is carried out *via* the use of a series of molecules derived from 4,5-dihydroxy-2,3-pentanedione (DPD) **3**, known as autoinducer-2 (AI-2) molecules, the structures of the *Salmonella typhimurium* and *Vibrio harveyi* AI-2s **4** and **5** are shown in Figure 3.1.[16,17]

3.3.2 Inhibitors of AHL Based Quorum Sensing as Biofilm Inhibitors

The role of AHL based quorum sensing in biofilm formation was discovered upon the observation that a *P. aeruginosa* mutant lacking the AHL synthase enzyme LasI forms biofilms with altered morphology that are sensitive to biocides. Exogenous addition of the AHL resulted in the formation of wild-type like biofilms that were resistant to biocides.[18] Native AHLs possess a number of issues that make them unsuitable for use as a therapeutic, such as instability and immunomodulatory activity.[19] This has led to the development of a number of synthetic AHL quorum sensing antagonists (Figure 3.2), many of which have been investigated as biofilm inhibitors, against *P. aeruginosa*[20] and several other Gram-negative bacteria.

Some of the most potent AHL analogues have been developed by the Blackwell group, for example analogues **6** and **7** inhibit biofilm formation by *P. aeruginosa* PAO1 at low micromolar concentrations,[21] and this group has also identified several synthetic AHL quorum sensing antagonists, including the

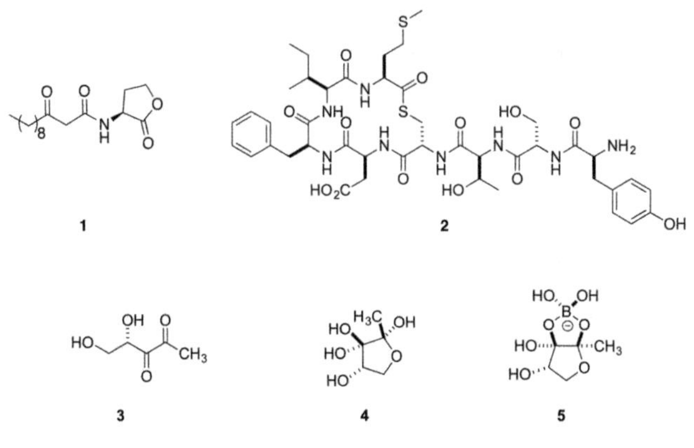

Figure 3.1 Native quorum sensing molecules.

halogen containing compounds **8** and **9**, that inhibit biofilm formation by the major nosocomial pathogen *Acinetobacter baumannii*.[22]

More extensive structural modifications to the native AHL scaffold have also been pursued, for example replacement of the amide moiety with a triazole group led to the identification of compounds **10** and **11**, which also inhibit *P. aeruginosa* biofilm formation.[23] The *N*-acyl cyclopentylamide analogue, C10-CPA **12** also inhibits biofilm formation by *P. aeruginosa*.[24] Compound **13** was identified from a series of covalent modifiers of LasR, the transcriptional regulator to which AHLs bind in *P. aeruginosa*. This compound selectively binds Cys 79 in the AHL binding pocket and inhibits quorum sensing and biofilm formation.[17]

One major hurdle with the use of native AHLs as quorum sensing or biofilm inhibitors *in vivo*, is a lack of stability due to the hydrolytic susceptibility of the lactone moiety. To circumvent this issue, a series of thiolactone AHL analogues was investigated by the Bassler group, leading to the identification of compound **14**, which inhibits *P. aeruginosa* biofilm formation, and has also been show to possess *in vivo* activity in a *Caenorhabditis elegans P. aeruginosa* infection assay.[25]

There have also been several classes of non-lactone compounds developed that interfere with AHL-based signaling and inhibit biofilm formation. Perhaps the most widely studied of these are the brominated furanones (Figure 3.3),[26–28] which are produced by the marine algae *Delisea pulchra*.

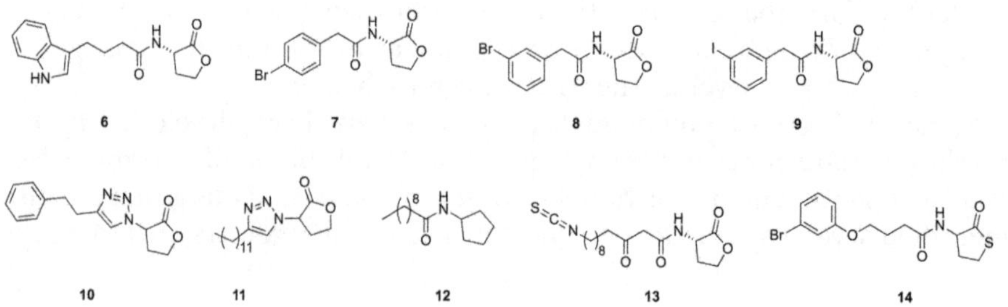

Figure 3.2 Synthetic AHL analogues.

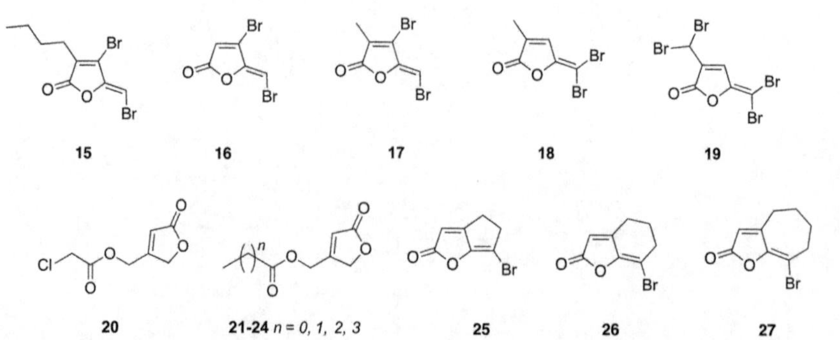

Figure 3.3 Brominated furanones.

The naturally occurring (5Z)-4-Bromo-5-(bromomethylene)-3-butyl-2(5H)-furanone **15** inhibits biofilm formation in *Escherichia coli* and *Bacillus subtilis*,[28,29] while the synthetic furanone C-30 **16** increased the susceptibility of *P. aeruginosa* within a biofilm to tobramycin and promoted clearance by the host immune response in a mouse pulmonary infection model.[30] This same compound also effected a significantly more rapid clearance of *P. aeruginosa* from silicone implants as compared to a placebo-treated group in a mouse foreign-body infection model.[31] Transcriptome analysis of furanone C-30 treated *P. aeruginosa* showed that this compound specifically targeted quorum sensing systems. Furanone C-30 retains activity against Gram-positive bacteria, inhibiting biofilm formation by both *Streptococcus intermedius* and the oral pathogen *Streptococcus mutans*.[32]

Other synthetic furanones with anti-biofilm activity include; **17**, **18**, and **19**, which inhibit biofilm formation by *E. coli*,[33] and the non-brominated furanones **20–24**, which inhibit *P. aeruginosa* biofilm formation.[34] Efforts to mute some of the more undesirable properties of the brominated furanone class of biofilm inhibitors, such as mammalian toxicity and low aqueous stability[35] are ongoing, and recently a class of bicyclic brominated furanones was reported that includes compounds **25–27**. These compounds were shown to act as quorum sensing antagonists and inhibit biofilm formation by *P. aeruginosa* and *E. coli* while exhibiting reduced mammalian cell toxicity.[36]

The use of enzymes to degrade AHLs as means of inhibiting quorum sensing is known as quorum quenching and is exploited by numerous organisms.[37] One of the first examples of quorum quenching described in the literature involved the inhibition of AHL activity in the plant pathogens *Erwinia carotovora* and *Agrobacterium tumefaciens* by expression of a lactonase gene from Bacillus sp. in host cells.[38,39] Following this, the hydrolase BpiB05, which was isolated from the soil metagenome, and subsequently produced recombinantly in *E. coli*, was shown to inhibit quorum sensing in an *A. tumefaciens* reporter strain carrying a traI-lacZ- fusion, and to reduce biofilm formation in addition to motility and pyocyanin synthesis in *P. aeruginosa*.[40] Another lactonase, SsoPox-I, which is an engineered variant of SsoPox produced by *Sulfolobus solfataricus*, inhibits quorum sensing and reduces biofilm formation in *P. aeruginosa* and was also active *in vivo*, reducing mortality and histological lung damage in a rat respiratory *P. aeruginosa* infection model.[41]

3.3.3 Inhibitors of AIP-Based Quorum Sensing as Biofilm Inhibitors

As mentioned earlier, Gram-positive bacteria use peptides, known as AIPs as signaling molecules.[42] In the most extensively studied Gram-positive quorum sensing system, that of *S. aureus*, the AIP **2** (Figure 3.1) binds the histidine kinase AgrC, leading to down-regulation of genes that encode adhesins required for biofilm formation.[43] Another peptide, the RNA-III inhibiting peptide (RIP) **28** (Figure 3.4), inhibits phosphorylation of a protein known as TRAP (target of RNA-III activating peptide), which leads to reduced biofilm

YSPWTNF-NH$_2$ YAPWTNF-NH$_2$

28 **29**

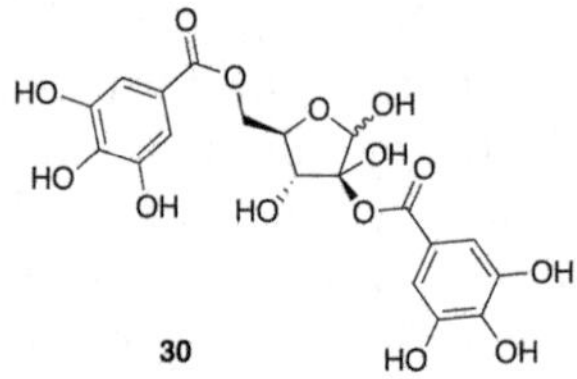

30

Figure 3.4 Inhibitors of AIP-based signaling.

formation.[44] RIP itself has been investigated as an anti-biofilm agent and has been shown to infections in multiple animal models.[45]

Several synthetic AIP analogues have developed and investigated as anti-virulence agents[46] including for the ability to inhibit biofilm production, and to enhance the effects of antibiotics *in vivo*. The synthetic RIP derivative FS3 **29** enhances the efficacy of tigecycline in a rat model of staphylococcal vascular graft infection.[47] A virtual screen of an RIP-based pharmacophore against a database of commercially available small-molecule compounds led to the identification of the non-peptide RNA-III inhibitor 2′,5-di-*O*-galloyl-D-hamamelose (hamamelitannin) **30**, which was shown to prevent *S. aureus* and *Staphylococcus epidermidis* mediated device associated infections *in vivo*.[48]

3.3.4 Inhibitors of AI-2 Based Quorum Sensing as Biofilm Inhibitors

The AI-2 quorum sensing system is utilized by many bacteria and employs a collection of molecules derived from DPD **3** (Figure 3.1) and known as AI-2. The synthase for DPD is conserved in the genome of over 55 bacterial species,[49] although the receptor from AI-2 has only been identified in Vibrio spp., in *Salmonella enterica* serovar Typhimurium and in *E. coli*.[50] The structures of AI-2 from *V. harveyi* (S-THMF borate **4**) and *S. typhimurium* (R-THMF **5**) are currently the only AI-2 structures that have been elucidated and are shown in Figure 3.1.[51]

DPD itself is unstable at high concentrations, however several synthetic AI-2 analogues have been developed to interfere with AI-2 signaling (Figure 3.5),[52] including the more stable DPD analogue Ac$_2$-DPD **31**, which inhibits biofilm formation by *Bacillus cereus* in addition to acting upon other AI-2 mediated behaviors, most likely a result of the release of DPD by *in situ* ester hydrolysis.[53] Other synthetic AI-2 analogues that have been developed include the AI-2 system agonists C4-alkoxy-5-hydroxy-2,3-pentanediones **32** and **33**, which activate the AI-2 pathway in *V. harveyi* more potently than DPD[54] and several C1-substituted DPD analogues including propyl-DPD **34** and butyl-DPD **35** that act as AI-2 system antagonists.[55] Isobutyl-DPD **36** has been shown to inhibit maturation of *E. coli* biofilms and, when administered in combination with gentamicin, to effect near complete clearance of pre-formed *E. coli* biofilms.[56] Another DPD analogue, 4-fluoro-5- hydroxypentane-2,3-dione **37**, has been reported to inhibit biofilm formation in *V. harveyi*.[57]

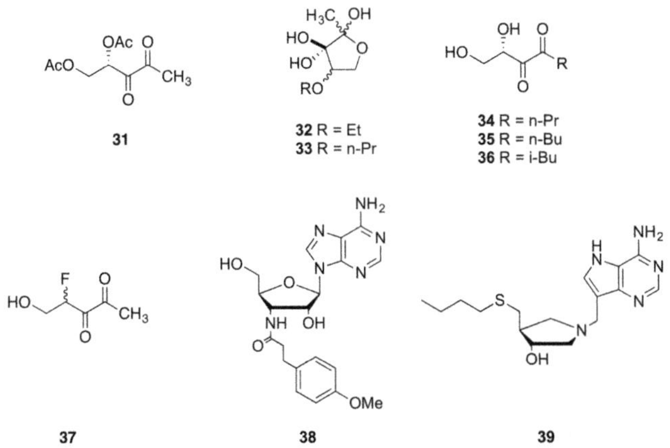

Figure 3.5 Inhibitors of AI-2 based signaling.

An alternative approach to developing compounds that inhibit biofilm formation by disrupting the AI-2 signaling pathway, is to prevent the biosynthesis of the AI-2 signal. This has been achieved using a series of nucleoside analogues, including the AI-2 system antagonist **38**, which affects biofilm formation in several *Vibrio* species.[58] Compound **39** inhibits 5′-methylthioadenosine nucleosidase (MTAN), which is involved in AI-2 biosynthesis, and was shown to inhibit both AI-2 production and biofilm formation in *E. coli* and *Vibrio cholerae* without affecting planktonic growth.[59]

3.4 Inhibition of Bacterial Signaling Pathways

Bacteria use internal signaling pathways to regulate numerous behaviors, including the formation of biofilms. Designing compounds to block or interfere with these pathways is another strategy to controlling biofilm formation.

3.4.1 Interference with c-di-GMP Signaling to Inhibit Biofilm Formation

One bacterial signaling molecule that is known to play a significant role in biofilm formation is bis-(3′5′)-cyclic di-guanylic acid (c-di-GMP) **40** (Figure 3.6). c-di-GMP is a ubiquitous second messenger bacterial signaling molecule that is formed and degraded by diguanylate cyclases (DGCs) and phosphodiesterases (PDEs) respectively in response to several environmental signals such as oxygen, light, and small molecules.[60,61] The design of compounds that interfere with c-di-GMP signaling is therefore another attractive target for the control of biofilm formation.[62]

Sulfathiazole **41** acts as a DGC inhibitor and has been shown to inhibit *E. coli* biofilm formation at low micromolar concentrations without significantly inhibiting bacterial growth,[63] while the benzimidazole **42**, which was

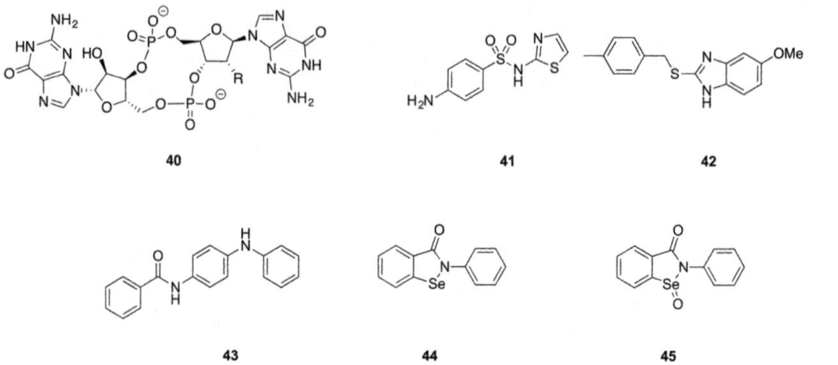

Figure 3.6 Inhibitors of c-di-GMP signaling.

identified as an inducer of a c-di-GMP-inducible transcriptional fusion in *V. cholerae*, has broad-spectrum biofilm inhibition activity, inhibiting biofilm formation by *P. aeruginosa*, *Klebsiella pneumoniae*, *Erwinia amylovora*, *Shigella boydii*, and *S. aureus*.[64] Other DCG inhibitors that prevent biofilm formation include compound **43**, which inhibits biofilm formation by *V. cholerae* and *P. aeruginosa*,[65] and ebselen **44** and ebselen oxide **45**, which reduce DGC activity by covalently modifying cysteine residues, which inhibit biofilm formation by *P. aeruginosa*.[66]

3.4.2 Inhibition of Indole Signaling Pathways to Prevent Biofilm Formation

Indole has been proposed as a universal intercellular signal molecule,[67] and is known to play a role in the control of many bacterial behaviors including biofilm formation[68] and is another attractive target for the design if compounds that prevent biofilm formation.[69] Many bacterial species, both those that produce indole and those that do not, will alter phenotypes in response to the presence of indole.[67] One of these phenotypes is biofilm formation. For example *E. coli*, which produces indole from the breakdown of tryptophan, will exhibit a decrease in biofilm formation in response to the presence of indole.[70] Conversely, for the non-indole producing *P. aeruginosa* indole effects an increase in biofilm formation.[68]

In addition to the parent molecule itself, the various metabolites of indole also have a significant impact on several bacterial behaviors including biofilm formation. For example, the oxidized derivatives 5-hydroxyindole **46** and 7-hydroxyindole **47** inhibit biofilm formation by enterohemorrhagic *E. coli* (EHEC), with 7-hydroxyindole **47** effecting a greater degree of inhibition than indole itself at the same concentration.[70] Indole-3-acetaldehyde, **48**, which is produced by the plant pathogen *Rhodococcus* sp. BFI 332, inhibits biofilm formation by EHEC, while the spent medium of *Rhodococcus* sp. BFI 332, has an inhibitory effect on biofilm formation by *S. aureus* and *S. epidermidis*.[71]

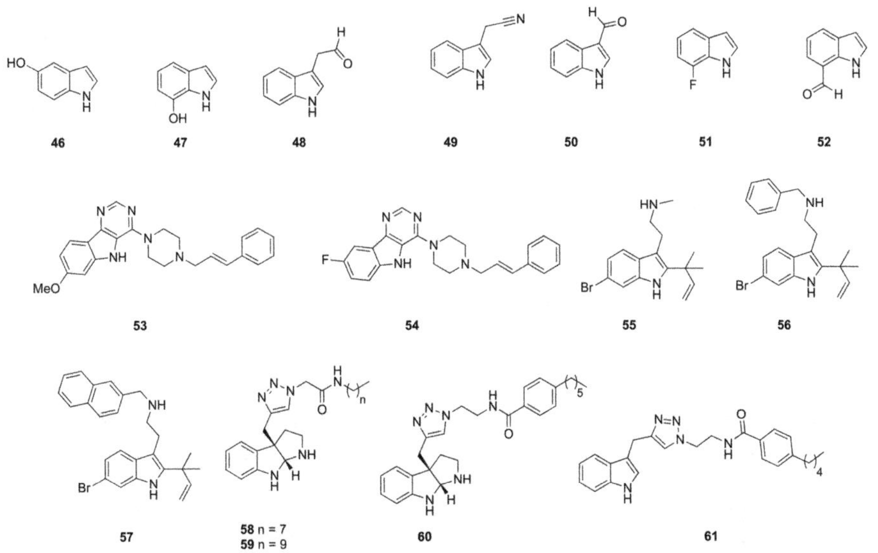

Figure 3.7 Indole derived biofilm inhibitors.

The indole derived plant metabolites 3-indolylacetonitrile (IAN) **49** and indole-3-carboxyaldehyde (I3CA) **50** reduce biofilm formation by *E. coli* O157:H7 to a greater degree than does indole at the same concentration, and weakly inhibit biofilm formation by *P. aeruginosa* in contrast to indole, which promotes biofilm formation by this bacterium.[72]

Several synthetic indole derivatives have been investigated for their anti-biofilm activity (Figure 3.7), with the simple synthetic indole derivatives 7-fluoroindole (7FI) **51**, and 7-formylindole **52** inhibiting biofilm formation by *P. aeruginosa*.[73] These simple synthetic derivatives, along with the naturally occurring indole analogues mentioned require high concentrations, often up to 1 mM, in order to exert their effects, which has led to the development of more complex indole containing compounds that elicit their effects at much lower concentrations. For example, the indole containing compound **53** was identified from a high-throughput screen for inhibitors of biofilm formation by *S. enterica* serovar Typhimurium, and a subsequent structure-activity relationship study led to the discovery of the more active 8-fluoro-4-[4-(3-phenyl-2-propen-1-yl)-1-piperazinyl]-5H-pyrimido[5,4-b] indole **54**, which inhibited biofilm formation at low micromolar concentrations.[74]

Another series of more complex indole derivatives that affect biofilm production, in addition to other behaviors, is based on the pyrroloindoline and indole containing secondary metabolites from *Flustra foliacea*, a North Sea bryzoan.[75] The natural product desformylflustrabromine (dFBr) **55** moderately inhibits biofilm formation by *E. coli* and *S. aureus* but exhibits microbicidal effects on planktonic growth of the bacteria.[76] Synthetic manipulation of the dFBr scaffold led to compounds **56** and **57**, which more potently inhibit biofilm formation by these two bacteria without concurrent microbicidal effects. Mechanistic studies suggest that the observed anti-biofilm activity in *E. coli* may be occurring through modulation of indole-based signaling pathways as the activity of these two compounds is dependent on the same factors as the activity of indole itself, *i.e.* temperature, the transcriptional regulator SdiA, and tryptophanase (TnaA).[76,77] Other flustramine derived small molecules have been reported to inhibit biofilm formation by various bacterial species including compound **58**, which inhibits biofilm formation by *A. baumannii* compound **59**, which inhibits biofilm formation by *E. coli*, and the Gram-positive acting compounds **60** and **61**, which inhibit biofilm formation by *S. aureus*.[78,79]

3.4.3 Inhibition of Two-Component Signal Transduction Systems (TCS) to Prevent Biofilm Formation

TCS are a class of regulatory systems that allow bacteria to sense and respond to changes in their environment and typically consist, as the name suggests, of two components - a histidine kinase sensor and a DNA-binding response regulator.[80] TCS regulate the expression of genes that control numerous behaviors, one of which is biofilm formation and maintenance,[81] making them attractive targets for the development of anti-biofilm compounds.

Several compounds have been developed that target either the histidine kinase, or the response regulator, and evaluated for their effects on various bacterial behaviors.[82] The histidine kinase inhibitor Walkmycin C **62** (Figure 3.8), which is produced by *Streptomyces* sp. strain MK632-100F11, inhibits the autophosphorylation activity of the *S. mutans* kinases VicK and CiaH, which play a role in sucrose-dependent biofilm formation, and also causes the formation of abnormal biofilms.[83,84] Carolacton **63** affects the expression of several TCS in *S. mutans* biofilm cells, and potently and selectively affects biofilm viability, killing bacterial cells that residing in a biofilm state at

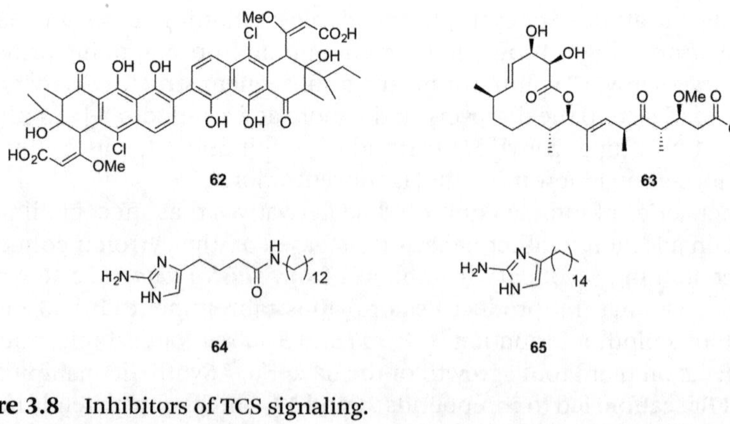

Figure 3.8 Inhibitors of TCS signaling.

nanomolar concentrations while exhibiting only minor microbicidal effects on planktonic bacteria.[85,86]

The 2-aminoimidazole (2-AI) class of small molecules, which are derived from the marine sponge alkaloids oroidin and bromoageliferin, are potent broad-spectrum inhibitors of biofilm formation and are discussed in Section 3.5. 2-AI **64**, inhibits and disperse biofilms in both *P. aeruginosa* and *A. baumannii*[87] and pull-down assays using a biotinylated analogue of this compound identified the response regulator BfmR, which is known to play a role in biofilm formation,[88] as the target for this class of compounds in *A. baumannii*.[89] The related 2-AI **65**, which inhibits biofilm formation by *S. mutans* and inhibits accumulation of *Porphyromonas gingivalis* on a substratum of *Streptococcus gordonii*, downregulates the histidine kinase ComD in *S. mutans*.[90,91]

3.4.4 Inhibition of Other Signaling Pathways to Prevent Biofilm Formation

Several bacteria use long chain fatty acids to mediate a number of cellular processes including gene expression and intracellular signaling.[92] One such fatty acid signaling molecule, *cis*-2-decenoic acid **66** (Figure 3.9), which is produced by *P. aeruginosa*, has been shown to induce the dispersion of established biofilms of several bacterial species including *E. coli*, *K. pneumoniae*, *Proteus mirabilis*, *Streptococcus pyogenes*, *B. subtilis*, and *S. aureus*.[93] *cis*-2-Decenoic acid has also been reported to enhance the effects of tobramycin and ciprofloxacin against *P. aeruginosa* biofilm cells.[94]

Another native bacterial fatty acid signaling molecule is *cis*-11-methyl-2-dodecenoic acid **67**, known as diffusible signal factor (DSF), which is produced by *Xanthomonas campestris* and has been shown to disaggregate cell flocs formed by *X. campestris*.[95] The *Burkholderia* diffusible signal factor (BDSF), *cis*-2-dodecenoic acid **68**, produced by *Burkholderia cenocepacia*, has been reported to both inhibit and disperse biofilms of *Francisella novicida*, potentially as a results of increased levels of (p)ppGpp,[96] and also to inhibit biofilm formation and adherence of the fungus *Candida albicans*.[97] The non-bacterial derived mitochondrial fatty acid oxidation intermediate palmitoyl-DL-carnitine **69** has been reported to prevent biofilm formation by *P. aeruginosa*, *E. coli*, and *Listeria monocytogenes* but does not disperse preformed biofilms.[98,99]

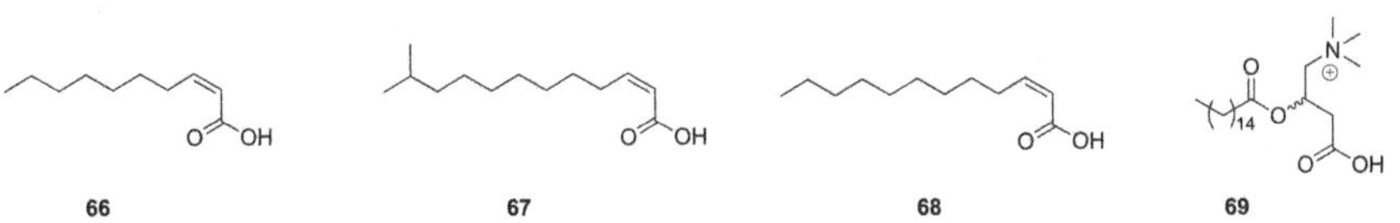

66 **67** **68** **69**

Figure 3.9 Fatty acid based inhibitors of bacterial signaling.

3.5 Identification of Natural Products and Analogues that Prevent Biofilm Formation

Many plants and marine organisms produce natural products that possess anti-biofilm activity as part of their defense systems against infection by microorganisms. These natural products have the potential either for therapeutic use themselves, or as scaffolds for the development of more viable molecules, much like as described for the indole containing products in Section 3.4.

3.5.1 Plant Natural Products and Analogues that Prevent Biofilm Formation

The disulfide allicin **70** (Figure 3.10), which is obtained from the garlic plant and has long been known to possess antibiotic properties,[100] was demonstrated to exhibit anti-biofilm activity. Garlic extracts have been shown to result in an increased sensitivity of *P. aeruginosa* biofilms to tobramycin *in vitro*, and also to clear *P. aeruginosa* lung infections *in vivo* when used prophylactically.[101] The dithianes **71** and **72** are thought to be responsible for this activity through modulation of LuxR quorum sensing systems.[102]

Several polyphenol compounds with anti-biofilm activity have been isolated from various plant sources. These include: ellagic acid **73** and its glycosylated

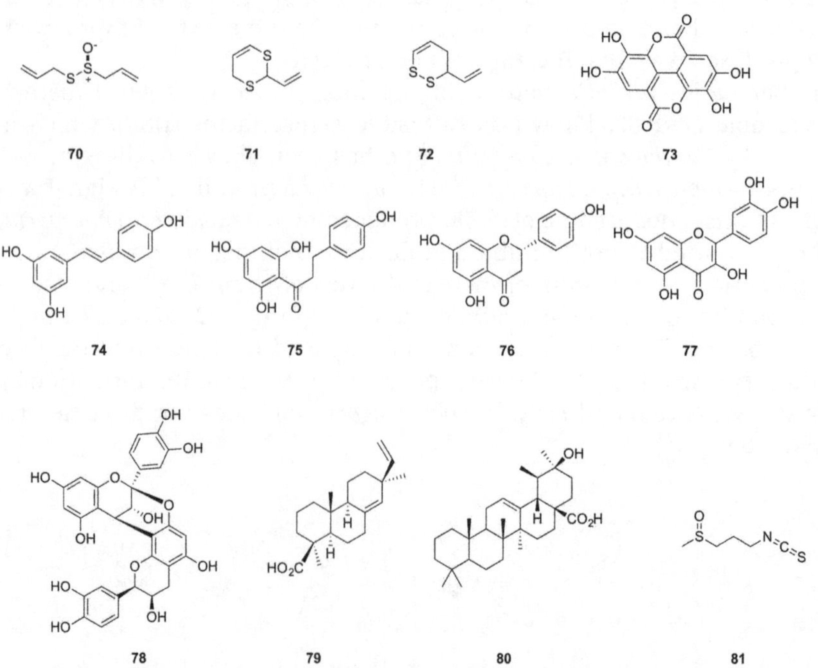

Figure 3.10 Plant natural products that prevent biofilm formation.

derivatives (extracted from *Rubus ulmifolius* (Elmleaf blackberry)) that inhibit *S. aureus* biofilm formation,[5] resveratrol **74**, (found in the skin of grapes and berries) that inhibits *V. cholerae* biofilm formation,[103] phloretin **75**, (found in the leaves of apple trees) that inhibits biofilm formation by EHEC potentially through repression of several AI-2 importer genes,[104] and naringenin **76** and quercetin **77** (found in citrus fruits), which inhibit *E. coli* and *V. harveyi* biofilm formation and have been shown to be antagonists of AHL and AI-2-mediated cell–cell signaling in *V. harveyi*.[105] Another class of plant-produced polyphenols are A-type proanthocyanidins (PACs) **78**, which are produced by cranberry plants and inhibit biofilm formation by *P. aeruginosa* and were also reported to potentiate the effects of gentamicin when tested in a *Galleria mellonella* model of infection.[106]

Other plant natural products that have shown anti-biofilm activity include the resin acid 4-epi-pimaric acid **79** (isolated from *Aralia cachemirica* L. (Araliaceae)) that inhibits *S. mutans* biofilm formation,[107] and the triterpene ursolic acid **80** (found in apple peel), which inhibits *E. coli* biofilm formation and has been shown to upregulate genes responsible for several biofilm related processes formation including chemotaxis and motility. The isothiocyanate iberin **81**, which is produced by horseradish, has been reported to inhibit biofilm formation by *P. aeruginosa* by inhibiting the Gac/Rsm quorum sensing network.[108]

3.5.2 Marine Natural Products and Analogues that Prevent Biofilm Formation

Many sponges and other marine organisms are known to produce compounds with anti-bacterial activities as defense against infection and biofouling.[109] As discussed earlier, the algae *D. pulchra* produces brominated furanones that have shown potent anti-biofilm activity. The marine sponge *Agelas coniferas* produces a series of 2-AI containing alkaloids, including oroidin **82** and bromoageliferin **83** (Figure 3.11),[110] that are believed to a play a role as a chemical anti-feeding defense mechanism. These alkaloids were first reported to inhibit biofilm formation by the Gram-negative marine α-proteobacterium *Rhodospirillum salexigens* in 1997,[111] while oroidin **82** was later reported to inhibit biofilm formation by *P. aeruginosa*.[112] The 2-AI scaffold has since been used to develop a series of simplified, readily accessible derivatives that exhibit anti-biofilm activity against a broad spectrum of bacteria and fungi. The structurally simple bromoageliferin analogue TAGE (*trans*-bromoageliferin) **84** inhibits and disperses *P. aeruginosa* biofilms,[113] while the di-brominated acylpyrrole containing analogue **85** is a more potent inhibitor of *P. aeruginosa* biofilms.[114] Several oroidin analogues are also potent anti-biofilm compounds including dihydrosventrin (DHS) **86**, which exhibits activity against *P. aeruginosa*, *A. baumannii*, and *Bordatella bronchiseptica*,[115] and the *para*-bromo phenyl derivative **87**, which exhibits increased anti-biofilm activity against *A. baumannii* as compared to DHS.[116]

The reverse amide class of 2-AI compounds, in which the directionality of amide bond is inverted from that in the natural products, contains a number of potent anti-biofilm compounds.[117] As mentioned earlier, 2-AI **64** (Figure 3.8), inhibits and disperse biofilms in both *P. aeruginosa* and *A. baumannii* and was used to identified the molecular target for this class of compounds. The related 2-AIs **88** and **89** (Figure 3.11) also potently inhibit biofilm, formation by *P. aeruginosa*, while the triazole containing compound **90** and other 2-aminoimidaozle/triazole conjugates (2-AITs) are potent inhibitors of biofilm formation by a broad-spectrum of bacteria including *P. aeruginosa*, *A. baumannii*, and *S. aureus*.[118,119] Compound **90** also suppresses resistance to several antibiotics in methicillin resistant *S. aureus* (MRSA) and

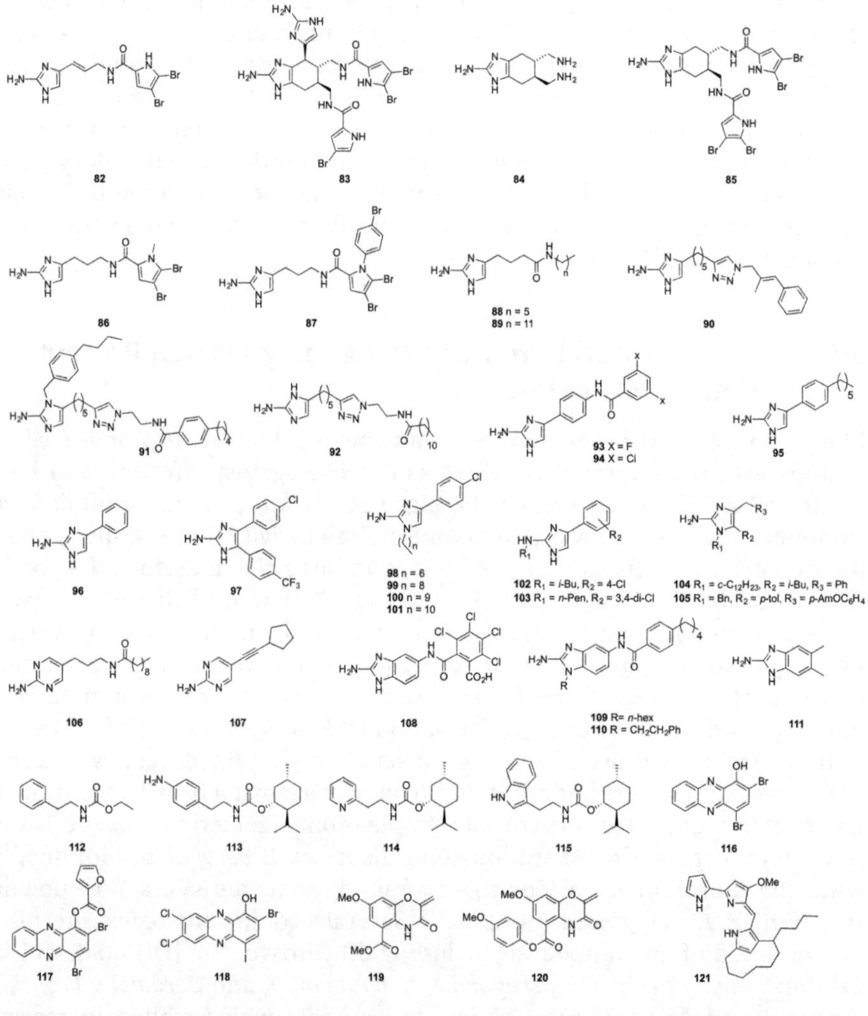

Figure 3.11 Marine natural products and analogues that prevent biofilm formation, and the bacterial derived natural product streptorubin B.

A. baumannii,[120] and the 2-AIT analogue **91** is a potent suppressor of oxacillin resistance in MRSA, *via* a mechanism that is dependent on the VraRS TCS.[121] The related triazole containing compound **92** inhibits biofilm formation by *S. mutans* and inhibited bacterial colonization and reduced the incidence of dental caries incidence *in vivo* in a rat infection model.[122]

Compound **93**, which belongs to the aryl 2-AI class of compounds was identified as a potent inhibitor of *E. coli* biofilm formation,[123] and this compound along with several related members of this class of 2-AI compounds including compounds **94** and **95**, suppresses antibiotic resistance in Gram–negative bacteria.[124–126] The mechanism for suppression of colistin resistance by compound **95** in *A. baumannii* was shown to be dependent on the PmrAB TCS.[125] A related series of aryl 2-AI compounds including **96–103** have been reported to exhibit anti-biofilm activity against the *P. aeruginosa* and the intestinal pathogen *S. typhimurium*,[74,127] as do the trisubstituted naamine alkaloids **104** and **105**, which are produced by the marine sponge *Leucetta chagosensis*.[128]

Compounds based upon 2-AI related heterocycles including 2-aminopyrimidine (2-AP) and 2-aminobenzimidazole (2-ABI) have also proven to be effective inhibitors of biofilm formation. For example the 2-APs **106** and **107** inhibit biofilm formation by *S. aureus*.[129] The 2-ABI **108** potently inhibits biofilm formation by Gram-positive bacteria, acting *via* a mechanism that was shown to inhibited by the presence of Zn(II), as do the related *N*1-substituted 2-ABIs **109** and **110**.[130,131] Another 2-ABI, 5,6-dimethylbenzimidazole (DMABI) **111**, potently inhibits biofilm formation by *P. aeruginosa* and moderately inhibits LasR based quorum sensing.[132]

The marine bacterium SCRC3P79 (Cytophaga sp.) produces ethyl *N*-(2-phenethyl) carbamate **112**, which possesses moderate anti-biofilm activity against *R. salexigens*,[111] in addition to several medically relevant pathogens. Analogue synthesis led to the identification of a number of compounds with increased activity, including the menthyl derived compounds **113–115**, which potently inhibit biofilm formation by several strains of *S. aureus*.[133]

The marine phenazine antibiotic **116** was recently used as a scaffold to develop a series of halogenated phenazines, including **117**, which has been reported to prevent biofilm formation by MRSA and to eradicate MRSA biofilms.[134] Subsequent analogue synthesis led to the identification of the potent anti-biofilm compound **118**, which eradicates biofilms of several Gram-positive bacterial species including MRSA, methicillin resistant *S. epidermidis* (MRSE), and vancomycin resistant enterococci (VRE).[135]

The auromomycin derived natural product **119** was identified from a screen of a 1248-member prefractionated marine natural-product library for biofilm inhibitory activity against *V. cholerae*. Compound **119** was shown to be a moderate inhibitor of *V. cholerae* biofilms, though it did not disperse pre-formed biofilms. The efficacy of compound **119**, however, was enhanced by the addition of sub-MIC levels of several antibiotics.[136] Development of the benzo[1,4]-oxazine scaffold led to the identification of compound **120**, which is a highly potent biofilm inhibitor against *V. cholereae*, and, unlike the parent compound, also potently disperses *V. cholerae* biofilms.[137]

3.5.3 Other Natural Products that Prevent Biofilm Formation

The prodiginine streptorubin B **121** (Figure 3.11) was identified from a screen of a library of actinomycete culture extracts for compounds with anti-biofilm activity. Streptorubin B is produced by *Streptomyces* sp. strain MC11024 and is a potent inhibitor of biofilm formation in *S. aureus*, exhibiting activity at concentrations as low as at 1 µg mL^{-1}.[138]

3.6 Antimicrobial Peptides (AMPs) that Prevent Biofilm Formation

AMPs are relatively small cationic amphipathic peptides produced by most organisms, both eukaryotic and prokaryotic, as part of the innate defense against invading pathogens.[139] AMPs are active against a broad spectrum of bacteria and, unlike most antibiotics, retain activity against metabolically inactive cells.[140] In addition to their antibiotic activities against planktonic bacteria, many AMPs have been shown to prevent biofilm formation, disperse preexisting biofilms and kill biofilm cells of several species of both Gram-positive and Gram-negative bacteria. Although the precise mechanisms of action of their anti-biofilm activity are not yet known, and there remain several obstacles to the clinical use of AMPS such as instability, host cell toxicity, and low bioavailability, they are promising agents for the development of anti-biofilm therapeutics, particularly for topical application in the treatment of wound infections in which biofilms play a significant role.[141]

One of the first reports of the anti-biofilm activity of an AMP was in 2008 when the human AMP LL-37 **122** (Figure 3.12), a member of the cathelicidin family of AMPs, was shown to exhibit anti-biofilm activity against *P. aeruginosa in vitro*, decreasing attachment, potently inhibiting the formation of biofilms at concentrations well below MIC levels, and reducing biofilm thickness of pre-formed biofilms. Although the exact mechanism of anti-biofilm

LL-37 **122**	LLGDFFRKSKEKIGKEFKRIVQRIKDFLRNLVPRTES
NA-CATH **123**	KRFKKFFKKLKNSVKKRAKKFFKKPKVIGVTFPF
SMAP-29 **124**	RGLRRLGRKIAHGVKKYGPTVLRIIRIAG
D-LL-37 **125**	*LLGDFFRKSKEKIGKEFKRIVQRIKDFLRNLVPRTES*
P10 **126**	LAREYKKIVEKLKRWLRQVLRTLR
P60.4Ac **127**	IGKEFKRIVERIKRFLRELVRPLR
17BIPHE2 **128**	GXKRLVQRLKDXLRNLV (X = biphenylalanine)
NACATH:ATRA1-ATRA1 **129**	KRFKKFFKKLKNSVKKRFKKFFKKLKVIGVTFPF
Novispirin-G10 **130**	KNLRRIIRKGIHIIKKYG
IDR-1018 **131**	VRLIVAVRIWRR

Figure 3.12 Sequences of peptides that prevent biofilm formation (italics indicates D-amino acids).

activity was not determined it was shown that LL-37 down-regulated a number of quorum sensing controlled genes including *lasI* and *rhlR* and also stimulated twitching motility.[142] Since this initial report, LL-37 has also been reported to inhibit biofilm formation by *A. baumannii*,[143] *S. aureus*[144] and *S. epidermidis*,[145] LL-37 exhibits *in vivo* activity, effecting a significantly prolonged survival time compared to untreated worms in a *G. mellonela P. aeruginosa* infection model.[146] The related cathelicidin NA-CATH **123** from the elapid snake *Naja atra* also inhibits biofilm formation by *S. aureus*, while the sheep cathelicidin SMAP-29 **124** inhibits biofilm formation by *P. aeruginosa* at sub MIC levels.[147]

Several synthetic AMP mimetics, analogues, and truncated peptides have been developed in attempts to improve the efficacy along with some of the less desirable properties of AMPS such low stability, toxicity against host cells and low bioavailability. The LL-37 mimetic D-LL-37 **125**, in which every amino acid was replaced with the corresponding D-amino acid to impart increased protease stability to the peptide, exhibits comparable anti-biofilm activity against *P. aeruginosa* and *S. aureus* to the native LL-37, displayed enhanced stability to trypsin degradation and was more effective upon survival of *P. aeruginosa* infected *G. mellonela* after 48 h.[144,146]

Several fragments or truncated versions of LL-37 have been developed that show increased anti-biofilm activity. For example, the fragments P10 **126** and P60.4Ac **127** are more efficient than LL-37 in eliminating biofilm associated bacteria in a thermally wounded human skin equivalent (HSE) MRSA infection model.[148] A series of peptides based upon GF-17 (residues 17-32 of LL-37) were designed with improved stability by incorporating non-natural amino acids. The most active peptide, 17BIPHE2 **128** exhibits resistance to several eukaryotic and bacterial proteases, inhibits *S. aureus* biofilm formation and disrupted pre-formed biofilms *in vitro* by *S. aureus*, and shows activity *in vivo*, preventing biofilm formation in a mouse model of catheter-associated infection.[149,150]

The NA-CATH analogue NACATH:ATRA1-ATRA1 **129**, in which the imperfect 11 amino acid repeat found in the native peptide was replaced with a perfect repeat, is a more potent inhibitor of *S. aureus* biofilm formation than NA-CATH.[144] Novispirin-G10 **130**, which is derived from SMAP-29, demonstrated activity in a rat burn wound *P. aeruginosa* infection model.[151]

The immunomodulatory peptide IDR (innate defense regulator) 1018 **131**, the sequence of which is based on the bovine neutrophil host defense peptide bactenecin Bac2a, was initially identified from a screen for immune modulators.[152] Peptide 1018 exhibits weak antimicrobial activity against planktonic bacteria but is a potent broad spectrum antibiofilm peptide, preventing biofilm formation, and eradicating pre-formed biofilms of numerous bacterial species including *P. aeruginosa*, *E. coli*, *A. baumannii*, *K. pneumoniae*, MRSA, *S. typhimurium* and *B. cenocepacia*. The authors posit that the mechanism of action of this peptide involves targeting the stringent response by binding to, and causing degradation of, the alarmones (p)ppGpp, an important signal in biofilm development.[153] This peptide also exhibited synergy with several antibiotics including ceftazidime, ciprofloxacin, imipenem, and tobramycin against biofilms of *P. aeruginosa*. Synergy was also observed between peptide 1018 and ceftazidime against biofilms of *A. baumannii, S. enterica*, and MRSA, and with tobramycin against biofilms of *E. coli, A. baumannii* and *K. pneumoniae*.[154]

3.7 Inhibition of Efflux to Prevent Biofilm Formation

The effect of the presence of efflux systems on biofilm formation has recently come into consideration, with a report that inactivation of any of the efflux systems of *S. enterica* serovar Typhimurium resulted in the inability to form a biofilm.[155] The authors posit that this phenomenon is due to transcriptional repression of curli biosynthesis genes and consequent inhibition of curli production. The use of efflux inhibitors as potential biofilm inhibitors was therefore investigated and three compounds with different mechanisms of efflux inhibition, chlorpromazine **132**, CCCP **133**, and PAβN **134** (Figure 3.13), were shown to effectively prevent biofilm formation in *E. coli, P. aeruginosa*, and *S. aureus*.[156]

3.8 Matrix Degradation to Prevent Biofilm Formation

While much of the research into ways to prevent biofilm formation has centered upon the identification of small molecules in a traditional drug-discovery like approach, the use of biologics, such as enzymes, as anti-biofilm entities (such as in the quorum quenching examples described earlier) has also been explored. The most common use of enzymes as a potential way to prevent biofilm formation is to degrade the biofilm matrix. This is an attractive approach to biofilm eradication as the matrix plays a vital role in biofilm integrity, forming the basis of the three-dimensional structure and accounting for up to 90% of the dry mass of the biofilm. The matrix also facilitates cell–cell communication, one of the hallmarks of the biofilm lifestyle. The biofilm matrix is composed of several extracellular polymeric substances

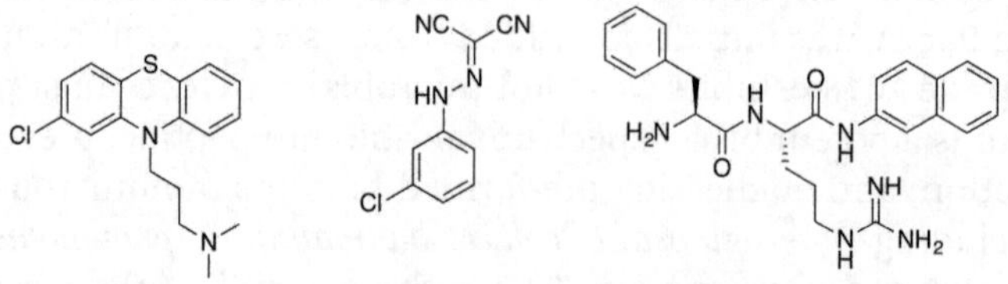

132 **133** **134**

Figure 3.13 Efflux pump inhibitors that prevent biofilm formation.

produced by the bacteria that include polysaccharides, proteins, lipids and nucleic acids. The secretion of enzymes such as glycosidases, proteases, and DNases that degrade the various components of the biofilm matrix is an innate phenomenon used by many bacteria,[157] and examples of endogenously produced matrix degrading enzymes include the *S. aureus* DNase thermonuclease, the glycoside hydrolase dispersin B, and the alginate lyase class of polysaccharide hydrolyzing enzymes. These enzymes and many others are used by the bacteria to initiate dispersion from the biofilm, which contributes to bacterial survival and disease transmission.[157] Based upon these activities, several of these matrix degrading enzymes have been investigated as potential therapeutic agents.

Dispersin B is a soluble glycoside hydrolase produced by the proteobacterium *Aggregatibacter actinomycetemcomitans* that degrades poly-*N*-acetylglucosamine (PGA), a major matrix component of several bacterial biofilms.[158] Dispersin B has been shown to inhibit the formation of biofilms by several medically relevant bacterial species including *S. aureus*, *S. epidermidis*, and *E. coli*, to disperse *S. epidermidis* and *E. coli* biofilms, and to sensitize *S. epidermidis* biofilm cells to the action of antimicrobials.[159–162] Dispersin B has demonstrated activity *in vivo*, lowering the rate of catheter colonization by *S. aureus* in combination with triclosan in a rabbit model of infection.[163]

Alginate lyases catalyze the degradation of the matrix component alginate, a copolymer of α-L-guluronate and β-D-mannuronate, and have been isolated from a wide range of organisms, including algae, marine invertebrates, and marine and terrestrial microorganisms.[164] Alginate lyase has been shown to enhance the microbicidal activity of aminoglycosides against *P. aeruginosa* biofilms *in vitro*.[165–167] Alginate lyase has also demonstrated *in vivo* efficacy, enhancing the effectiveness of amikacin in clearing *P. aeruginosa* in a rabbit model of endocarditis.[168]

The use of DNase for the prevention of biofilm formation by a variety of bacteria has also been investigated and it has been shown that biofilms formed in the presence of DNases exhibit reduced biomass and decreased antibiotic tolerance.[169] NucB is an extracellular DNase produced by *Bacillus licheniformis* that has been shown to effect rapid biofilm dispersal against a range of bacteria including *B. subtilis*, *E. coli*, and *Micrococcus luteus*.[168] Recombinant human DNase I (rhDNase), which is also known as dornase alfa and marketed as Pulmozyme by Genentech, inhibits and disperses *S. aureus* biofilms *in vitro* and has been shown to increase the susceptibility of *S. aureus* biofilm cells to several antimicrobials including chlorhexidine gluconate and povidone iodine, as well as to increase the susceptibility of *P. aeruginosa* biofilm cells to aminoglycosides.[166,170] rhDNase has also exhibited *in vivo* activity, increasing the efficacy of tobramycin upon *S. aureus*-infected *C. elegans*.[170] Pulmozyme is marketed for the treatment of pulmonary disease in cystic fibrosis (CF) patients,[171] and it has been shown that administration of Pulmozyme leads to reduced demand for antibiotics and improved lung function in CF patients.[172]

Several proteases have been investigated as anti-biofilm agents including the serine protease Esp from *S. epidermidis*, which was shown to both inhibit and eradicate preformed biofilms of *S. aureus* as well as enhance the susceptibility of *S. aureus* biofilms to the antimicrobial peptide human beta-defensin 2 (hBD2). Esp is also active *in vivo*, eliminating human nasal colonization by *S. aureus*.[173] Another serine protease Proteinase K, which is produced by the fungus *Tritirachium album* Limber, exhibits anti-biofilm activity against *S. aureus*,[174] while the metalloprotease serratopeptidase (SPEP) from *Serratia marcescens*, a widely used an anti-inflammatory therapeutic, has shown biofilm inhibitory activity against *P. aeruginosa*, *S. epidermidis* and *Listeria monocytogenes*.[175,176] This enzyme enhances the activity of ofloxacin against biofilms of *P. aeruginosa* and *S. epidermidis*.[176]

3.9 Conclusions

Adoption of the biofilm phenotype by pathogenic bacteria contributes significantly to the recalcitrance of bacterial infections to standard antimicrobial treatments. Prevention of biofilm formation is therefore an attractive strategy in the ongoing fight against drug resistant infection. We have highlighted many of the numerous strategies that have been investigated for the prevention of biofilm formation, including inhibition of bacterial communication and signaling pathways, the identification of natural products with anti-biofilm activity, the use of antimicrobial peptides that prevent biofilm formation, and the use of matrix degrading enzymes to prevent and eradicate bacterial biofilms. The development of therapeutics that effectively prevent biofilm formation would be a valuable addition to the current armory of anti-bacterial agents, potentially allowing conventional antibiotics to regain activity.

References

1. J. W. Costerton, P. S. Stewart and E. P. Greenberg, Bacterial biofilms: A common cause of persistent infections, *Science*, 1999, **284**(5418), 1318–1322.
2. N. M. Oliveira, E. Martinez-Garcia, J. Xavier, W. M. Durham, R. Kolter and W. Kim, *et al.*, Biofilm Formation As a Response to Ecological Competition, *PLoS Biol.*, 2015, **13**(7), e1002191.
3. G. A. O'Toole and P. S. Stewart, Biofilms strike back, *Nat. Biotechnol.*, 2005, **23**(11), 1378–1379.
4. D. J. Musk and P. J. Hergenrother, Chemical countermeasures for the control of bacterial biofilms: Effective compounds and promising targets, *Curr. Med. Chem.*, 2006, **13**(18), 2163–2177.
5. C. L. Quave, M. Estevez-Carmona, C. M. Compadre, G. Hobby, H. Hendrickson and K. E. Beenken, *et al.*, Ellagic acid derivatives from Rubus ulmifolius inhibit Staphylococcus aureus biofilm formation and improve response to antibiotics, *PLoS One*, 2012, **7**(1), e28737.

6. M. Shirtliff and J. G. Leid, *The Role of Biofilms in Device-Related Infections*, Springer, 2008.
7. M. E. Davey and G. A. O'Toole, Microbial biofilms: from ecology to molecular genetics, *Microbiol. Mol. Biol. Rev.*, 2000, **64**(4), 847–867.
8. T. F. C. Mah and G. A. O'Toole, Mechanisms of biofilm resistance to antimicrobial agents, *Trends Microbiol.*, 2001, **9**(1), 34–39.
9. H. C. Flemming, T. R. Neu and D. J. Wozniak, The EPS matrix: the "house of biofilm cells", *J. Bacteriol.*, 2007, **189**(22), 7945–7947.
10. A. L. Spoering and K. Lewis, Biofilms and planktonic cells of Pseudomonas aeruginosa have similar resistance to killing by antimicrobials, *J. Bacteriol.*, 2001, **183**(23), 6746–6751.
11. A. R. Coates and Y. Hu, Targeting non-multiplying organisms as a way to develop novel antimicrobials, *Trends Pharmacol. Sci.*, 2008, **29**(3), 143–150.
12. R. Saginur, M. Stdenis, W. Ferris, S. D. Aaron, F. Chan and C. Lee, *et al.*, Multiple combination bactericidal testing of staphylococcal biofilms from implant-associated infections, *Antimicrob. Agents Chemother.*, 2006, **50**(1), 55–61.
13. N. Hoiby, T. Bjarnsholt, M. Givskov, S. Molin and O. Ciofu, Antibiotic resistance of bacterial biofilms, *Int. J. Antimicrob. Agents*, 2010, **35**(4), 322–332.
14. J. B. Kaplan, Antibiotic-induced biofilm formation, *Int. J. Artif. Organs*, 2011, **34**(9), 737–751.
15. A. Camilli and B. L. Bassler, Bacterial small-molecule signaling pathways, *Science*, 2006, **311**(5764), 1113–1116.
16. Y. Irie and M. R. Parsek, Quorum sensing and microbial biofilms, *Curr. Top. Microbiol. Immunol.*, 2008, **322**, 67–84.
17. N. Amara, R. Mashiach, D. Amar, P. Krief, S. A. Spieser and M. J. Bottomley, *et al.*, Covalent inhibition of bacterial quorum sensing, *J. Am. Chem. Soc.*, 2009, **131**(30), 10610–10619.
18. D. G. Davies, M. R. Parsek, J. P. Pearson, B. H. Iglewski, J. W. Costerton and E. P. Greenberg, The involvement of cell-to-cell signals in the development of a bacterial biofilm, *Science*, 1998, **280**(5361), 295–298.
19. E. A. Yates, B. Philipp, C. Buckley, S. Atkinson, S. R. Chhabra and R. E. Sockett, *et al.*, N-acylhomoserine Lactones undergo lactonolysis in a pH-, temperature-, and acyl chain length-dependent manner during growth of Yersinia pseudotuberculosis and Pseudomonas aeruginosa, *Infect. Immun.*, 2002, **70**(10), 5635–5646.
20. M. E. Mattmann and H. E. Blackwell, Small molecules that modulate quorum sensing and control virulence in Pseudomonas aeruginosa, *J. Org. Chem.*, 2010, **75**(20), 6737–6746.
21. G. D. Geske, R. J. Wezeman, A. P. Siegel and H. E. Blackwell, Small molecule inhibitors of bacterial quorum sensing and biofilm formation, *J. Am. Chem. Soc.*, 2005, **127**(37), 12762–12763.
22. D. M. Stacy, M. A. Welsh, P. N. Rather and H. E. Blackwell, Attenuation of Quorum Sensing in the Pathogen Acinetobacter baumannii Using Non-native N-Acyl Homoserine Lactones, *ACS Chem. Biol.*, 2012, 7(10), 1719–1728.

23. G. Brackman, M. Risseeuw, S. Celen, P. Cos, L. Maes, H. J. Nelis, S. Van Calenbergh and T. Coenye, Synthesis and evaluation of the quorum sensing inhibitory effect of substituted triazolyldihydrofuranones, *Bioorg. Med. Chem.*, 2012, **20**, 4737–4743.
24. T. Ishida, T. Ikeda, N. Takiguchi, A. Kuroda, H. Ohtake and J. Kato, Inhibition of quorum sensing in Pseudomonas aeruginosa by N-acyl cyclopentylamides, *Appl. Environ. Microbiol.*, 2007, **73**(10), 3183–3188.
25. C. T. O'Loughlin, L. C. Miller, A. Siryaporn, K. Drescher, M. F. Semmelhack and B. L. Bassler, A quorum-sensing inhibitor blocks Pseudomonas aeruginosa virulence and biofilm formation, *Proc. Natl. Acad. Sci. U. S. A.*, 2013, **110**(44), 17981–17986.
26. M. Hentzer, K. Riedel, T. B. Rasmussen, A. Heydorn, J. B. Andersen and M. R. Parsek, *et al.*, Inhibition of quorum sensing in Pseudomonas aeruginosa biofilm bacteria by a halogenated furanone compound, *Microbiology*, 2002, **148**, 87–102.
27. M. Manefield, T. B. Rasmussen, M. Henzter, J. B. Andersen, P. Steinberg and S. Kjelleberg, *et al.*, Halogenated furanones inhibit quorum sensing through accelerated LuxR turnover, *Microbiology*, 2002, **148**, 1119–1127.
28. D. Ren, J. J. Sims and T. K. Wood, Inhibition of biofilm formation and swarming of Escherichia coli by (5Z)-4-bromo-5-(bromomethylene)-3-butyl-2(5H)-furanone, *Environ. Microbiol.*, 2001, **3**(11), 731–736.
29. D. Ren, J. J. Sims and T. K. Wood, Inhibition of biofilm formation and swarming of Bacillus subtilis by (5Z)-4-bromo-5-(bromomethylene)-3-butyl-2(5H)-furanone, *Lett. Appl. Microbiol.*, 2002, **34**(4), 293–299.
30. M. Hentzer, H. Wu, J. B. Andersen, K. Riedel, T. B. Rasmussen and N. Bagge, *et al.*, Attenuation of Pseudomonas aeruginosa virulence by quorum sensing inhibitors, *EMBO J.*, 2003, **22**(15), 3803–3815.
31. L. D. Christensen, C. Moser, P. O. Jensen, T. B. Rasmussen, L. Christophersen and S. Kjelleberg, *et al.*, Impact of Pseudomonas aeruginosa quorum sensing on biofilm persistence in an *in vivo* intraperitoneal foreign-body infection model, *Microbiology*, 2007, **153**(pt 7), 2312–2320.
32. J. Lonn-Stensrud, F. C. Petersen, T. Benneche and A. A. Scheie, Synthetic bromated furanone inhibits autoinducer-2-mediated communication and biofilm formation in oral streptococci, *Oral Microbiol. Immunol.*, 2007, **22**(5), 340–346.
33. Y. Han, S. Hou, K. A. Simon, D. Ren and Y. Y. Luk, Identifying the important structural elements of brominated furanones for inhibiting biofilm formation by Escherichia coli, *Bioorg. Med. Chem. Lett.*, 2008, **18**(3), 1006–1010.
34. C. Kim, J. Kim, H. Y. Park, H. J. Park, J. H. Lee and C. K. Kim, *et al.*, Furanone derivatives as quorum-sensing antagonists of Pseudomonas aeruginosa, *Appl. Microbiol. Biotechnol.*, 2008, **80**(1), 37–47.
35. M. Hentzer and M. Givskov, Pharmacological inhibition of quorum sensing for the treatment of chronic bacterial infections, *J. Clin. Invest.*, 2003, **112**(9), 1300–1307.

36. S. Yang, O. A. Abdel-Razek, F. Cheng, D. Bandyopadhyay, G. S. Shetye and G. Wang, *et al.,* Bicyclic brominated furanones: a new class of quorum sensing modulators that inhibit bacterial biofilm formation, *Bioorg. Med. Chem.*, 2014, **22**(4), 1313–1317.
37. Y. H. Dong and L. H. Zhang, Quorum sensing and quorum-quenching enzymes, *J. Microbiol.*, 2005, **43**, 101–109.
38. Y. H. Dong, J. L. Xu, X. Z. Li and L. H. Zhang, AiiA, an enzyme that inactivates the acylhomoserine lactone quorum-sensing signal and attenuates the virulence of Erwinia carotovora, *Proc. Natl. Acad. Sci. U. S. A.*, 2000, **97**(7), 3526–3531.
39. Y. H. Dong, L. H. Wang, J. L. Xu, H. B. Zhang, X. F. Zhang and L. H. Zhang, Quenching quorum-sensing-dependent bacterial infection by an N-acyl homoserine lactonase, *Nature*, 2001, **411**(6839), 813–817.
40. P. Bijtenhoorn, C. Schipper, C. Hornung, M. Quitschau, S. Grond and N. Weiland, *et al.,* BpiB05, a novel metagenome-derived hydrolase acting on N-acylhomoserine lactones, *J. Biotechnol.*, 2011, **155**(1), 86–94.
41. S. Hraiech, J. Hiblot, J. Lafleur, H. Lepidi, L. Papazian and J. M. Rolain, *et al.,* Inhaled lactonase reduces Pseudomonas aeruginosa quorum sensing and mortality in rat pneumonia, *PLoS One*, 2014, **9**(10), e107125.
42. G. J. Lyon and R. P. Novick, Peptide signaling in Staphylococcus aureus and other Gram-positive bacteria, *Peptides*, 2004, **25**(9), 1389–1403.
43. C. Bordi and S. de Bentzmann, Hacking into bacterial biofilms: a new therapeutic challenge, *Ann. Intensive Care*, 2011, **1**(1), 19.
44. C. A. Fux, P. Stoodley, L. Hall-Stoodley and J. W. Costerton, Bacterial biofilms: a diagnostic and therapeutic challenge, *Expert Rev. Anti-Infect. Ther.*, 2003, **1**(4), 667–683.
45. A. Giacometti, O. Cirioni, Y. Gov, R. Ghiselli, M. S. Del Prete and F. Mocchegiani, *et al.,* RNA III inhibiting peptide inhibits *in vivo* biofilm formation by drug-resistant Staphylococcus aureus, *Antimicrob. Agents Chemother.*, 2003, **47**(6), 1979–1983.
46. C. P. Gordon, P. Williams and W. C. Chan, Attenuating Staphylococcus aureus virulence gene regulation: a medicinal chemistry perspective, *J. Med. Chem.*, 2013, **56**(4), 1389–1404.
47. O. Simonetti, O. Cirioni, F. Mocchegiani, I. Cacciatore, C. Silvestri and L. Baldassarre, *et al.,* The efficacy of the quorum sensing inhibitor FS8 and tigecycline in preventing prosthesis biofilm in an animal model of staphylococcal infection, *Int. J. Mol. Sci.*, 2013, **14**(8), 16321–16332.
48. M. D. Kiran, N. V. Adikesavan, O. Cirioni, A. Giacometti, C. Silvestri and G. Scalise, *et al.,* Discovery of a quorum-sensing inhibitor of drug-resistant staphylococcal infections by structure-based virtual screening, *Mol. Pharmacol.*, 2008, **73**(5), 1578–1586.
49. C. M. Waters and B. L. Bassler, Quorum sensing: cell-to-cell communication in bacteria, *Annu. Rev. Cell Dev. Biol.*, 2005, **21**, 319–346.
50. G. Brackman and T. Coenye, Quorum sensing inhibitors as anti-biofilm agents, *Curr. Pharm. Des.*, 2015, **21**(1), 5–11.

51. S. T. Miller, K. B. Xavier, S. R. Campagna, M. E. Taga, M. F. Semmelhack and B. L. Bassler, *et al.,* Salmonella typhimurium recognizes a chemically distinct form of the bacterial quorum-sensing signal AI-2, *Mol. Cell*, 2004, **15**(5), 677–687.
52. M. Guo, S. Gamby, Y. Zheng and H. O. Sintim, Small molecule inhibitors of AI-2 signaling in bacteria: state-of-the-art and future perspectives for anti-quorum sensing agents, *Int. J. Mol. Sci.*, 2013, **14**(9), 17694–17728.
53. M. Frezza, L. Soulere, D. Balestrino, M. Gohar, C. Deshayes and Y. Queneau, *et al.,* Ac2-DPD, the bis-(O)-acetylated derivative of 4,5-dihydroxy-2,3-pentanedione (DPD) is a convenient stable precursor of bacterial quorum sensing autoinducer AI-2, *Bioorg. Med. Chem. Lett.*, 2007, **17**(5), 1428–1431.
54. K. Tsuchikama, J. Zhu, C. A. Lowery, G. F. Kaufmann and K. D. Janda, C4-alkoxy-HPD: a potent class of synthetic modulators surpassing nature in AI-2 quorum sensing, *J. Am. Chem. Soc.*, 2012, **134**(33), 13562–13564.
55. C. A. Lowery, J. Park, G. F. Kaufmann and K. D. Janda, An unexpected switch in the modulation of AI-2-based quorum sensing discovered through synthetic 4,5-dihydroxy-2,3-pentanedione analogues, *J. Am. Chem. Soc.*, 2008, **130**(29), 9200–9201.
56. V. Roy, M. T. Meyer, J. A. Smith, S. Gamby, H. O. Sintim and R. Ghodssi, *et al.,* AI-2 analogs and antibiotics: a synergistic approach to reduce bacterial biofilms, *Appl. Microbiol. Biotechnol.*, 2013, **97**(6), 2627–2638.
57. M. Kadirvel, F. Fanimarvasti, S. Forbes, A. McBain, J. M. Gardiner and G. D. Brown, *et al.,* Inhibition of quorum sensing and biofilm formation in Vibrio harveyi by 4-fluoro-DPD; a novel potent inhibitor of signalling, *Chem. Commun.*, 2014, **50**(39), 5000–5002.
58. G. Brackman, S. Celen, K. Baruah, P. Bossier, S. Van Calenbergh and H. J. Nelis, *et al.,* AI-2 quorum-sensing inhibitors affect the starvation response and reduce virulence in several Vibrio species, most likely by interfering with LuxPQ, *Microbiology*, 2009, **155**(pt 12), 4114–4122.
59. J. A. Gutierrez, T. Crowder, A. Rinaldo-Matthis, M. C. Ho, S. C. Almo and V. L. Schramm, Transition state analogs of 5′-methylthioadenosine nucleosidase disrupt quorum sensing, *Nat. Chem. Biol.*, 2009, **5**(4), 251–257.
60. H. Yan and W. Chen, 3',5'-Cyclic diguanylic acid: a small nucleotide that makes big impacts, *Chem. Soc. Rev.*, 2010, **39**(8), 2914–2924.
61. R. Tamayo, J. T. Pratt and A. Camilli, Roles of cyclic diguanylate in the regulation of bacterial pathogenesis, *Annu. Rev. Microbiol.*, 2007, **61**, 131–148.
62. D. L. Caly, D. Bellini, M. A. Walsh, J. M. Dow and R. P. Ryan, Targeting cyclic di-GMP signalling: a strategy to control biofilm formation? *Curr. Pharm. Des.*, 2015, **21**(1), 12–24.
63. D. Antoniani, P. Bocci, A. Maciag, N. Raffaelli and P. Landini, Monitoring of diguanylate cyclase activity and of cyclic-di-GMP biosynthesis by whole-cell assays suitable for high-throughput screening of biofilm inhibitors, *Appl. Microbiol. Biotechnol.*, 2010, **85**(4), 1095–1104.

64. K. Sambanthamoorthy, A. A. Gokhale, W. Lao, V. Parashar, M. B. Neiditch and M. F. Semmelhack, *et al.*, Identification of a novel benzimidazole that inhibits bacterial biofilm formation in a broad-spectrum manner, *Antimicrob. Agents Chemother.*, 2011, **55**(9), 4369–4378.
65. K. Sambanthamoorthy, R. E. Sloup, V. Parashar, J. M. Smith, E. E. Kim and M. F. Semmelhack, *et al.*, Identification of small molecules that antagonize diguanylate cyclase enzymes to inhibit biofilm formation, *Antimicrob. Agents Chemother.*, 2012, **56**(10), 5202–5211.
66. O. J. Lieberman, M. W. Orr, Y. Wang and V. T. Lee, High-throughput screening using the differential radial capillary action of ligand assay identifies ebselen as an inhibitor of diguanylate cyclases, *ACS Chem. Biol.*, 2014, **9**(1), 183–192.
67. J. Lee and J. H. Lee, Indole as an intercellular signal in microbial communities, *FEMS Microbiol. Rev.*, 2010, **34**(4), 426–444.
68. J. Lee, A. Jayaraman and T. K. Wood, Indole is an inter-species biofilm signal mediated by SdiA, *BMC Microbiol.*, 2007, **7**, 42.
69. R. J. Melander, M. J. Minvielle and C. Melander, Controlling bacterial behavior with indole-containing natural products and derivatives, *Tetrahedron*, 2014, **70**(37), 6363–6372.
70. J. Lee, T. Bansal, A. Jayaraman, W. E. Bentley and T. K. Wood, Enterohemorrhagic Escherichia coli biofilms are inhibited by 7-hydroxyindole and stimulated by isatin, *Appl. Environ. Microbiol.*, 2007, **73**(13), 4100–4109.
71. T. K. Wood, J. Lee, X. S. Zhang, M. Hegde, W. E. Bentley and A. Jayaraman, Indole cell signaling occurs primarily at low temperatures in Escherichia coli, *ISME J.*, 2008, **2**(10), 1007–1023.
72. J. H. Lee, M. H. Cho and J. Lee, 3-indolylacetonitrile decreases Escherichia coli O157:H7 biofilm formation and Pseudomonas aeruginosa virulence, *Environ. Microbiol.*, 2011, **13**(1), 62–73.
73. J. H. Lee, Y. G. Kim, M. H. Cho, J. A. Kim and J. Lee, 7-fluoroindole as an antivirulence compound against Pseudomonas aeruginosa, *FEMS Microbiol. Lett.*, 2012, **329**(1), 36–44.
74. S. C. Robijns, B. De Pauw, B. Loosen, A. Marchand, P. Chaltin and S. C. De Keersmaecker, *et al.*, Identification and characterization of 4-[4-(3-phenyl-2-propen-1-yl)-1-piperazinyl]-5H-pyrimido[5,4-b]indole derivatives as Salmonella biofilm inhibitors, *FEMS Immunol. Med. Microbiol.*, 2012, **65**(2), 390–394.
75. L. Peters, G. M. Konig, A. D. Wright, R. Pukall, E. Stackebrandt and L. Eberl, *et al.*, Secondary metabolites of Flustra foliacea and their influence on bacteria, *Appl. Environ. Microbiol.*, 2003, **69**(6), 3469–3475.
76. C. A. Bunders, M. J. Minvielle, R. J. Worthington, M. Ortiz, J. Cavanagh and C. Melander, Intercepting bacterial indole signaling with flustramine derivatives, *J. Am. Chem. Soc.*, 2011, **133**(50), 20160–20163.
77. M. J. Minvielle, K. Eguren and C. Melander, Highly active modulators of indole signaling alter pathogenic behaviors in Gram-negative and Gram-positive bacteria, *Chemistry*, 2013, **19**(51), 17595–17602.

78. C. Bunders, J. Cavanagh and C. Melander, Flustramine inspired synthesis and biological evaluation of pyrroloindoline triazole amides as novel inhibitors of bacterial biofilms, *Org. Biomol. Chem.*, 2011, **9**(15), 5476–5481.
79. M. J. Minvielle, C. A. Bunders and C. Melander, Indole/triazole conjugates are selective inhibitors and inducers of bacterial biofilms, *MedChemComm*, 2013, **4**(6), 916–919.
80. A. M. Stock, V. L. Robinson and P. N. Goudreau, Two-component signal transduction, *Annu. Rev. Biochem.*, 2000, **69**, 183–215.
81. Y. Gotoh, Y. Eguchi, T. Watanabe, S. Okamoto, A. Doi and R. Utsumi, Two-component signal transduction as potential drug targets in pathogenic bacteria, *Curr. Opin. Microbiol.*, 2010, **13**(2), 232–239.
82. R. J. Worthington, M. S. Blackledge and C. Melander, Small-molecule inhibition of bacterial two-component systems to combat antibiotic resistance and virulence, *Future Med. Chem.*, 2013, **5**(11), 1265–1284.
83. Y. Eguchi, N. Kubo, H. Matsunaga, M. Igarashi and R. Utsumi, Development of an antivirulence drug against Streptococcus mutans: repression of biofilm formation, acid tolerance, and competence by a histidine kinase inhibitor, walkmycin C, *Antimicrob. Agents Chemother.*, 2011, **55**(4), 1475–1484.
84. F. Qi, J. Merritt, R. Lux and W. Shi, Inactivation of the ciaH Gene in Streptococcus mutans diminishes mutacin production and competence development, alters sucrose-dependent biofilm formation, and reduces stress tolerance, *Infect. Immun.*, 2004, **72**(8), 4895–4899.
85. M. Reck, K. Rutz, B. Kunze, J. Tomasch, S. K. Surapaneni and S. Schulz, *et al.,* The biofilm inhibitor carolacton disturbs membrane integrity and cell division of Streptococcus mutans through the serine/threonine protein kinase PknB, *J. Bacteriol.*, 2011, **193**(20), 5692–5706.
86. B. Kunze, M. Reck, A. Dotsch, A. Lemme, D. Schummer and H. Irschik, *et al.,* Damage of Streptococcus mutans biofilms by carolacton, a secondary metabolite from the myxobacterium Sorangium cellulosum, *BMC Microbiol.*, 2010, **10**, 199.
87. T. E. Ballard, J. J. Richards, A. L. Wolfe and C. Melander, Synthesis and Antibiofilm Activity of a Second-Generation Reverse-Amide Oroidin Library: A Structure-Activity Relationship Study, *Chem.–Eur. J.*, 2008, **14**(34), 10745–10761.
88. A. P. Tomaras, M. J. Flagler, C. W. Dorsey, J. A. Gaddy and L. A. Actis, Characterization of a two-component regulatory system from Acinetobacter baumannii that controls biofilm formation and cellular morphology, *Microbiology*, 2008, **154**, 3398–3409.
89. R. J. Thompson, B. G. Bobay, S. D. Stowe, A. L. Olson, L. Peng and Z. Su, *et al.,* Identification of BfmR, a response regulator involved in biofilm development, as a target for a 2-Aminoimidazole-based antibiofilm agent, *Biochemistry*, 2012, **51**(49), 9776–9778.
90. C. Liu, R. J. Worthington, C. Melander and H. Wu, A new small molecule specifically inhibits the cariogenic bacterium Streptococcus

mutans in multispecies biofilms, *Antimicrob. Agents Chemother.*, 2011, **55**(6), 2679–2687.
91. C. J. Wright, H. Wu, R. J. Melander, C. Melander and J. Lamont, Disruption of heterotypic community development by Porphyromonas gingivalis with small molecule inhibitors, *Mol. Oral. Microbiol.*, 2014, **29**(5), 185–193.
92. C. C. Dirusso and P. N. Black, Bacterial long chain fatty acid transport: gateway to a fatty acid-responsive signaling system, *J. Biol. Chem.*, 2004, **279**(48), 49563–49566.
93. D. G. Davies and C. N. H. Marques, A Fatty Acid Messenger Is Responsible for Inducing Dispersion in Microbial Biofilms, *J. Bacteriol.*, 2009, **191**(5), 1393–1403.
94. C. N. Marques, D. G. Davies and K. Sauer, Control of Biofilms with the Fatty Acid Signaling Molecule cis-2-Decenoic Acid, *Pharmaceuticals (Basel)*, 2015, **8**(4), 816–835.
95. J. M. Dow, L. Crossman, K. Findlay, Y. Q. He, J. X. Feng and J. L. Tang, Biofilm dispersal in Xanthomonas campestris is controlled by cell-cell signaling and is required for full virulence to plants, *Proc. Natl. Acad. Sci. U. S. A.*, 2003, **100**(19), 10995–11000.
96. S. N. Dean, M. C. Chung and M. L. van Hoek, Burkholderia Diffusible Signal Factor Signals to Francisella novicida To Disperse Biofilm and Increase Siderophore Production, *Appl. Environ. Microbiol.*, 2015, **81**(20), 7057–7066.
97. J. Tian, L. X. Weng, Y. Q. Zhang and L. H. Wang, BDSF inhibits Candida albicans adherence to urinary catheters, *Microb. Pathog.*, 2013, **64**, 33–38.
98. I. B. Wenderska, M. Chong, J. McNulty, G. D. Wright and L. L. Burrows, Palmitoyl-DL-carnitine is a multitarget inhibitor of Pseudomonas aeruginosa biofilm development, *ChemBioChem*, 2011, **12**(18), 2759–2766.
99. U. T. Nguyen, I. B. Wenderska, M. A. Chong, K. Koteva, G. D. Wright and L. L. Burrows, Small-molecule modulators of Listeria monocytogenes biofilm development, *Appl. Environ. Microbiol.*, 2012, **78**(5), 1454–1465.
100. A. J. Slusarenko, A. Patel and D. Portz, Control of plant diseases by natural products: Allicin from garlic as a case study, *Eur. J. Plant Pathol.*, 2008, **121**(3), 313–322.
101. T. Bjarnsholt, P. O. Jensen, T. B. Rasmussen, L. Christophersen, H. Calum and M. Hentzer, *et al.*, Garlic blocks quorum sensing and promotes rapid clearing of pulmonary Pseudomonas aeruginosa infections, *Microbiology*, 2005, **151**, 3873–3880.
102. T. Persson, T. H. Hansen, T. B. Rasmussen, M. E. Skinderso, M. Givskov and J. Nielsen, Rational design and synthesis of new quorum-sensing inhibitors derived from acylated homoserine lactones and natural products from garlic, *Org. Biomol. Chem.*, 2005, **3**(2), 253–262.
103. N. Augustine, A. K. Goel, K. C. Sivakumar, R. Ajay Kumar and S. Thomas, Resveratrol - A potential inhibitor of biofilm formation in Vibrio cholerae, *Phytomedicine*, 2014, **21**(3), 286–289.

104. J. H. Lee, S. C. Regmi, J. A. Kim, M. H. Cho, H. Yun and C. S. Lee, *et al.*, Apple flavonoid phloretin inhibits Escherichia coli O157:H7 biofilm formation and ameliorates colon inflammation in rats, *Infect. Immun.*, 2011, **79**(12), 4819–4827.
105. A. Vikram, G. K. Jayaprakasha, P. R. Jesudhasan, S. D. Pillai and B. S. Patil, Suppression of bacterial cell-cell signalling, biofilm formation and type III secretion system by citrus flavonoids, *J. Appl. Microbiol.*, 2010, **109**(2), 515–527.
106. R. K. Ulrey, S. M. Barksdale, W. D. Zhou and M. L. van Hoek, Cranberry proanthocyanidins have anti-biofilm properties against Pseudomonas aeruginosa, *BMC Complementary Altern. Med.*, 2014, **14**, 499.
107. F. Ali, P. L. Sangwan, S. Koul, A. Pandey, S. Bani and S. T. Abdullah, *et al.*, 4-epi-Pimaric acid: a phytomolecule as a potent antibacterial and anti-biofilm agent for oral cavity pathogens, *Eur. J. Clin. Microbiol.*, 2012, **31**(2), 149–159.
108. S. Y. Tan, Y. Liu, S. L. Chua, R. M. Vejborg, T. H. Jakobsen and S. C. Chew, *et al.*, Comparative systems biology analysis to study the mode of action of the isothiocyanate compound Iberin on Pseudomonas aeruginosa, *Antimicrob. Agents Chemother.*, 2014, **58**(11), 6648–6659.
109. S. D. Stowe, J. J. Richards, A. T. Tucker, R. Thompson, C. Melander and J. Cavanagh, Anti-biofilm compounds derived from marine sponges, *Mar. Drugs*, 2011, **9**(10), 2010–2035.
110. J. C. Braekman, D. Daloze, C. Stoller and R. W. M. Vansoest, Chemotaxonomy of Agelas (Porifera, Demospongiae), *Biochem. Syst. Ecol.*, 1992, **20**(5), 417–431.
111. A. Yamada, H. Kitamura, K. Yamaguchi, S. Fukuzawa, C. Kamijima and K. Yazawa, *et al.*, Development of chemical substances regulating biofilm formation, *Bull. Chem. Soc. Jpn.*, 1997, **70**(12), 3061–3069.
112. J. J. Richards, T. E. Ballard, R. W. Huigens and C. Melander, Synthesis and Screening of an Oroidin Library Against Pseudomonas aeruginosa Biofilms, *ChemBioChem*, 2008, **9**(8), 1267–1279.
113. R. W. Huigens, J. J. Richards, G. Parise, T. E. Ballard, W. Zeng and R. Deora, *et al.*, Inhibition of Pseudomonas aeruginosa biofilm formation with bromoageliferin analogues, *J. Am. Chem. Soc.*, 2007, **129**(22), 6966–6967.
114. R. W. Huigens, L. Ma, C. Gambino, A. Basso, P. D. R. Moeller and J. Cavanagh, *et al.*, Control of Bacterial Biofilms with Marine Alkaloid Derivatives, *Mol. BioSyst.*, 2008, **4**(6), 614–621.
115. J. J. Richards, R. W. Huigens, T. E. Ballard, A. Basso, J. Cavanagh and C. Melander, Inhibition and Dispersion of Proteobacterial Biofilms, *Chem. Commun.*, 2008, (14), 1698–1700.
116. J. J. Richards, C. S. Reed and C. Melander, Effects of N-pyrrole substitution on the anti-biofilm activities of oroidin derivatives against Acinetobacter baumannii, *Bioorg. Med. Chem. Lett.*, 2008, **18**(15), 4325–4327.

117. J. J. Richards, T. E. Ballard and C. Melander, Inhibition and Dispersion of Pseudomonas aeruginosa Biofilms with Reverse Amide 2-Aminoimidazole Oroidin Analogues, *Org. Biomol. Chem.*, 2008, **6**(8), 1356–1363.
118. S. A. Rogers and C. Melander, Construction and Screening of a 2-Aminoimidazole Library Identifies a Small Molecule Capable of Inhibiting and Dispersing Biofilms Across Bacterial Order, Class, and Phylum, *Angew. Chem., Int. Ed.*, 2008, **47**(28), 5229–5231.
119. S. A. Rogers, J. D. Bero and C. Melander, Chemical Synthesis and Biological Screening of 2-Aminoimidazole-Based Bacterial and Fungal Antibiofilm Agents, *ChemBioChem*, 2010, **11**(3), 396–410.
120. S. A. Rogers, R. W. Huigens 3rd, J. Cavanagh and C. Melander, Synergistic effects between conventional antibiotics and 2-aminoimidazole-derived antibiofilm agents, *Antimicrob. Agents Chemother.*, 2010, **54**(5), 2112–2118.
121. T. L. Harris, R. J. Worthington and C. Melander, Potent Small-Molecule Suppression of Oxacillin Resistance in Methicillin-Resistant Staphylococcus aureus, *Angew Chem. Int. Ed. Engl.*, 2012, **51**(45), 11254–11257.
122. W. Pan, M. Fan, H. Wu, C. Melander and C. Liu, A new small molecule inhibits Streptococcus mutans biofilms *invitro* and *in vivo*, *J. Appl. Microbiol.*, 2015, **119**(5), 1403–1411.
123. C. A. Bunders, J. J. Richards and C. Melander, Identification of aryl 2-aminoimidazoles as biofilm inhibitors in Gram-negative bacteria, *Bioorg. Med. Chem. Lett.*, 2010, **20**(12), 3797–3800.
124. R. J. Worthington, M. S. Blackledge and C. Melander, Small Molecule Inhibition of Bacterial Two-Component Systems to Combat Antibiotic Resistance and Virulence, *Future Med. Chem.*, 2013, **5**(11), 1265–1284.
125. T. L. Harris, R. J. Worthington, L. E. Hittle, D. V. Zurawski, R. K. Ernst and C. Melander, Small Molecule Downregulation of PmrAB Reverses Lipid A Modification and Breaks Colistin Resistance, *ACS Chem. Biol.*, 2014, **9**(1), 122–127.
126. C. M. Brackett, R. J. Melander, I. H. An, A. Krishnamurthy, R. J. Thompson and J. Cavanagh, *et al.*, Small-Molecule Suppression of beta-Lactam Resistance in Multidrug-Resistant Gram-Negative Pathogens, *J. Med. Chem.*, 2014, **57**(17), 7450–7458.
127. H. P. Steenackers, D. S. Ermolat'ev, B. Savaliya, A. De Weerdt, D. De Coster and A. Shah, *et al.*, Structure-activity relationship of 4(5)-aryl-2-amino-1H-imidazoles, N1-substituted 2-aminoimidazoles and imidazo[1,2-a] pyrimidinium salts as inhibitors of biofilm formation by Salmonella typhimurium and Pseudomonas aeruginosa, *J. Med. Chem.*, 2011, **54**(2), 472–484.
128. D. S. Ermolat'ev, J. B. Bariwal, H. P. Steenackers, S. C. De Keersmaecker and E. V. Van der Eycken, Concise and diversity-oriented route toward polysubstituted 2-aminoimidazole alkaloids and their analogues, *Angew. Chem., Int. Ed. Engl.*, 2010, **49**(49), 9465–9468.

129. E. A. Lindsey, R. J. Worthington, C. Alcaraz and C. Melander, 2-Aminopyrimidine as a novel scaffold for biofilm modulation, *Org. Biomol. Chem.*, 2012, **10**(13), 2552–2561.
130. S. A. Rogers, R. W. Huigens and C. A. Melander, 2-Aminobenzimidazole that Inhibits and Disperses Gram-Positive Biofilms Through a Zinc-Dependent Mechanism, *J. Am. Chem. Soc.*, 2009, **131**(29), 9868–9869.
131. E. A. Lindsey, C. M. Brackett, T. Mullikin, C. Alcaraz and C. Melander, The discovery of N-1 substituted 2-aminobenzimidazoles as zinc-dependent S. aureus biofilm inhibitors, *MedChemComm*, 2012, **3**(11), 1462–1465.
132. R. Frei, A. S. Breitbach and H. E. Blackwell, 2-Aminobenzimidazole derivatives strongly inhibit and disperse Pseudomonas aeruginosa biofilms, *Angew. Chem., Int. Ed. Engl.*, 2012, **51**(21), 5226–5229.
133. S. A. Rogers, D. C. Whitehead, T. Mullikin and C. Melander, Synthesis and bacterial biofilm inhibition studies of ethyl N-(2-phenethyl) carbamate derivatives, *Org. Biomol. Chem.*, 2010, **8**(17), 3857–3859.
134. A. T. Garrison, F. Bai, Y. Abouelhassan, N. G. Paciaroni, S. G. Jin and R. W. Huigens, Bromophenazine derivatives with potent inhibition, dispersion and eradication activities against Staphylococcus aureus biofilms, *RSC Adv.*, 2015, **5**(2), 1120–1124.
135. A. T. Garrison, Y. Abouelhassan, D. Kallifidas, F. Bai, M. Ukhanova and V. Mai, *et al.*, Halogenated Phenazines that Potently Eradicate Biofilms, MRSA Persister Cells in Non-Biofilm Cultures, and Mycobacterium tuberculosis, *Angew. Chem., Int. Ed. Engl.*, 2015, 14819–14823.
136. K. C. Peach, A. T. Cheng, A. G. Oliver, F. H. Yildiz and R. G. Linington, Discovery and biological characterization of the auromomycin chromophore as an inhibitor of biofilm formation in Vibrio cholerae, *ChemBioChem*, 2013, **14**(16), 2209–2215.
137. C. J. Warner, A. T. Cheng, F. H. Yildiz and R. G. Linington, Development of benzo[1,4]oxazines as biofilm inhibitors and dispersal agents against Vibrio cholerae, *Chem. Commun.*, 2015, **51**(7), 1305–1308.
138. N. Suzuki, N. Ohtaguro, Y. Yoshida, M. Hirai, H. Matsuo and Y. Yamada, *et al.*, A Compound Inhibits Biofilm Formation of Staphylococcus aureus from Streptomyces, *Biol. Pharm. Bull.*, 2015, **38**(6), 889–892.
139. K. V. Reddy, R. D. Yedery and C. Aranha, Antimicrobial peptides: premises and promises, *Int. J. Antimicrob. Agents*, 2004, **24**(6), 536–547.
140. M. Di Luca, G. Maccari and R. Nifosi, Treatment of microbial biofilms in the post-antibiotic era: prophylactic and therapeutic use of antimicrobial peptides and their design by bioinformatics tools, *Pathog. Dis.*, 2014, **70**(3), 257–270.
141. N. Strempel, J. Strehmel and J. Overhage, Potential application of antimicrobial peptides in the treatment of bacterial biofilm infections, *Curr. Pharm. Des.*, 2015, **21**(1), 67–84.
142. J. Overhage, A. Campisano, M. Bains, E. C. Torfs, B. H. Rehm and R. E. Hancock, Human host defense peptide LL-37 prevents bacterial biofilm formation, *Infect. Immun.*, 2008, **76**(9), 4176–4182.

143. X. Feng, K. Sambanthamoorthy, T. Palys and C. Paranavitana, The human antimicrobial peptide LL-37 and its fragments possess both antimicrobial and antibiofilm activities against multidrug-resistant Acinetobacter baumannii, *Peptides*, 2013, **49**, 131–137.
144. S. N. Dean, B. M. Bishop and M. L. van Hoek, Natural and synthetic cathelicidin peptides with anti-microbial and anti-biofilm activity against Staphylococcus aureus, *BMC Microbiol.*, 2011, **11**, 114.
145. E. Hell, C. G. Giske, A. Nelson, U. Romling and G. Marchini, Human cathelicidin peptide LL37 inhibits both attachment capability and biofilm formation of Staphylococcus epidermidis, *Lett. Appl. Microbiol.*, 2010, **50**(2), 211–215.
146. S. N. Dean, B. M. Bishop and M. L. van Hoek, Susceptibility of Pseudomonas aeruginosa Biofilm to Alpha-Helical Peptides: D-enantiomer of LL-37, *Front. Microbiol.*, 2011, **2**, 128.
147. A. Pompilio, M. Scocchi, S. Pomponio, F. Guida, A. Di Primio and E. Fiscarelli, *et al.,* Antibacterial and anti-biofilm effects of cathelicidin peptides against pathogens isolated from cystic fibrosis patients, *Peptides*, 2011, **32**(9), 1807–1814.
148. E. M. Haisma, A. de Breij, H. Chan, J. T. van Dissel, J. W. Drijfhout and P. S. Hiemstra, *et al.,* LL-37-derived peptides eradicate multidrug-resistant Staphylococcus aureus from thermally wounded human skin equivalents, *Antimicrob. Agents Chemother.*, 2014, **58**(8), 4411–4419.
149. G. Wang, M. L. Hanke, B. Mishra, T. Lushnikova, C. E. Heim and V. Chittezham Thomas, *et al.,* Transformation of human cathelicidin LL-37 into selective, stable, and potent antimicrobial compounds, *ACS Chem. Biol.*, 2014, **9**(9), 1997–2002.
150. R. M. G. Biswajit Mishra, K. Lau, T. Lushnikova and G. Wang, Anti-Staphylococcal Biofilm Effects of Human Cathelicidin Peptides, *ACS Med. Chem. Lett.*, 2016, **7**(1), 117–121.
151. L. Steinstraesser, B. F. Tack, A. J. Waring, T. Hong, L. M. Boo and M. H. Fan, *et al.,* Activity of novispirin G10 against Pseudomonas aeruginosa *in vitro* and in infected burns, *Antimicrob. Agents Chemother.*, 2002, **46**(6), 1837–1844.
152. S. C. Mansour, C. de la Fuente-Nunez and R. E. Hancock, Peptide IDR-1018: modulating the immune system and targeting bacterial biofilms to treat antibiotic-resistant bacterial infections, *J. Pept. Sci.*, 2015, **21**(5), 323–329.
153. C. de la Fuente-Nunez, F. Reffuveille, E. F. Haney, S. K. Straus and R. E. Hancock, Broad-spectrum anti-biofilm peptide that targets a cellular stress response, *PLoS Pathog.*, 2014, **10**(5), e1004152.
154. F. Reffuveille, C. de la Fuente-Nunez, S. Mansour and R. E. Hancock, A broad-spectrum antibiofilm peptide enhances antibiotic action against bacterial biofilms, *Antimicrob. Agents Chemother.*, 2014, **58**(9), 5363–5371.
155. S. Baugh, A. S. Ekanayaka, L. J. Piddock and M. A. Webber, Loss of or inhibition of all multidrug resistance efflux pumps of Salmonella enterica serovar Typhimurium results in impaired ability to form a biofilm, *J. Antimicrob. Chemother.*, 2012, **67**(10), 2409–2417.

156. S. Baugh, C. R. Phillips, A. S. Ekanayaka, L. J. Piddock and M. A. Webber, Inhibition of multidrug efflux as a strategy to prevent biofilm formation, *J. Antimicrob. Chemother.*, 2014, **69**(3), 673–681.
157. J. B. Kaplan, Biofilm dispersal: mechanisms, clinical implications, and potential therapeutic uses, *J. Dent. Res.*, 2010, **89**(3), 205–218.
158. N. Ramasubbu, L. M. Thomas, C. Ragunath and J. B. Kaplan, Structural analysis of dispersin B, a biofilm-releasing glycoside hydrolase from the periodontopathogen Actinobacillus actinomycetemcomitans, *J. Mol. Biol.*, 2005, **349**(3), 475–486.
159. E. A. Izano, M. A. Amarante, W. B. Kher and J. B. Kaplan, Differential roles of poly-N-acetylglucosamine surface polysaccharide and extracellular DNA in Staphylococcus aureus and Staphylococcus epidermidis biofilms, *Appl. Environ. Microbiol.*, 2008, **74**(2), 470–476.
160. G. Donelli, I. Francolini, D. Romoli, E. Guaglianone, A. Piozzi and C. Ragunath, *et al.*, Synergistic activity of dispersin B and cefamandole nafate in inhibition of staphylococcal biofilm growth on polyurethanes, *Antimicrob. Agents Chemother.*, 2007, **51**(8), 2733–2740.
161. J. B. Kaplan, C. Ragunath, K. Velliyagounder, D. H. Fine and N. Ramasubbu, Enzymatic detachment of Staphylococcus epidermidis biofilms, *Antimicrob. Agents Chemother.*, 2004, **48**(7), 2633–2636.
162. Y. Itoh, X. Wang, B. J. Hinnebusch, J. F. Preston 3rd and T. Romeo, Depolymerization of beta-1,6-N-acetyl-D-glucosamine disrupts the integrity of diverse bacterial biofilms, *J. Bacteriol.*, 2005, **187**(1), 382–387.
163. R. O. Darouiche, M. D. Mansouri, P. V. Gawande and S. Madhyastha, Antimicrobial and antibiofilm efficacy of triclosan and DispersinB combination, *J. Antimicrob. Chemother.*, 2009, **64**(1), 88–93.
164. T. Y. Wong, L. A. Preston and N. L. Schiller, ALGINATE LYASE: review of major sources and enzyme characteristics, structure-function analysis, biological roles, and applications, *Annu. Rev. Microbiol.*, 2000, **54**, 289–340.
165. M. A. Alkawash, J. S. Soothill and N. L. Schiller, Alginate lyase enhances antibiotic killing of mucoid Pseudomonas aeruginosa in biofilms, *APMIS*, 2006, **114**(2), 131–138.
166. M. Alipour, Z. E. Suntres and A. Omri, Importance of DNase and alginate lyase for enhancing free and liposome encapsulated aminoglycoside activity against Pseudomonas aeruginosa, *J. Antimicrob. Chemother.*, 2009, **64**(2), 317–325.
167. J. W. Lamppa and K. E. Griswold, Alginate lyase exhibits catalysis-independent biofilm dispersion and antibiotic synergy, *Antimicrob. Agents Chemother.*, 2013, **57**(1), 137–145.
168. A. S. Bayer, S. Park, M. C. Ramos, C. C. Nast, F. Eftekhar and N. L. Schiller, Effects of alginase on the natural history and antibiotic therapy of experimental endocarditis caused by mucoid Pseudomonas aeruginosa, *Infect. Immun.*, 1992, **60**(10), 3979–3985.
169. V. V. Tetz and G. V. Tetz, Effect of extracellular DNA destruction by DNase I on characteristics of forming biofilms, *DNA Cell Biol.*, 2010, **29**(8), 399–405.

170. J. B. Kaplan, K. LoVetri, S. T. Cardona, S. Madhyastha, I. Sadovskaya and S. Jabbouri, *et al.*, Recombinant human DNase I decreases biofilm and increases antimicrobial susceptibility in staphylococci, *J. Antibiot. (Tokyo)*, 2012, **65**(2), 73–77.
171. G. Parsiegla, C. Noguere, L. Santell, R. A. Lazarus and Y. Bourne, The structure of human DNase I bound to magnesium and phosphate ions points to a catalytic mechanism common to members of the DNase I-like superfamily, *Biochemistry*, 2012, **51**(51), 10250–10258.
172. B. Frederiksen, T. Pressler, A. Hansen, C. Koch and N. Hoiby, Effect of aerosolized rhDNase (Pulmozyme) on pulmonary colonization in patients with cystic fibrosis, *Acta Paediatr.*, 2006, **95**(9), 1070–1074.
173. T. Iwase, Y. Uehara, H. Shinji, A. Tajima, H. Seo and K. Takada, *et al.*, Staphylococcus epidermidis Esp inhibits Staphylococcus aureus biofilm formation and nasal colonization, *Nature*, 2010, **465**(7296), 346–349.
174. J. H. Park, J. H. Lee, M. H. Cho, M. Herzberg and J. Lee, Acceleration of protease effect on Staphylococcus aureus biofilm dispersal, *FEMS Microbiol. Lett.*, 2012, **335**(1), 31–38.
175. C. Longhi, G. L. Scoarughi, F. Poggiali, A. Cellini, A. Carpentieri and L. Seganti, *et al.*, Protease treatment affects both invasion ability and biofilm formation in Listeria monocytogenes, *Microb. Pathog.*, 2008, **45**(1), 45–52.
176. L. Selan, F. Berlutti, C. Passariello, M. R. Comodi-Ballanti and M. C. Thaller, Proteolytic enzymes: a new treatment strategy for prosthetic infections? *Antimicrob. Agents Chemother.*, 1993, **37**(12), 2618–2621.

CHAPTER 4

Narrow Spectrum Antibacterial Agents

LAUREN A. SPAGNUOLO[a], PETER J. TONGE[a] AND STEWART L. FISHER*[b]

[a]Department of Chemistry and Institute for Chemical Biology and Drug Discovery, Stony Brook University, Stony Brook, NY 11794, USA; [b]C4 Therapeutics, Inc., 675 W. Kendall Street, Cambridge, MA 02142, USA
*E-mail: stewfisher@SLFisherConsulting.com

4.1 Introduction

The advent of the antibiotic era, heralded by the discoveries of arsphenamine by Ehrlich,[1] prontosil by Domagk[2] and penicillin G by Fleming[3] is now considered one of the most important events in the history of human health. The widespread, empirical use of drugs inspired by these compounds to treat a variety of infections, many of which were previously characterized by high rates of mortality and morbidity, transformed infectious diseases clinical practice and spurred intense efforts to identify new, broader spectrum antibacterial agents. The subsequent four decades of antibacterial discovery was extremely fruitful, and the vast majority of current antibacterial therapies can be traced to this "Golden Age of Antibacterial Discovery".[4] The impact of these discoveries cannot be over-emphasized, since comparisons to the "pre-antibiotic era" indicate dramatic improvements in all of the measures of an advanced society, from childhood and maternal childbirth mortality rates to extended average lifespans. Further, while these direct effects of antibiotic

Drug Discovery Series No. 58
Antibiotic Drug Discovery: New Targets and Molecular Entities
Edited by Steven M. Firestine and Troy Lister

Published by the Royal Society of Chemistry, www.rsc.org

therapy are readily apparent, it can be argued that these agents have enabled many of the advances of modern medicine such as organ transplantation, elective orthopedic surgeries and therapeutic options for life-threatening ailments that require immunomodulation, such as most current oncology therapies.[5] Thus, it is clear that antibiotics have both transformed and enabled modern medical practice.

With few exceptions, the drugs comprising the modern arsenal of antibacterial agents exhibit activity across a range of pathogens. The extent of coverage across the collection of known pathogenic species has provided a basis for a rough classification as either broad or narrow spectrum. While this classification arose from a relative comparison between generations of agents within an antibacterial class, this distinction faded over time with the continued discovery of new classes and subsequent generations of compounds that varied in potency across the range of pathogenic bacterial organisms. Like the adage, "How long is a piece of string?", the debate of narrow *versus* broad spectrum is highly contextual and dependent on the clinical indication and the underlying microbiological standard of care. For example, cephalosporins are typically considered broad-spectrum agents, particularly when compared to the comparatively narrower spectrum of the natural penicillins. However, cephalosporins can also be considered narrow-spectrum agents relative to subsequent generations of β-lactam antibiotics when activity against intracellular pathogens, anaerobes and non-fermenting organisms is considered.[6] For these reasons, the classification on antibacterial spectrum remains controversial and some have called for redefining these classifications based on other key factors, such as clinical indication and drug distribution to targeted organs, or even on the limitations in spectrum coverage, either through inherent or acquired resistance mechanisms.[7]

Despite this debate, it is clear that, from a clinical infectious diseases perspective, broader spectrum coverage facilitates an empirical approach to antimicrobial therapy. This is particularly evident for serious infections, where clinical outcomes are directly linked to the elapsed time between case presentation and appropriate antibiotic intervention. For example, high mortality rates are associated with bacterial ventilator-acquired pneumonia, but these rates doubled when adequate antimicrobial therapies were not administered within the first 24 h.[8,9] Since the microbiological diagnosis often requires both culture and resistance phenotyping—a process that can take up to 48 h—initial therapeutic decisions are based on clinical assessments and a broad-spectrum antibiotic regimen that is designed to cover the range of suspected pathogens.[10] Following microbiological identification, appropriate narrow spectrum therapy can be initiated—a process known as "de-escalation". Therefore, three factors dominate clinical practice and favor the use of the broadest spectrum agents for first line therapy: a relatively long delay for accurate diagnosis, the requirement for rapid therapeutic intervention and the historical abundance of effective, safe broad spectrum antibiotics. These factors have driven infectious disease clinical practice since the antibiotic era inception and continue to this day.

However, while this strategy remains the predominant recommended clinical practice, there are a number of emerging factors that threaten the viability this approach. First and foremost, resistance to the existing clinical agents has risen steadily over time. Resistant mechanisms have been identified for all antibacterial classes, regardless of discovery origin (natural product, synthetic lead) and are often observed at or even before commercialization. Further, it is now clear that antibiotic resistance mechanisms are widespread,[11] readily transferred across species, and have ancient origins.[12] Critically, strains containing multiple resistance mechanisms for the most serious pathogens have emerged and pan-resistant organisms have also surfaced. A recent survey of infectious disease clinicians revealed that a majority (>85%) had encountered at least one untreatable case resulting from a pan-resistant pathogen in the past year and that many (>35%) of these infections resulted in mortality or significant clinical complications.[13] This data, along with a growing body of surveillance evidence, indicates that multi-drug resistant strains are increasing at alarming rates.

Even as these data emerge, the number of new antibacterial agent approvals has steadily declined over the past decade and, looking forward, the clinical development pipeline is nearly bare.[14] The lack of new drugs can be traced to both technical and strategic factors. First, the discovery of new antibacterial agents has proved to be extraordinarily difficult, particularly for broad spectrum or Gram-negative-targeted therapies. The factors underpinning these challenges are multifactorial, but it is clear that antibacterial agents require a different chemical space than host-based therapies[15,16] and that bacterial cellular membranes present formidable barriers to compound penetration.[17] Further, despite the historical success, the technical challenges are elevated when in the context of broad spectrum agents. First, the available target space is more limited, since the vast genomic diversity across bacterial pathogens places strict limits on the number of viable drug targets and physiological pathways.[18,19] Second, broad-spectrum antibacterial agents often require elevated exposure requirements to ensure coverage of the least susceptible pathogen, resulting in limited therapeutic indices. These technical challenges, along with concerns in the regulatory approval process and effective reimbursement strategies for novel antibacterial agents, have led to a steady decay in pharmaceutical company research investment in this area over the past two decades.[20] Unlike other disease areas, there are several factors that limit the return on investment. Antibacterial therapies often require short-term dosing for effective therapy, but pricing is often based on a sales volume model with limited premiums. Further, new agents are typically reserved following approval to ensure that the utility of these agents is preserved for the most serious infections. As a result, the uptake volume of new agents is typically low and prolonged, leading to significantly lower net present value (NPV) calculations relative to other therapeutic areas. Finally, there is growing evidence that broad-spectrum treatments result in dramatic and largely irreversible damage to the intestinal microbiome.[21] These effects are most well documented in cases of *Clostridium difficile*-induced diarrhea[22] following broad-spectrum therapy, but links between microbiome dysbioses and severe, chronic diseases such as Type I diabetes,[23] inflammatory bowel disease[24,25] and colon cancer[26,27] are growing. To combat these risks, there is increasing pressure to develop narrow spectrum agents that target key pathogens but spare the host microbiome.

The combined forces of emergent pan-drug resistant pathogens, a weak development pipeline and diminished investment in new antibacterial agents have resulted in a global healthcare crisis.[28] Recently, the United States Executive Branch declared that antimicrobial resistance presents a national security threat and an Executive Order has been issued that outlines a series of actions to address this problem.[29] Importantly, the development of narrow spectrum agents is now seen as a means to address the most critical pathogens where the unmet medical need is high and growing. Since these targeted therapies are anticipated to demonstrate significant benefit to patients, proposals for a streamlined clinical trial process[30,31] and enhanced reimbursement rates[32] for these types of agents are gaining traction. Further, since these therapies will likely require pairing with high sensitivity and specificity clinical diagnostics to be fully effective first-line treatments, interest has surged in the development of rapid molecular diagnostics. For example, the United Kingdom recently announced a Longitude Prize Challenge[33]—named after the last challenge of this type over 300 years ago, which was based on the development of a standardized global positional grid system to improve marine navigation—for the development of a novel and specific antibacterial clinical diagnostic. Taken together, there is a resurgent interest in the discovery of narrow spectrum antibacterial agents, particularly against Gram-negative pathogens. This review will outline recent progress in the discovery of narrow spectrum agents that, at the highest level, includes agents with spectrums defined by Gram stain classification. Two general approaches are considered, based on small molecule compound library source: natural products and synthetic small molecules. Given the breadth of research in this area, a comprehensive review covering early target validation work through marketed agents is beyond the scope of this report. Rather, this review will highlight promising discovery project examples that have demonstrated proof of concept through microbiological activity and *in vivo* efficacy, and for those cases that have advanced into development, the path to the clinic is discussed. Marketed agents with narrow spectrum activity such as daptomycin[34] and linezolid[35] are not covered, since these have been the subject of extensive reviews to date. It is noted that there is potential overlap with other chapters in this series, including the discovery of agents against *C. dificile* (Chapter 1), agents targeting *M. tuberculosis* (Chapter 7) and the Gram-negative LPS biosynthesis target LpxC (Chapter 5); interested readers are directed to those reports for narrow spectrum antibacterial examples in these areas.

4.2 Natural Products

Given the productivity of the "Golden Age" of antibiotic discovery, it is not surprising that many present-day narrow spectrum antibacterial programs are based on natural product-derived leads. These programs vary in scope, but many are based on legacy scaffolds that, given a strong bias for broad spectrum activity, were overlooked at the time of initial isolation. As a case in point, Wyeth initiated an internal review of their natural product screening efforts at the turn of the millennium to re-assess compounds from their previous 50 years of isolation and characterization. Among the most promising was the mannopeptimycins, a series of glycopeptide antibacterials (**1**, Figure 4.1). These derivatives were first isolated as a complex (AC98) from *Streptomyces hygroscopicus* by American Cyanamid in 1958.[36] Upon further purification, the complex was found to contain five derivatives that differ in substitution on the terminal mannose ring, which was subsequently found to be the dominant feature for biological activity.[37] In general, these derivatives have adequate potency against Gram-positive organisms, including methicillin-resistant *Staphylococcus aureus* and vancomycin-resistant *Enterococcus faecium*. Several of these analogs exhibited good efficacies in a murine lethal infection model, suggesting that this series had potential for further optimization.

The mannopeptimycins inhibit the cell wall biosynthetic pathway by directly binding to the advanced peptidoglycan biosynthesis intermediate, Lipid II.[38] While this is mechanistically similar to the vancomycin glycopeptides, these molecules have a distinct, but as yet undetermined, binding motif and no cross-resistance is observed between these classes.[36] Mannopeptimycins show weak, non-specific binding to lipoteichoic acids and this may aid in concentrating the drug to the bacterial cell wall surface. As expected from this mechanism, resistance mutants proved near impossible to isolate despite extensive efforts using a variety of techniques.[36]

Several parallel strategies were employed to improve the activity and pharmacokinetics of the series including selective chemical semi-synthesis, directed biosynthesis and biosynthetic engineering. Despite the chemical complexity of the mannopeptimycins, a variety of acetal and ketal derivatives were successfully prepared using relatively simple reaction and chromatography steps starting with the parent molecule.[39–41] Of these, symmetrical and hindered ketals were found to be optimal for both microbiological activity and pharmacokinetic profile.[41] Among these modifications, replacement of the core 2-methyl-phenylalanine residue with cyclohexyl-alanine through addition of this metabolite during fermentation was found to be the most productive in overall activity improvement.[41] Using a combination of these two approaches, AC98-6446 (**1**, Figure 4.1) was identified as the lead molecule for pre-clinical development.[42]

AC98-6446 has excellent microbiological activity that is comparable or more potent than clinical comparators against key Gram-positive pathogens including Staphylococcal spp (including MRSA), Streptococcal spp (including penicillin-resistant *Streptococcus pneumoniae*) and Enterococcus spp

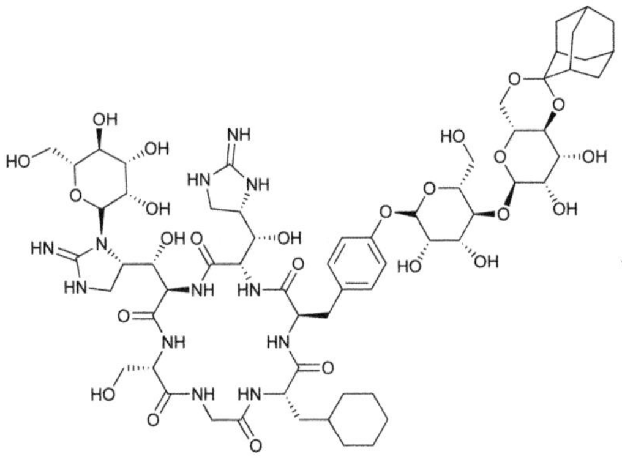

1
(AC98-6446)

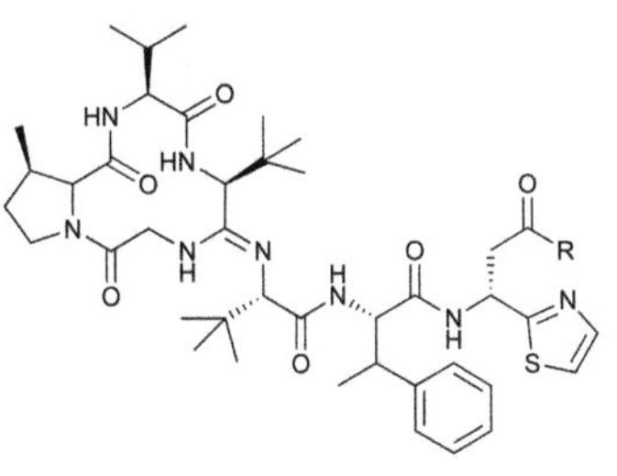

2a R = OMe (Bottromycin A_2)
2b R = $NHNH_2$

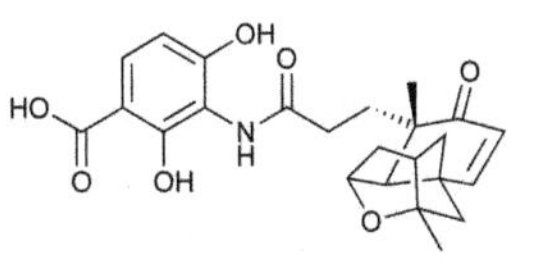

3
(Platensimycin)

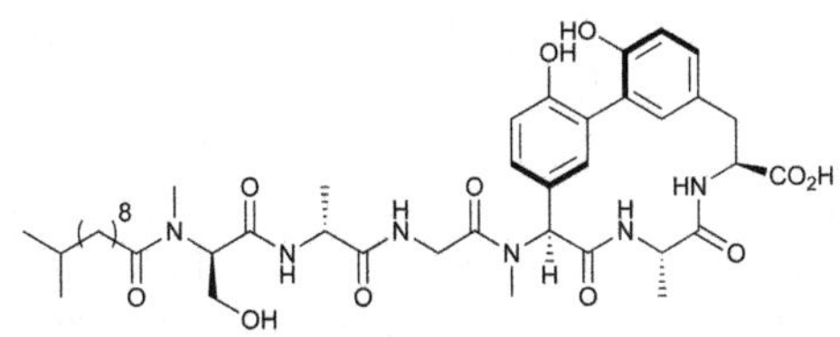

4
(Arylomycin A_2)

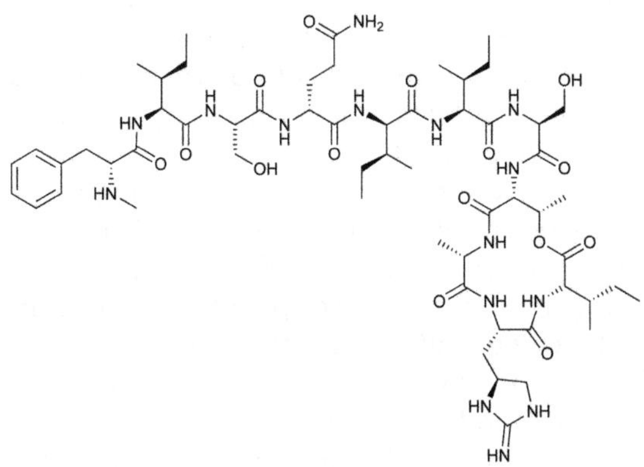

5
(Teixobactin)

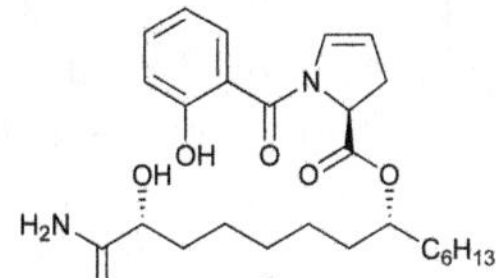

6
(Promysalin)

Figure 4.1 Representative analogs of narrow spectrum, natural product-derived antibacterial classes.

(incuding VRE).[43] In contrast to vancomycin, this derivative exhibits concentration dependent bactericidal activity against most species. AC98-6446 exhibits excellent *in vivo* activity against clinically resistant Gram-positive pathogens in both murine acute lethal and bacterial thigh infection models, as well as a rat endocarditis infection model.[44] In all cases, AC98-6446 was at least five-fold more efficacious than vancomycin and significant reductions in overall bacterial load were noted. The pharmacokinetics of the drug was assessed in four species using *iv* administration and overall exposure was dependent on the metabolic clearance rate. As a result, higher species (monkey and dog) had the longest plasma $t_{1/2}$ and highest exposure. The increased interest in the mannopeptimycin scaffold has sparked recent total synthesis efforts[45,46] and identification of enzymes responsible for the synthesis of the hexapeptide core and other chemical modifications like mannosylation, isovalerylation, hydroxylation, and methylation.[47–49] These advances now enable a combination approach utilizing total synthesis, semi-synthesis, and chemoenzymatic synthesis that with further drive the potential for this unique class of antibacterials.

The bottromycins represent another series of "rediscovered" natural product, peptide-based antibiotics. First isolated in 1957 from *Streptococcus bottropensis*,[50] the active component, Bottromycin A2 (**2a**, Figure 4.1), had good activity against Gram-positive organisms.[51] Over time, a number of derivatives have been isolated that vary in substitution at the central proline residue[52,53] and isolation of the closely related analog chandramycin[54] has expanded the naturally-occurring family members. Structural characterization of this series proved very challenging; while the initial reports supported a linear peptide structure, it is now clear that these compounds have a cyclic peptide backbone with a novel amidine linkage.[55] In all cases, the ester moiety was found to be a key element for microbiological activity and hydrolysis to the parent acid resulted in loss of activity.[53] Despite the good microbiological activity and relative metabolic stability in serum and mouse liver homogenates, Bottromycin A2 was not effective in murine experimental models of Streptococcal and Staphylococcal infections.[56] In contrast, a hydrazide derivative prepared by direct reaction with the parent ester, was efficacious in the murine Staphylococcal and Streptococcal infection models even though it exhibits a significantly weaker microbiological potency profile (**2b**, Figure 4.1).[56] This differential in activity was attributed to the significantly higher serum exposure relative to Bottromycin A2. Further, this analog was well tolerated in single injection (LD_{50}: 1000 mg kg^{-1}, intraperitoneal injection) and repeat dosing (200 mg kg^{-1}, 30 days qd subcutaneous administration).

The bottromycins inhibit protein translation, but utilize a unique mode of action that is characterized by binding to the 50S subunit and blocking the binding of tRNA to the acceptor site[57–60] causing inhibition of protein translation at the elongation phase and accumulation of aminoacyl-tRNAs. As expected from this novel mechanism, a narrow distribution of MIC values was obtained with a collection of methicillin-resistant *S. aureus* strains with no evidence of resistant subpopulations.[61]

Based on the early success of the hydrazide analog and the relative ease of the semi-synthetic preparations, most of the optimization efforts in this series of antibacterials have focused on further derivatization of the ester moiety. A large variety of amides, ureas and hydrazides have been prepared using the parent bottromycin ester as the starting material.[52,53,62] In most cases, the microbiological profile is maintained but improvements in the metabolic stability have been noted. The most promising analogs have excellent Gram-positive activity including coverage of MRSA and VRE clinical isolates. Recent reports on the total synthesis of Bottromycin A2 suggests that modifications throughout the scaffold may now be accessible.[55,63] Recently, proteins involved in the biosynthetic pathway of Bottromycin biosynthesis have been elucidated and characterized upon the identification of the gene clusters from *S. bottropensis* and *S. scabies*.[64,65] These results stimulate the need for a full elucidation of the mechanism for bottromycin biosynthesis, which will aid in the generation of bottromycin libraries with improved properties. A novel mechanism for the amidine ring formation has been proposed for the biosynthetic pathway in that the amidine-ring synthesis, unconventional D-amino acids and β-methylated residues comprising the bottromycin scaffold are of ribosomal origin.[64–66] The authors note that these findings, while clearly relevant to secondary metabolite biosynthesis, have broader implications for understanding nature's ability to afford complex and diverse mature proteins upon post-translational modification. As a result, the continued development of the bottromycin class, including recent work on the Bottromycin D analog,[67] remains an area of active interest with far reaching implications.

Researchers at Merck identified a novel antibacterial class, exemplified by platensimycin[68] (See **3**, Figure 4.1), using a novel, array-based siRNA phenotypic screen of *S. aureus*.[69] These compounds target fatty acid biosynthesis and, through a series of elegant biochemical studies, were shown to exhibit a unique mode of inhibition. Platensimycin specifically bound to the acylated form of the fatty acid condensing enzyme, FabF, thereby trapping the enzyme mid-way through the catalytic cycle. Platensimycin has good activity across Gram-positive species, such as Staphylococci, Streptococci and Enterococcus spp. In contrast, Gram-negative pathogens were not inhibited due to poor intracellular penetration. Platensimycin demonstrated *in vivo* efficacy in a *S. aureus* murine infection model; however, continuous infusion was required to achieve sufficient exposure, suggesting liabilities in the pharmacokinetic profile of this compound. The total synthesis of platensimycin has been reported,[70] along with the related analog platencin,[71,72] thereby enabling chemical exploration of these scaffolds[73,74] beyond the array of naturally occurring members of this class and semisynthetic derivatives.[75–78]

In 2002, the Arylomycin class of lipopeptide antibacterials was identified using standard isolation and characterization approaches.[79,80] These novel hexapeptide scaffolds contain a C-terminal biaryl-bridged macrocycle comprised of *N*-methylphenylglycine, glycine and tyrosine with a variety of alkyl chain variants capping the N-terminal, D-amino acid-containing tail (See **4**, Figure 4.1). Two major subgroups were observed, which differed only in the nitrosylation of the tyrosine residue. The microbiological activity of these compounds was very narrow, with very limited Gram-positive activity and no effect on the growth of Gram-negative organisms. The related lipoglycopeptide class was identified soon after, using a high throughput screen of recombinant bacterial signal peptidase I against a collection of natural product extracts.[81] Signal peptidase I (SPase I) is an essential, highly conserved serine protease involved in the processing of outer-membrane and secreted proteins, and cleaves the signal leader peptide sequence upon export through the cytoplasmic membrane. The lipoglycopeptides, which differ only in the glycosylation of the tyrosine phenol, demonstrated potent inhibition of *E. coli* SPase I and less potent activity was observed against the *S. pneumoniae* enzyme. While the weak microbiological activity of these compounds in *S. pneumoniae* tracked with the enzyme potency, activity was only observed in LPS- and efflux pump-deficient strains of *E. coli*, suggesting that the spectrum was limited by poor outer membrane permeation. The preparation of Arylomycin A2 by total synthesis confirmed these results and expanded profiling revealed potent, selective cellular activity against *Staphylococcus epidermidis*, but limited activity against *S. aureus*.[82] A molecular basis of this narrow spectrum was revealed through subsequent selection experiments, which demonstrated that a single mutation, S29P was sufficient to confer high level resistance to the drug.[83] In most bacterial species, proline is conserved at this corresponding position, thereby conferring a genotypic basis for limited spectrum of activity. From a structural and biochemical perspective, binding studies on the S29P *S. epidermidis* enzyme have demonstrated reduced affinity for arylomycin derivatives and strains encoding this mutation show reduced growth inhibition susceptibility. In contrast, purified recombinant enzymes and strains from Gram-positive and Gram-negative species encoding the corresponding P29S mutation (*S. epidermidis* numbering) show enhanced binding and activity in cell growth assays. This genotype:phenotype correlation has been rationalized at the molecular level through analysis of the *E. coli* enzyme crystal structure with bound Arylomycin A2.[84] While the biaryl-bridged macrocycle warhead binds in the active site through a network of key contacts with the catalytic machinery, the N-terminal peptide tail forms several H-bond interactions with protein backbone amides in the substrate-binding channel. The conversion of S29 to proline—the only naturally-occurring amino acid without a main chain amide H-bond donor—removes a critical drug:backbone amide hydrogen bond interaction required for potent binding. This insight has catalyzed a range of efforts to expand the potency and spectrum of the series, including modifications to the lipid tail and N-terminal peptide residue replacements.[85,86] These efforts have had limited success to date but have demonstrated that aromatic substitutions in the lipid chain and aliphatic replacements of the N-terminal residues are tolerated, particularly for activity against wild-type *S. aureus* strains. While the S29P genotype:phenotype correlation has been generally highly predictive,[87] surprising activity has been observed against wild-type strains of *Yerisinia pestis*, an organism that naturally encodes the S29P resistance marker.[83] Recent physiological studies indicate that the enhanced activity of arylomycins against *Y. pestis* is due to an increased dependence on SPase activity to support high levels of protein secretion required under physiological conditions.[88] These results suggest that the arylomycin microbiological activity spectrum is attributed to a complex array of factors including genotypic, permeation and physiologic target vulnerability.

In a recent and intriguing high profile report, the novel narrow spectrum antibacterial teixobactin (See **5**, Figure 4.1) was identified using an innovative soil-implant chip culture system.[89] This approach significantly improves the growth recovery rate of soil-based organisms, thereby enabling the isolation of extracts from novel producing organisms. Using this method, teixobactin was identified from extracts produced from *Elefthera terrae* by screening for activity against *S. aureus*. Good to exceptional bactericidal activity was observed across a range of Gram-positive pathogens, including highly resistant isolates. Remarkably, attempts to select *S. aureus* resistant mutants, even by sustained serial culture passage over 27 days, were unsuccessful, suggesting that the physiological target was not a protein. Physiological pathway inhibition analyses suggested that the compound targets late steps of cell wall biosynthesis, and that biochemical assays of cell wall biosynthesis were inhibited when initiated with Lipid I, Lipid II and undecaprenyl pyrophosphate. Further, supplementation of compound treated cultures with purified Lipid II prevented inhibition of growth by teixobactin and protected lipid intermediates from extraction from the cell membrane when incubated in two-fold stoichiometric excess. Taken together, these results suggest that teixobactin binds cell wall biosynthesis intermediates in a manner analogous to the glycopeptide antibiotics such as vancomycin. From a pharmacodynamic perspective, teixobactin demonstrated favorable pharmacokinetic and protein binding profiles in mice and a robust efficacy response was observed in both a septicemia and mouse thigh *S. aureus* and *S. pneumoniae* infection models.

There are very few examples of *Pseudomonas* species-producing natural products on the market. Mupirocin, a polyketide, is one exception.[90] Yet, the rhizosphere—composed of the space adjacent to plant roots—is rich in *Pseudomonas* species and *Pseudomonas* species-producing natural products. Secondary metabolites in the rhizosphere are loosely grouped into four major classes, encompassing lipopeptides (daptomycin, for example), siderophores (such as pyoverdine), antibiotics/antifungals, and quorum-sensing molecules, albeit these groupings are not mutually exclusive.[91] However, understanding their multifaceted roles within the microbiome can provide critical insights into how microorganisms communicate and co-exist with competing organisms, offering a glance at how pathogens thrive. *Pseudomonas putida*, strain RW10S1, was first observed having a virulence phenotype against other *Pseudomonas* species upon its isolation from a rice plant in Sri Lanka in 1992.[92] The isolation of promysalin (See **6**, Figure 4.1) and discovery that this natural product scaffold contains *pro*line, *my*ristic acid, and *sal*icylic acid motifs, afforded promysalin its name and prompted a wave of research aimed at devising synthetic routes, assigning absolute stereochemistry, developing SAR, and determining the mode of action and natural biosynthetic pathway for this natural product.[90,93–96] In an attempt to reannotate the biosynthetic gene clusters outlined previously,[90] Steele and co-workers successfully employed bioinformatics analysis to assign the absolute stereochemistry of the three chiral centers of promysalin.[95] Further, both diverted and convergent total synthesis efforts have produced a diverse array of analogs and enabled synthesis of gram-scale quantities of the antagonistic molecule.[93–96] Promysalin was found to inhibit the growth of *Pseudomonas aeruginosa* at low μM concentrations, disperse biofilm formation, and inhibit production of the siderophore, pyoverdine.[95] Further, the lack of a virulence phenotype on Gram-positive organisms grants vast implications both in agriculture and in the clinic. Interestingly, promysalin behaves like a siderophore but unlike other siderophores, whose production is turned on in a low-iron environment, production of promysalin depends on regulation of the Gac/Rsm system, which is known to regulate communication in γ-proteobacteria,[90] thus hinting at a more complex antagonistic role of this molecule. Recently, it was found that promysalin binds iron and scavenges iron from competitor organisms, and that the iron-binding motif is largely responsible for its narrow spectrum activity.[96] Work to further determine the intricate role promysalin plays in the rhizosphere is ongoing but it is hypothesized that promysalin may inhibit siderophore-transport pathways.[96] Studies to determine the target(s) of promysalin are also being investigated. Taken together, promysalin shows great promise as a narrow spectrum antibiotic against Gram-negative pathogens, specifically the clinically relevant, *Pseudomonas aeruginosa*.

In summary, it is clear that approaches that leverage natural products remain productive, particularly for Gram-positive-directed agents. While the historical success rate for the identification of natural product-derived agents against Gram-negative pathogens remains challenging, the combined use of genomics, metabolomics and synthetic biology to engineer novel secondary metabolite producers has the potential to yield novel antibacterial agents. A recent analysis of the chemical potential of these approaches suggests that nature encodes a diverse set of untapped chemical transformations which, if successfully unlocked and understood, could be leveraged to expand the natural product chemical space even further.[97] Since these efforts are only augmented by the continued discovery of novel producing strains, coupled with genomic and metabolomic computational analysis producer pathways,[98] it is clear that the potential of natural products remains high and growing.

4.3 Synthetic and Target-Based Approaches

While there has been significant debate regarding the productivity and value of target based, high throughput screening approaches for antibacterial drug discovery,[99] there are now a number of examples where this approach has been effective in finding novel leads for narrow spectrum applications. In fact, it has been argued that the requirement for broad spectrum activity has been a factor in the overall attrition rates for project progression.[17] To date, several targets with broad spectrum potential have yielded narrow spectrum clinical candidates including peptide deformylase (PDF), the fatty acid biosynthesis enzyme FabI and FtsZ, a tubulin-like GTPase.

Peptide deformylase is a metalloenzyme, which relies on coordination to Fe^{2+} to effect the removal of the N-terminal formyl group of the leading methionine as a nascent polypeptide chain emerges from the ribosome after translation. The protein maturation event catalyzed by PDF affords this target with broad spectrum potential. Versicor was the first to identify a PDF inhibitor from high throughput screening and their efforts found that actinonin, a hydroxamic acid natural product produced from Streptomyces ssp targets PDF.[100] Actinonin was found to display exquisite on-target activity inside the cell whereby over-expression and reduced expression models showed recovery of growth and increased susceptibility, respectively.[101] Poor *in vivo* properties plagued the advancement of actinonin. However, through their PDF screening initiatives, GSK developed the clinical candidate, GSK1322322 (7, Figure 4.2), which triggers catastrophic cellular consequences specifically on *S. aureus* that were not matched by either moxifloxacin or linezolid.

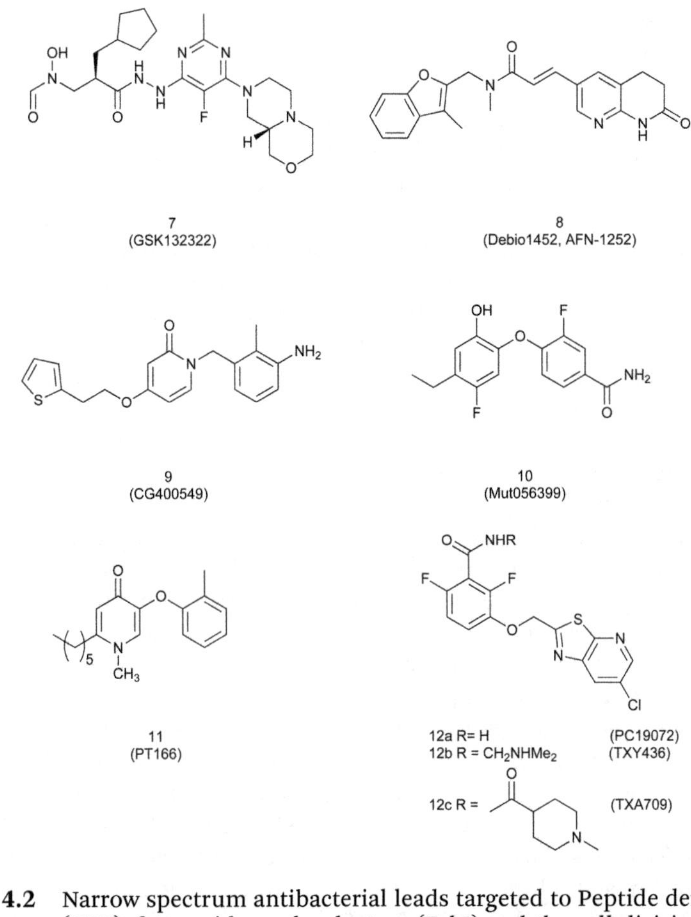

Figure 4.2 Narrow spectrum antibacterial leads targeted to Peptide deformylase (PDF), fatty acid enoyl reductase (FabI) and the cell division enzyme FtsZ.

GSK1322322 inhibits growth for 6 h of 91% of 100 random clinical isolates of *S aureus* at 1/8-fold to 1/32-fold MIC concentrations.[102]

In addition, to achieve similar killing capacity, concentrations near the MIC were required in other representative Gram-positive and Gram-negative organisms like *S. pneumoniae* and *H. influenzae*, respectively, demonstrating the potential for GSK1322322 as a narrow spectrum agent. GSK1322322 is a hydrazinopyrimidine that has completed Phase IIa clinical trials meeting the end-point goals for safety and efficacy in patients with ABSSSI, reducing lesion sizes by more than 20% after 72 h in 96% of cases compared to 100% of cases treated with linezolid.[103] The authors indicate that additional Phase II studies are planned to evaluate minimal dosing requirements. Interest in PDF as a drug target has prompted a phase I study investigating the effects of GSK1322322 on the gut microbiome.[104] The design of the study was innovative in that the microbiome was assessed by NGS (next-generation sequencing) of stool samples both pre-treatment and post-treatment with placebo, orally administered GSK1322322, *iv*-administered GSK1322322, and a combination of oral-*iv* GSK1322322 to offer a more complete picture for changes that occur in response to the drug. Not surprising, a decrease in the counts of some bacterial taxa was observed in the oral-*iv* treated patients compared to the *iv*-only or placebo treated patients; however, a decrease in bacterial diversity was also observed. Due to the increased interest in disseminating the effects of antibiotics on the homeostasis of the gut microbiome, this first in class study raises awareness to issues that should be considered perhaps earlier in the drug pipeline.

Although GSK1322322 continues to show great promise for treatment of drug-susceptible and drug resistant *S. aureus*, its activity against other pathogens including *S. pneumoniae* and *H. influenzae* should not be under-estimated as GSK1322322 has entered Phase III clinical trials in Europe *via* the Innovative Medicines Initiative COMBACTE for ABSSSI and CABP infections.[14] Recently, Dubois *et al.* evaluated GSK1322322 against strains of *Legionella pneumophila*, a Gram-negative pathogen, and the causative agent of Legionnaires' Disease, which has received great mainstream news coverage in recent months.[105]

Fatty acid enoyl reductase (FabI), the final rate-limiting enzyme for fatty acid elongation in the FASII pathway that catalyzes the formation of acyl-ACP from enoyl-ACP, is the target of isoniazid, a front-line drug used to treat *Mycobacterium tuberculosis*. Despite the clinical utility against the *M. tuberculosis* FabI enzyme, no drugs are approved for Gram-negative or Gram-positive infections to date. The FASII pathway is a validated and essential drug target for Gram-negative pathogens because the biosynthesis of Lipid A relies on utilization of β-hydroxyl fatty acid components that cannot be provided by exogenous fatty acids.[106] Although some Gram-positive organisms can scavenge fatty acids from the host and bypass the FAS II pathway,[107] this is not the case for *Staphylococcus aureus*.[108,109] Two clinical candidates targeting FabI have completed Phase II clinical trials with human efficacy. Debio1452 (formally AFN-1252, **8**, Figure 4.2) and CG400549 (**9**, Figure 4.2) with naphthyridinone and 2-pyridone scaffolds, respectively, are narrow spectrum agents against

S. aureus.[14,110] Debio1452 has demonstrated potential against a wide array of Staphylococcal clinical isolates without activity against other Gram-positive or Gram-negative pathogens.[111,112] In a mouse model of infection with lethal septicemia, an oral dose of 1 mg kg^{-1} of AFN-1252 provided 100% survival.[113] In the same study, the frequency of spontaneous resistance was very low.

The discovery that triclosan inhibits FabI across a range of pathogens caused a wave of extensive SAR drug discovery efforts to optimize the diphenyl ether scaffold and capitalize on tuning its broad spectrum potential in the pursuit of narrow spectrum antibiotics.[114] MUT056399 (**10**, Figure 4.2), based on the diphenyl ether scaffold, is a promising triclosan derivative, displaying *in vivo* data against murine models of systemic Staph infections, including MRSA and VRSA. MUT056399 completed Phase I clinical trial in 2010.[115,116] MUT056399 differs from Debio1452 and CG400549 in that it also displays activity across important Gram-negative species. In addition to the clinical candidate that has emerged, enzyme-guided drug discovery efforts have focused on optimizing the slow-binding kinetics of diphenyl ether inhibitor binding to FabI from different organisms. By rationally optimizing the residence time of the drug-FabI complex, Tonge *et al.* have developed inhibitors that are tuned to the needs of the active sites of specific homologs of FabI.[117–119] Similarly, Kisker *et al.* designed a 4-pyridone compound, PT166 (**11**, Figure 4.2), based on CG400549 and guided by insights obtained from X-ray crystallography data, that afforded an expanded spectrum of activity to include Gram-negative pathogens like *Francisella tularenis* and Mycobacteria spp.[120] This target-directed approach offers the potential for narrow spectrum identification of agents against specific pathogens of interest.

FtsZ is a tubulin-like GTPase responsible for cytokinesis. Inhibitors of FtsZ act by increasing self-polymerization potential, effectively stabilizing FtsZ polymers and preventing the formation of the Z ring structure required for cell division. Although FtsZ is a conserved bacterial protein essential for cell division and has broad spectrum potential, efforts have largely focused of studies against Staphylococcal spp. The lead compound PC190723 (**12a**, Figure 4.2), discovered through a fragment-based SAR medicinal chemistry program,[121] prompted research initiatives aimed at improving the poor drug-like properties of the core hydrophobic benzamide-thiazolopyridine scaffold. TXY541, a prodrug of PC190723, significantly improved properties of the scaffold by offering a protonated piperidine ring in place of the benzamide.[122] *In vitro* and *in vivo* studies of TXY541 demonstrated promising activity against Staphylococci spp and *Bacillus subtilis* but poor activity against other important Gram-positive organisms and no activity against Gram-negative pathogens. Likewise, TXY436 (**12b**, Figure 4.2) and TXA709 (**12c**, Figure 4.2), further optimized prodrugs of PC190723, showed good activity in an oral dosing mouse model of systemic infection in both MSSA and MRSA infection whereas no efficacy was demonstrated with PC190723, with TXA709 demonstrating lower exposure requirements.[123,124] The basis for the narrow spectrum Gram-positive activity was elucidated through structural homology model building of the enzyme across bacterial species, and it was discovered that there is an intrinsic difference in the inhibitor binding site that is unique to the susceptible species.[125] In addition, MD simulations coupled with cavity searching of the FtsZ binding pocket illustrate marked differences in the energetic landscapes of FtsZ from *S. aureus*, *M. tuberculosis*,[126] and *P. aeruginosa*.[127] However, it was found that chemical inhibition of the RND efflux pumps renders *Klebsiella pneumoniae*, *Acinetobacter baumannii*, and *E. coli* strains all susceptible to the prodrug, TXY436,[128] suggesting that the observed narrow spectrum results from both target binding specificity and limited compound penetration.

Genetic studies have also suggested that FtsZ may be a particularly attractive target for combination-based approaches. A knock-down approach of 245 genes was undertaken and susceptibility of MRSA to β-lactams was measured and FtsZ was identified as a target with synergistic potential.[129] To this end, it was found that PC190723-resistant mutants of MRSA were hypersensitive to β-lactams in a mouse thigh infection model, suggesting that the resistance mechanisms invoked mutually exclusive fitness costs. Further, fluorescence microscopy analyses of PC190723-treated cells indicated that PBP2, the target of β-lactams, and FtsZ are delocalized from the cell division septum, thereby demonstrating a mechanistic basis for the observed *in vitro* and *in vivo* synergy.

Bacterial DNA ligase (LigA), an essential enzyme involved in joining Okazaki DNA fragments formed by DNA polymerase on the lagging strand of chromosomal DNA, has been a focus of numerous medicinal chemistry campaigns. The first reported inhibitors for this enzyme were a series of pyridochromanone analogs (**13**, Figure 4.3) that exhibited reasonable inhibition *in vitro* against both the *E. coli* and *S. pneumoniae* enzymes[130] and acted as competitive inhibitors to the natural substrate, nicatinamide adenine dinucleotide (NAD^+).

However, the spectrum of microbiological activity observed for this series was limited to Gram-positive organisms, due to poor cell penetration across the Gram-negative outer membrane. The compounds demonstrated rapid bactericidal activity against *S. aureus* and were not active against the human bacterial ligase enzyme. Similarly, a series of adenosine inhibitors were identified using an HTS-based approach (**14**, Figure 4.3), and these compounds demonstrated good *in vitro* enzyme inhibition against LigA enzymes from both Gram-positive and Gram-negative organisms, but a microbiological spectrum that was limited to *S. aureus*, Streptococcal spp and *H. influenzae*.[131] The mode of action of these molecules was confirmed to be *via* DNA ligase inhibition through a range of techniques, including elevated MIC values in LigA overexpression strains, inhibition of thymidine incorporation in *de novo* DNA synthesis and the isolation of LigA-mediated resistant mutants. Importantly, a representative analog in this series (**14b**, Figure 4.3) was profiled for *in vivo* efficacy using murine thigh and lung infection models of *S. aureus* and *S. pneumoniae*, respectively. A clear dose dependent effect was observed in both models and no adverse reactions were observed, but achieving sufficient drug exposure required co-administration of the metabolic modifier

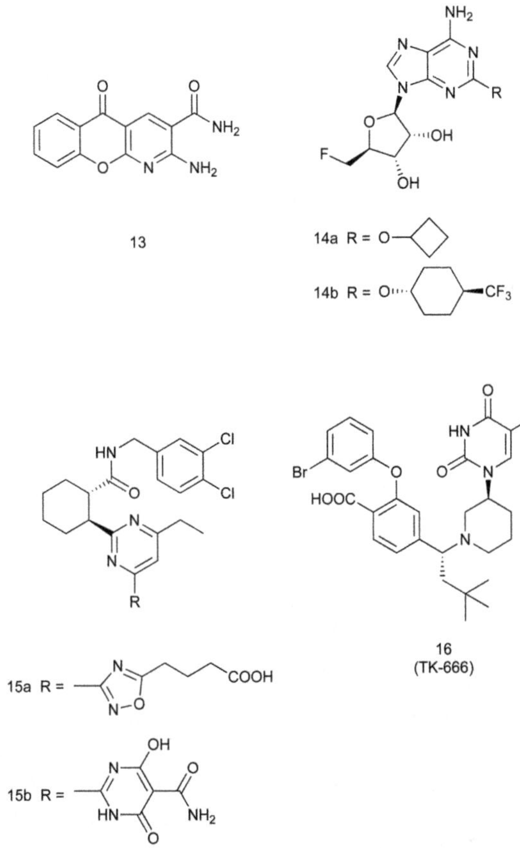

Figure 4.3 Narrow spectrum antibacterial leads targeting bacterial NAD$^+$-dependent DNA ligase, 4′-phosphopantethiene adenylyltransferase (PPAT) and thymidylate monophosphate kinase (TMK).

aminobenzotriazole to block Phase I metabolism. This series of analogs were further optimized for both microbiological activity[132] and improved pharmacokinetic profiles[133] and these efforts resulted in advanced analogs with improved potency, but retained the original narrow spectrum activity. Interestingly, recent studies on DNA Ligase indicate that this target is vulnerable to the development of resistance,[134] an outcome attributed to cellular tolerance for mutations that diminish enzyme activity but favor native NAD$^+$ substrate occupancy over competitive inhibitor binding. These data suggest that NAD$^+$ competitive inhibitors will require exquisite potency, along with high and sustained exposures, to avoid the emergence of resistance.

Enzymes involved in producing essential cellular metabolites have also been successfully targeted using synthetic strategies. For example, a series of cycloalkyl pyrimidine (**15**, Figure 4.3) inhibitors of 4′-phosphopantethiene adenylyltransferase (PPAT) were discovered using an *in vitro* high throughput screen.[135] This enzyme catalyzes a critical step in the biosynthesis of Coenzyme A, an essential cofactor involved in a wide variety of cellular functions. These inhibitors were found to have good *in vitro* activity against enzymes selected from pathogens across the bacterial spectrum and, with the aid of structure-based design, optimized compounds were identified with exquisite potency. Biochemical studies revealed that the inhibitors were competitive with phosphopantethiene, but dependent upon ATP binding. Interestingly, cellular inhibition activity required exceeding a potent *in vitro* inhibition threshold (IC_{50} < 20 nM) and measurable microbiological activity was limited to Gram-positive pathogens, but included multidrug resistant strains such as MRSA. This series demonstrated bacteriostatic activity and a moderate, but acceptable, concentration dependent spontaneous frequency of resistance emergence. Advanced analogs (**15a, 15b**) demonstrated efficacy in lung and thigh murine mouse efficacy models for *S. pneumoniae* and *S. aureus*, but co-dosing with aminobenzotriazole was required to achieve sufficient exposure. In both cases, a delayed onset of inhibition was observed, suggesting that depletion of cellular Coenzyme A pools is required prior to growth stasis. The authors noted that while optimization of microbiological potency was readily achieved, challenges were noted in the development of potent inhibitors with balanced drug-like properties.

Thymidylate kinase (TMK), the enzyme that catalyzes the production of thymidine diphosphate from thymidine monophosphate and ATP, is an essential component of DNA biosynthesis and is therefore considered a high value antibacterial target. Multiple labs have reported leads against this target,[136,137] and the most promising series reported to date was discovered using a rational-design, substrate-based approach (**16**, Figure 4.3).[137] This series, based on a thymidine core scaffold, was assembled using a fragment-build based effort that successively analyzed and optimized a series of linker-ring systems, informed by inhibitor co-crystal structural studies. The optimized compounds were found to utilize a novel, induced-fit mechanism and occupy a cryptic binding region that was not obvious from the apo or substrate-bound forms. This unique binding mode is dependent on a diastereomeric requirement to ensure that the molecule adopts an optimal three-dimensional binding conformation. Despite initial differences in potency between the *S. pneumoniae* and *S. aureus* enzymes, optimization efforts resulted in analogs with picomolar affinity for both species. This biochemical potency translated into excellent microbiological activity across Streptococcal, Staphylococcal and Enterococcus spp and subsequent mechanistic studies confirmed that this activity was driven by TMK inhibition.[138] These inhibitors demonstrated rapid bactericidal activity and low frequencies of spontaneous resistance emergence. Finally, despite requiring relatively high doses with co-administration of aminobenzotriazole, robust responses were observed in efficacy studies using a *S. aureus* mouse thigh infection model with advanced compounds (**16**, Figure 4.3).

In contrast to the target-based approaches outlined above, a survey of peptidomimetic-based libraries using phenotypic cell growth inhibition identified a lead series with microbiological activity for Pseudomonas spp., including potent activity against *P. aeruginosa*.[139] The spectrum of these agents is extremely narrow, as exemplified by POL7080, which is inactive against other bacterial species. Using a forward genetic screen and photoaffinity approaches, LptD was identified as the cellular target of these agents, thereby providing a rationale for the stereoselective and specific activity. LptD is an essential outer-membrane protein involved in membrane biogenesis, and inhibition or genetic deletion results in internal accumulation of membrane material and formation of filaments prior to cell death. This series exhibit slow bactericidal effects and render the cells sensitive to antibacterial agents and detergents, suggesting that combination therapies would be synergistic. POL7080 demonstrated potent efficacy as a monotherapy in a mouse septicemia model that was significantly more active than the aminoglycoside control.

4.4 Future Prospects

Narrow spectrum agents can decrease the frequency of occurrence of pan-resistance and are less likely to perturb the homeostasis of the gut microbiome. Clearly, future antibacterial drug discovery initiatives will require careful consideration of these issues that have inundated the progression of novel agents into the marketplace. While it is clear that both target-based and phenotypic screens have produced novel narrow spectrum leads, there is a current emphasis on developing novel phenotypic screening methods for narrow spectrum or even pathogen-specific lead discovery. As an example of this approach, a bioluminescent signaling assay has been reported that screens for compounds that cause an attenuation in flux through the FASII pathway.[140] By nature of design, this is a narrow spectrum screening approach, as promoter induction occurs after sensing a perturbation in FASII flux in the engineered *P. aeruginosa* strain resulting in a gain-of-signal response. Because penetration of drugs across the membranes of Gram-negative pathogens is an area of intense study, this whole-cell screening approach positively identifies only those compounds that can accumulate in the *P. aeruginosa* cytosol. This example highlights the power of creative phenotypic assay design, and is representative of a general strategic shift that emphasizes narrow spectrum screening approaches. These "unconventional screening approaches"[141] include phenotypic approaches to target physiological pathways that are conditionally essential under conditions designed to reflect a nutrient-limited *in vivo* disease state, or impact virulence factor production required for bacterial survival in the host. This wider perspective has prompted a review of target essentiality. It is becoming clear that that the number and composition of targets deemed essential for growth is dependent on the species and growth conditions and that growth on laboratory media only partially reflects the selection pressures under *in vivo* conditions.[142–144] As a result, the classical target selection process based on genetic essentiality and conservation in complex laboratory growth media across bacterial species represents the most limiting and conservative selection criteria. Current strategies utilizing minimal media conditions that more closely mimic *in vivo* growth, such as metabolic suppression techniques[145] or inhibiting virulence factor production, have identified unique and narrow spectrum leads. For example, Romero *et al.* performed a screening of inhibitors targeting biofilms of *B. subtillis*.[146] In this imaging-based screen, the pellicles of *B. subtillis* form at an air–water interface and are characterized by their wrinkle-like appearance. A benzoquinone derivative and parthinolide were identified by their ability to cause dissolution of amyloid fibers supporting the extracellular matrix of the biofilm and therefore an observed change in the bacterial colony phenotype. Taken together, these approaches have demonstrated that the druggable target space may be much larger than previously thought, but the optimization and development of leads from these approaches will require alternative development microbiological strategies.

Screening approaches that consider combination therapies are also gaining momentum, with the expectation that these efforts will result in novel therapies with decreased risks for resistance emergence. As noted above, the observed synergy between FtsZ inhibitors and β-lactams in MRSA indicates that combination therapeutic approaches may help preserve the existing antibacterial clinical agents. These approaches may be particularly suited to narrow spectrum applications. A recent combinatorial screen of the FDA approved agents against *S. aureus*, *E. coli* and *P. aeruginosa* revealed that a distinct set of active combinations was identified for each organism. Further, this screen identified compounds that acted as adjunct therapies—compounds with minimal intrinsic activity were found to exhibit marked stimulation of antibacterial activity in the cognate pair compound. Taken together, these data suggest that combination approaches may have significant advantages over single agent approaches.

Upon writing this chapter, it was clear that the vast majority of examples targeted Gram-positive organisms. This bias was not unexpected, given similar trends in the overall antibacterial drug discovery pipeline,[14] but does reflect the significant challenge of overcoming the permeation barrier presented by the Gram-negative outer membrane. It should be emphasized that there is tremendous value in basic science efforts to study transport properties of antibiotics across the cell wall of Gram-negative pathogens. For instance, the studies of porins and their interaction with biomolecules to establish lessons learned that suggest drug scaffolds with engineered uptake groups will prove imperative to the design of narrow spectrum antibiotics targeting Gram-negative species. A recent report suggests that rational approaches can be used to engineer carbapenem analogs with enhanced passage through porins.[147] It is hoped that these early advances can be extended broadly, but these efforts will require coordinated, inter-collaborative efforts between Pharma and academics to achieve wide-ranging success.

4.5 Conclusions

The large number of advanced lead molecules outlined in this chapter provide strong evidence that the identification of narrow spectrum agents is not only feasible, but is likely to result in higher success rates than historical broad spectrum approaches. This conclusion provides much needed optimism for the future, given the rising unmet medical need for novel antibacterial agents and the emerging concerns of the long-term consequences of antibacterial-induced disruptions to microbiome homeostasis. It is clear that many challenges remain for the full realization of narrow spectrum agents to address the most pressing bacterial infections, including strategies to identify suitable leads against Gram-negative pathogens, the required changes in regulatory practice for effective clinical trialing and approval of these agents, and the development of rapid diagnostics for appropriate therapeutic intervention. However, these challenges can only be met with a robust pipeline of novel agents. Given the relative success rates evidenced in both phenotypic and target-based approaches to produce natural products and synthetic leads, there is reason to believe that a robust pipeline of agents can be assembled given a concerted and focused international effort.

References

1. K. J. Williams, *J. R. Soc. Med.*, 2009, **102**, 343–348.
2. R. I. Aminov, *Front. Microbiol.*, 2010, **1**, 134.
3. A. Fleming, *Br. J. Exp. Pharmacol.*, 1929, **10**, 226–236.
4. C. T. Walsh, *Antibiotics: Actions, Origins, Resistance*, ASM Press, Washington, DC, 2003.
5. B. Spellberg, M. Blaser, R. J. Guidos, H. W. Boucher, J. S. Bradley, B. I. Eisenstein, D. Gerding, R. Lynfield, L. B. Reller, J. Rex, D. Schwartz, E. Septimus, F. C. Tenover and D. N. Gilbert, *Clin. Infect. Dis.*, 2011, **52**(suppl. 5), S397–S428.
6. K. B. Holten and E. M. Onusko, *Am. Fam. Physician*, 2000, **62**, 611–620.
7. J. Acar, *Clin. Microbiol. Infect.*, 1997, **3**, 395–396.
8. C. M. Luna, P. Vujacich, M. S. Niederman, C. Vay, C. Gherardi, J. Matera and E. C. Jolly, *Chest*, 1996, **111**, 676–685.
9. M. Kollef, *Drugs*, 2003, **63**, 2157–2168.
10. S. Deresinski, *Clin. Infect. Dis.*, 2007, **45**(suppl. 3), S177–S183.
11. G. D. Wright, *Nat. Rev. Microbiol.*, 2007, **5**, 175–186.
12. V. M. D'Costa, C. E. King, L. Kalan, M. Morar, W. W. L. Sung, C. Schwarz, D. Froese, G. Zazula, F. Calmels, R. Debruyne, G. B. Golding, H. N. Poinar and G. D. Wright, *Nature*, 2011, **477**, 457–461.
13. F. Elliot, *ConsumersUnion*, 2014, pp. 1–7.
14. M. J. Pucci and K. Bush, *Clin. Microbiol. Rev.*, 2013, **26**, 792–821.
15. R. O'Shea and H. E. Moser, *J. Med. Chem.*, 2008, **51**, 2871–2878.
16. D. G. Brown, T. L. May-Dracka, M. M. Gagnon and R. Tommasi, *J. Med. Chem.*, 2014, **57**, 10144–10161.

17. R. Tommasi, D. G. Brown, G. K. Walkup, J. I. Manchester and A. A. Miller, *Nat. Rev. Drug Discovery*, 2015, 1–14.
18. D. Bumann, *Curr. Opin. Microbiol.*, 2008, **11**, 387–392.
19. L. L. Silver, *Nat. Rev. Drug Discovery*, 2007, **6**, 41–55.
20. E. Power, *Clin. Microbiol. Infect.*, 2006, **12**, 25–34.
21. C. A. Lopez, D. D. Kingsbury, E. M. Velazquez and A. J. Bäumler, *Cell Host Microbe*, 2014, **16**, 156–163.
22. M. C. Rea, A. Dobson, O. O'Sullivan, F. Crispie, F. Fouhy, P. D. Cotter, F. Shanahan, B. Kiely, C. Hill and R. P. Ross, *Proc. Natl. Acad. Sci. U. S. A.*, 2011, **108**, 4639–4644.
23. A. D. Kostic, D. Gevers, M. Knip, R. J. Xavier, H. La, S. Oikarinen, H. J. M. Harmsen, M. C. De Goffau, G. Welling, K. Alahuhta, T. Korhonen and S. M. Virtanen, *Cell Host Microbe*, 2015, **17**, 1–14.
24. D. Gevers, S. Kugathasan, L. A. Denson, Y. Vázquez-Baeza, W. Van Treuren, B. Ren, E. Schwager, D. Knights, S. J. Song, M. Yassour, X. C. Morgan, A. D. Kostic, C. Luo, A. González, D. McDonald, Y. Haberman, T. Walters, S. Baker, J. Rosh, M. Stephens, M. Heyman, J. Markowitz, R. Baldassano, A. Griffiths, F. Sylvester, D. Mack, S. Kim, W. Crandall, J. Hyams, C. Huttenhower, R. Knight and R. J. Xavier, *Cell Host Microbe*, 2014, **15**, 382–392.
25. A. D. Kostic, R. J. Xavier and D. Gevers, *Gastroenterology*, 2014, 1–11.
26. A. N. McCoy, F. Araújo-Pérez, A. Azcárate-Peril, J. J. Yeh, R. S. Sandler and T. O. Keku, *PLoS One*, 2013, **8**, e53653.
27. I. Okayasu, *Pathol. Int.*, 2012, **62**, 368–380.
28. J. O'Neill, *Rev. Antimicrob. Resist.*, 2015, 1–18.
29. *United States Executive Order – Combating Antibiotic-Resistant Bacteria | The White House*, http://www.whitehouse.gov/the-press-office/2014/09/18/executive-order-combating-antibiotic-resistant-bacteria.
30. J. H. Rex, B. I. Eisenstein, J. Alder, M. Goldberger, R. Meyer, A. Dane, I. Friedland, C. Knirsch, W. R. Sanhai, J. Tomayko, C. Lancaster and J. Jackson, *Lancet Infect. Dis.*, 2013, **13**, 269–275.
31. J. H. Rex, M. Goldberger, B. I. Eisenstein and C. Harney, *Ann. N. Y. Acad. Sci.*, 2014, **1323**, 11–21.
32. B. Spellberg and J. H. Rex, *Nat. Rev. Drug Discovery*, 2013, **12**, 963–964.
33. M. Rees, *Nature*, 2014, **509**, 401.
34. B. I. Eisenstein, F. B. Oleson and R. H. Baltz, *Clin. Infect. Dis.*, 2010, **50**(suppl. 1), S10–S15.
35. C. W. Ford, G. E. Zurenko and M. R. Barbachyn, *Curr. Drug Targets: Infect. Disord.*, 2001, **1**, 181–199.
36. M. P. Singh, P. J. Petersen, W. J. Weiss, J. E. Janso, S. W. Luckman, E. B. Lenoy, P. A. Bradford, R. T. Testa and M. Greenstein, *Antimicrob. Agents Chemother.*, 2003, **47**, 62–69.
37. H. He, R. T. Williamson, B. Shen, E. I. Graziani, H. Y. Yang, S. M. Sakya, P. J. Petersen and G. T. Carter, *J. Am. Chem. Soc.*, 2002, 9729–9736.
38. A. Ruzin, G. Singh, A. Severin, Y. Yang, R. G. Dushin, A. G. Sutherland, A. Minnick, M. Greenstein, M. K. May, D. M. Shlaes and P. A. Bradford, *Antimicrob. Agents Chemother.*, 2004, **48**, 728–738.

39. P.-E. Sum, D. How, N. Torres, P. J. Petersen, J. Ashcroft, E. I. Graziani, F. E. Koehn and T. S. Mansour, *Bioorg. Med. Chem. Lett.*, 2003, **13**, 2805–2808.
40. P.-E. Sum, D. How, N. Torres, H. Newman, P. J. Petersen and T. S. Mansour, *Bioorg. Med. Chem. Lett.*, 2003, **13**, 2607–2610.
41. R. G. Dushin, T. Z. Wang, P. E. Sum, H. He, A. G. Sutherland, J. S. Ashcroft, E. I. Graziani, F. E. Koehn, P. A. Bradford, P. J. Petersen, K. L. Wheless, D. How, N. Torres, E. B. Lenoy, W. J. Weiss, S. A. Lang, S. J. Projan, D. M. Shlaes and T. S. Mansour, *J. Med. Chem.*, 2004, **47**, 3487–3490.
42. H. He, *Appl. Microbiol. Biotechnol.*, 2005, **67**, 444–452.
43. P. J. Petersen, T. Z. Wang, R. G. Dushin and P. A. Bradford, *Antimicrob. Agents Chemother.*, 2004, **48**, 739–746.
44. W. J. Weiss, T. Murphy, E. Lenoy and M. Young, *Antimicrob. Agents Chemother.*, 2004, **48**, 1708–1712.
45. S. Fuse, H. Koinuma, A. Kimbara, M. Izumikawa, Y. Mifune, H. He, K. Shin-ya, T. Takahashi and T. Doi, *J. Am. Chem. Soc.*, 2014, **136**, 12011–12017.
46. S. N. Fischer, C. J. Schworer and M. Oberthur, *Synthesis (Stuttg.)*, 2014, **46**, 2234–2240.
47. Y. T. Huang, S. Y. Lyu, P. H. Chuang, N. S. Hsu, Y. S. Li, H. C. Chan, C. J. Huang, Y. C. Liu, C. J. Wu, W. Bin Yang and T. L. Li, *ChemBioChem*, 2009, **10**, 2480–2487.
48. H. He, B. Shen, P. J. Petersen, W. J. Weiss, H. Y. Yang, T.-Z. Wang, R. G. Dushin, F. E. Koehn and G. T. Carter, *Bioorg. Med. Chem. Lett.*, 2004, **14**, 279–282.
49. N. A. Magarvey, B. Haltli, M. He, M. Greenstein and J. A. Hucul, *Antimicrob. Agents Chemother.*, 2006, **50**, 2167–2177.
50. J. M. Waisvisz, M. G. van der Hoeven, J. van Peppen and W. C. M. Zwennis, *J. Am. Chem. Soc.*, 1957, **79**, 4520–4521.
51. S. Nakamura, T. Yajima, Y. Lin and H. Umezawa, *J. Antibiot. (Tokyo)*, 1967, **20**, 1–5.
52. S. Nakamura, S. Omura, T. Nishimura, N. Tanaka and H. Umezawa, *J. Antibiot. (Tokyo)*, 1967, **20**, 162–166.
53. W. J. Miller, L. Chaiet, G. Rasmussen, B. Christensen, J. Hannah, A. K. Miller and F. J. Wolf, *J. Med. Chem.*, 1968, **11**, 746–749.
54. S. K. Singh and S. Gurusiddaiah, *Antimicrob. Agents Chemother.*, 1984, **26**, 394–400.
55. H. Shimamura, H. Gouda, K. Nagai, T. Hirose, M. Ichioka, Y. Furuya, Y. Kobayashi, S. Hirono, T. Sunazuka and S. Ōmura, *Angew. Chem., Int. Ed.*, 2009, **48**, 914–917.
56. N. Tanaka, K. Sashikata, H. Yamaguchi and H. Umezawa, *J. Biochem.*, 1966, **60**, 405–410.
57. T. Otaka and A. Kaji, *J. Biol. Chem.*, 1976, **251**, 2299–2306.
58. T. Otaka and A. Kaji, *FEBS Lett.*, 1981, **123**, 173–176.
59. T. Otaka and A. Kaji, *FEBS Lett.*, 1983, **153**, 53–59.
60. T. Kinoshita and N. Tanaka, *J. Antibiot.*, 1970, **23**, 311–312.
61. S.-I. Sowa, N. Masumi, Y. Inouye, S. Nakamura, Y. Takesue and T. Yokoyama, *Hiroshima J. Med. Sci.*, 1992, **41**, 79–85.

62. Y. Kobayashi, M. Ichioka, T. Hirose, K. Nagai, A. Matsumoto, H. Matsui, H. Hanaki, R. Masuma, Y. Takahashi, S. Ōmura and T. Sunazuka, *Bioorg. Med. Chem. Lett.*, 2010, **20**, 6116–6120.
63. S. Ackermann, H.-G. Lerchen, D. Häbich, A. Ullrich and U. Kazmaier, *Beilstein J. Org. Chem.*, 2012, **8**, 1652–1656.
64. W. J. K. Crone, F. J. Leeper and A. W. Truman, *Chem. Sci.*, 2012, **3**, 3516.
65. L. Huo, S. Rachid, M. Stadler, S. C. Wenzel and R. Muller, *Chem. Biol.*, 2012, **19**, 1278–1287.
66. C. T. Walsh, *ACS Chem. Biol.*, 2014, **9**, 1653–1661.
67. Y. Hou, M. D. B. Tianero, J. C. Kwan, T. P. Wyche, C. R. Michel, G. A. Ellis, E. Vazquez-Rivera, D. R. Braun, W. E. Rose, E. W. Schmidt and T. S. Bugni, *Org. Lett.*, 2012, **14**, 5050–5053.
68. J. Wang, S. M. Soisson, K. Young, W. Shoop, S. Kodali, A. Galgoci, R. Painter, G. Parthasarathy, Y. S. Tang, R. Cummings, S. Ha, K. Dorso, M. Motyl, H. Jayasuriya, J. Ondeyka, K. Herath, C. Zhang, L. Hernandez, J. Allocco, A. Basilio, J. R. Tormo, O. Genilloud, F. Vicente, F. Pelaez, L. Colwell, S. H. Lee, B. Michael, T. Felcetto, C. Gill, L. L. Silver, J. D. Hermes, K. Bartizal, J. Barrett, D. Schmatz, J. W. Becker, D. Cully and S. B. Singh, *Nature*, 2006, **441**, 358–361.
69. S. B. Singh, J. W. Phillips and J. Wang, *Curr. Opin. Drug Discovery Dev.*, 2007, **10**, 160–166.
70. K. C. Nicolaou, D. J. Edmonds, A. Li and G. S. Tria, *Angew. Chem., Int. Ed.*, 2007, **46**, 3942–3945.
71. J. Wang, S. Kodali, S. H. Lee, A. Galgoci, R. Painter, K. Dorso, F. Racine, M. Motyl, L. Hernandez, E. Tinney, S. L. Colletti, K. Herath, R. Cummings, O. Salazar, I. González, A. Basilio, F. Vicente, O. Genilloud, F. Pelaez, H. Jayasuriya, K. Young, D. F. Cully and S. B. Singh, *Proc. Natl. Acad. Sci. U. S. A.*, 2007, **104**, 7612–7616.
72. D. C. J. Waalboer, M. C. Schaapman, F. L. Van Delft and F. P. J. T. Rutjes, *Tetrahedron Lett.*, 2008, **47**, 6576–6578.
73. K. C. Nicolaou, A. F. Stepan, T. Lister, A. Li, A. Montero, G. Scott, C. I. Turner, Y. Tang, J. Wang, R. M. Denton and J. David, *J. Am. Chem. Soc.*, 2008, **130**, 13110–13119.
74. K. C. Nicolaou, A. Li, D. J. Edmonds, G. S. Tria and S. P. Ellery, *J. Am. Chem. Soc.*, 2009, **131**, 16905–16918.
75. Z. Yu, M. J. Smanski, R. M. Peterson, K. Marchillo, D. Andes, S. R. Rajski and B. Shen, *Org. Lett.*, 2010, **12**, 1744–1747.
76. K. B. Herath, C. Zhang, H. Jayasuriya, J. G. Ondeyka, D. L. Zink, B. Burgess, J. Wang and S. B. Singh, *Org. Lett.*, 2008, **10**, 1699–1702.
77. C. Zhang, J. Ondeyka, D. L. Zink, B. Burgess, J. Wang and S. B. Singh, *Chem. Commun. (Camb.)*, 2008, 5034–5036.
78. S. B. Singh, H. Jayasuriya, K. B. Herath, C. Zhang, J. G. Ondeyka, D. L. Zink, G. Parthasarathy, J. W. Becker and S. M. Soisson, *Tetrahedron Lett.*, 2008, **49**, 3648–3651.
79. A. Höltzel, D. G. Schmid, G. J. Nicholson, S. Stevanovic, J. Schimana, K. Gebhardt, H.-P. Fiedler and G. Jung, *J. Antibiot. (Tokyo)*, 2002, **55**, 571–577.

80. J. Schimana, K. Gebhardt, A. Holtzel, D. G. Schmid, R. Sussmuth, J. Muller, R. Pukall and H.-P. Fieldler, *J. Antibiot.*, 2002, **55**, 565–570.
81. P. Kulanthaivel, A. J. Kreuzman, M. A. Strege, M. D. Belvo, T. A. Smitka, M. Clemens, J. B. Swartling, K. L. Minton, F. Zheng, E. L. Angleton, D. Mullen, L. N. Jungheim, V. J. Klimkowski, T. I. Nicas, R. C. Thompson and S. Bin Peng, *J. Biol. Chem.*, 2004, **279**, 36250–36258.
82. T. C. Roberts, P. A. Smith, R. T. Cirz and F. E. Romesberg, *J. Am. Chem. Soc.*, 2007, **129**, 15830–15838.
83. P. A. Smith, T. C. Roberts and F. E. Romesberg, *Chem. Biol.*, 2010, **17**, 1223–1231.
84. M. Paetzel, J. J. Goodall, M. Kania, R. E. Dalbey and M. G. P. Page, *J. Biol. Chem.*, 2004, **279**, 30781–30790.
85. T. C. Roberts, M. A. Schallenberger, J. Liu, P. A. Smith and F. E. Romesberg, *J. Med. Chem.*, 2011, **54**, 4954–4963.
86. J. Liu, P. A. Smith, D. B. Steed and F. Romesberg, *Bioorg. Med. Chem. Lett.*, 2013, **23**, 5654–5659.
87. P. A. Smith, M. E. Powers, T. C. Roberts and F. E. Romesberg, *Antimicrob. Agents Chemother.*, 2011, **55**, 1130–1134.
88. D. B. Steed, J. Liu, E. Wasbrough, L. Miller, S. Halasohoris, J. Miller, B. Somerville, J. R. Hershfield and F. E. Romesberg, *Antimicrob. Agents Chemother.*, 2015, **59**, 3887–3898.
89. L. L. Ling, T. Schneider, A. J. Peoples, A. L. Spoering, I. Engels, B. P. Conlon, A. Mueller, T. F. Schäberle, D. E. Hughes, S. Epstein, M. Jones, L. Lazarides, V. A. Steadman, D. R. Cohen, C. R. Felix, K. A. Fetterman, W. P. Millett, A. G. Nitti, A. M. Zullo, C. Chen and K. Lewis, *Nature*, 2015, **517**, 455–459.
90. W. Li, P. Estrada-De Los Santos, S. Matthijs, G.-L. Xie, R. Busson, P. Cornelis, J. Rozenski and R. De Mot, *Chem. Biol.*, 2011, **18**, 1320–1330.
91. C. E. Keohane, A. D. Steele and W. M. Wuest, *Synlett*, 2015, **26**, 2739–2744.
92. K. Vlassak, L. Van Holm, L. Duchateau, J. Vanderleyden and R. De Mot, *Plant Soil*, 1992, **145**, 51–63.
93. R. D. Kaduskar, A. A. Dhavan, S. Dallavalle, L. Scaglioni and L. Musso, *Tetrahedron*, 2016, **72**, 2034–2041.
94. K. W. Knouse and W. M. Wuest, *J. Antibiot.*, 2016, **69**, 337–339.
95. A. D. Steele, K. W. Knouse, C. E. Keohane and W. M. Wuest, *J. Am. Chem. Soc.*, 2015, **137**, 7314–7317.
96. A. D. Steele, C. E. Keohane, K. W. Knouse, S. E. Rossiter, S. J. Williams and W. M. Wuest, *J. Am. Chem. Soc.*, 2016, **138**, 5833–5836.
97. C. T. Walsh, *Nat. Chem. Biol.*, 2015, **11**, 620–624.
98. M. H. Medema and M. A. Fischbach, *Nat. Chem. Biol.*, 2015, **11**, 639–648.
99. D. J. Payne, M. N. Gwynn, D. J. Holmes and D. L. Pompliano, *Nat. Rev. Drug Discovery*, 2007, **6**, 29–40.
100. A. S. Waller and J. M. Clements, *Curr. Opin. Drug Discovery Dev.*, 2002, **5**, 785–792.
101. C. M. Apfel, H. Locher, S. Evers, B. Takács, C. Hubschwerlen, W. Pirson, G. P. Malcolm, W. Keck and M. G. P. Page, *Antimicrob. Agents Chemother.*, 2001, **45**, 1058–1064.

102. D. Butler, D. Chen, K. O'Dwyer, T. Lewandowski, K. Aubart and M. Zalacain, *Antimicrob. Agents Chemother.*, 2014, **58**, 290–296.
103. R. Corey, O. J. Naderer, W. D. O'Riordan, E. Dumont, L. S. Jones, M. Kurtinecz and J. Z. Zhu, *Antimicrob. Agents Chemother.*, 2014, **58**, 6518–6527.
104. S. Arat, A. Spivak, S. Van Horn, E. Thomas, C. Traini, G. Sathe, G. P. Livi, K. Ingraham, L. Jones, K. Aubart, D. J. Holmes, O. Naderer and J. R. Brown, *Antimicrob. Agents Chemother.*, 2015, **59**, 1182–1192.
105. J. Dubois, M. Dubois, J.-F. Martel, K. Aubart and D. Butler, *Antimicrob. Agents Chemother.*, 2015, **59**, 707–710.
106. J. Yao and C. O. Rock, *J. Biol. Chem.*, 2015, **290**, 5940–5946.
107. S. Brinster, G. Lamberet, B. Staels, P. Trieu-Cuot, A. Gruss and C. Poyart, *Nature*, 2009, **458**, 83–86.
108. W. Balemans, N. Lounis, R. Gilissen, J. Guillemont, K. Simmen, K. Andries and A. Koul, *Nature*, 2010, **463**, E3.
109. J. B. Parsons, M. W. Frank, C. Subramanian, P. Saenkham and C. O. Rock, *Proc. Natl. Acad. Sci.*, 2011, **108**, 15378–15383.
110. J. H. Yum, C. K. Kim, D. Yong, K. Lee, Y. Chong, C. M. Kim, J. M. Kim, S. Ro and J. M. Cho, *Antimicrob. Agents Chemother.*, 2007, **51**, 2591–2593.
111. J. A. Karlowsky, N. Kaplan, B. Hafkin, D. J. Hoban and G. G. Zhanel, *Antimicrob. Agents Chemother.*, 2009, **53**, 3544–3548.
112. R. K. Flamm, P. R. Rhomberg, N. Kaplan, R. N. Jones and D. J. Farrell, *Antimicrob. Agents Chemother.*, 2015, **59**, 2583–2587.
113. N. Kaplan, M. Albert, D. Awrey, E. Bardouniotis, J. Berman, T. Clarke, M. Dorsey, B. Hafkin, J. Ramnauth, V. Romanov, M. B. Schmid, R. Thalakada, J. Yethon and H. W. Pauls, *Antimicrob. Agents Chemother.*, 2012, **56**, 5865–5874.
114. H. Lu and P. J. Tonge, *Acc. Chem. Res.*, 2008, **41**, 11–20.
115. S. Escaich, L. Prouvensier, M. Saccomani, L. Durant, M. Oxoby, V. Gerusz, F. Moreau, V. Vongsouthi, K. Maher, I. Morrissey and C. Soulama-Mouze, *Antimicrob. Agents Chemother.*, 2011, **55**, 4692–4697.
116. V. Gerusz, A. Denis, F. Faivre, Y. Bonvin, M. Oxoby, S. Briet, G. LeFralliec, C. Oliveira, N. Desroy, C. Raymond, L. Peltier, F. Moreau, S. Escaich, V. Vongsouthi, S. Floquet, E. Drocourt, A. Walton, L. Prouvensier, M. Saccomani, L. Durant, J. M. Genevard, V. Sam-Sambo and C. Soulama-Mouze, *J. Med. Chem.*, 2012, **55**, 9914–9928.
117. H.-J. Li, C.-T. Lai, P. Pan, W. Yu, N. Liu, G. R. Bommineni, M. Garcia-Diaz, C. Simmerling and P. J. Tonge, *ACS Chem. Biol.*, 2014, **9**, 986–993.
118. A. Chang, J. Schiebel, W. Yu, G. R. Bommineni, P. Pan, M. V. Baxter, A. Khanna, C. A. Sotriffer, C. Kisker and P. J. Tonge, *Biochemistry*, 2013, **52**, 4217–4228.
119. J. Schiebel, A. Chang, B. Merget, G. R. Bommineni, W. Yu, L. A. Spagnuolo, M. V. Baxter, M. Tareilus, P. J. Tonge, C. Kisker and C. A. Sotriffer, *Biochemistry*, 2015, **54**, 1943–1955.
120. J. Schiebel, A. Chang, S. Shah, Y. Lu, L. Liu, P. Pan, M. W. Hirschbeck, M. Tareilus, S. Eltschkner, W. Yu, J. E. Cummings, S. E. Knudson,

G. R. Bommineni, S. G. Walker, R. A. Slayden, C. A. Sotriffer, P. J. Tonge and C. Kisker, *J. Biol. Chem.*, 2014, **289**, 15987–16005.

121. D. J. Haydon, N. R. Stokes, R. Ure, G. Galbraith, J. M. Bennett, D. R. Brown, P. J. Baker, V. V. Barynin, D. W. Rice, S. E. Sedelnikova, J. R. Heal, J. M. Sheridan, S. T. Aiwale, P. K. Chauhan, A. Srivastava, A. Taneja, I. Collins, J. Errington and L. G. Czaplewski, *Science*, 2008, **321**, 1673–1675.
122. M. Kaul, L. Mark, Y. Zhang, A. K. Parhi, E. J. Lavoie and D. S. Pilch, *Biochem. Pharmacol.*, 2013, **86**, 1699–1707.
123. M. Kaul, L. Mark, Y. Zhang, A. K. Parhi, Y. L. Lyu, J. Pawlak, S. Saravolatz, L. D. Saravolatz, M. P. Weinstein, E. J. LaVoie and D. S. Pilch, *Antimicrob. Agents Chemother.*, 2015, **59**, 4845–4855.
124. M. Kaul, L. Mark, Y. Zhang, A. K. Parhi, E. J. LaVoie and D. S. Pilch, *Antimicrob. Agents Chemother.*, 2013, **57**, 5860–5869.
125. M. Kaul, Y. Zhang, A. K. Parhi, E. J. Lavoie, S. Tuske, E. Arnold, J. E. Kerrigan and D. S. Pilch, *Biochimie*, 2013, **95**, 1880–1887.
126. S. E. Knudson, D. Awasthi, K. Kumar, A. Carreau, L. Goullieux, S. Lagrange, H. Vermet, I. Ojima and R. A. Slayden, *J. Antimicrob. Chemother.*, 2015, **70**, 3070–3073.
127. A. Miguel, J. Hsin, T. Liu, G. Tang, R. B. Altman and K. C. Huang, *PLoS Comput. Biol.*, 2015, **11**, e1004117.
128. M. Kaul, Y. Zhang, A. K. Parhi, E. J. Lavoie and D. S. Pilch, *Biochem. Pharmacol.*, 2014, **89**, 321–328.
129. C. M. Tan, A. G. Therien, J. Lu, S. H. Lee, A. Caron, C. J. Gill, C. Lebeau-Jacob, L. Benton-Perdomo, J. M. Monteiro, P. M. Pereira, N. L. Elsen, J. Wu, K. Deschamps, M. Petcu, S. Wong, E. Daigneault, S. Kramer, L. Liang, E. Maxwell, D. Claveau, J. Vaillancourt, K. Skorey, J. Tam, H. Wang, T. C. Meredith, S. Sillaots, L. Wang-Jarantow, Y. Ramtohul, E. Langlois, F. Landry, J. C. Reid, G. Parthasarathy, S. Sharma, A. Baryshnikova, K. J. Lumb, M. G. Pinho, S. M. Soisson and T. Roemer, *Sci. Transl. Med.*, 2012, **4**, 126ra35.
130. H. Brötz-Oesterhelt, I. Knezevic, S. Bartel, T. Lampe, U. Warnecke-Eberz, K. Ziegelbauer, D. Häbich and H. Labischinski, *J. Biol. Chem.*, 2003, **278**, 39435–39442.
131. S. D. Mills, A. E. Eakin, E. T. Buurman, J. V. Newman, N. Gao, H. Huynh, K. D. Johnson, S. Lahiri, A. B. Shapiro, G. K. Walkup, W. Yang and S. S. Stokes, *Antimicrob. Agents Chemother.*, 2011, **55**, 1088–1096.
132. S. S. Stokes, H. Huynh, M. Gowravaram, R. Albert, M. Cavero-Tomas, B. Chen, J. Harang, J. T. Loch, M. Lu, G. B. Mullen, S. Zhao, C.-F. Liu and S. D. Mills, *Bioorg. Med. Chem. Lett.*, 2011, **21**, 4556–4560.
133. S. S. Stokes, M. Gowravaram, H. Huynh, M. Lu, G. B. Mullen, B. Chen, R. Albert, T. J. O'Shea, M. T. Rooney, H. Hu, J. V. Newman and S. D. Mills, *Bioorg. Med. Chem. Lett.*, 2012, **22**, 85–89.
134. S. D. Podos, J. A. Thanassi and M. J. Pucci, *Antimicrob. Agents Chemother.*, 2012, **56**, 4095–4102.
135. B. L. M. de Jonge, G. K. Walkup, S. D. Lahiri, H. Huynh, G. Neckermann, L. Utley, T. J. Nash, J. Brock, M. San Martin, A. Kutschke, M. Johnstone,

V. Laganas, L. Hajec, R.-F. Gu, H. Ni, B. Chen, K. Hutchings, E. Holt, D. McKinney, N. Gao, S. Livchak and J. Thresher, *Antimicrob. Agents Chemother.*, 2013, **57**, 6005–6015.
136. J. Y. Choi, M. S. Plummer, J. Starr, C. R. Desbonnet, H. Soutter, J. Chang, J. R. Miller, K. Dillman, A. A. Miller and W. R. Roush, *J. Med. Chem.*, 2012, **55**, 852–870.
137. G. Martínez-Botella, J. N. Breen, J. E. S. Duffy, J. Dumas, B. Geng, I. K. Gowers, O. M. Green, S. Guler, M. F. Hentemann, F. A. Hernandez-Juan, D. Joseph-McCarthy, S. Kawatkar, N. A. Larsen, O. Lazari, J. T. Loch, J. A. Macritchie, A. R. McKenzie, J. V. Newman, N. B. Olivier, L. G. Otterson, A. P. Owens, J. Read, D. W. Sheppard and T. A. Keating, *J. Med. Chem.*, 2012, **55**, 10010–10021.
138. T. A. Keating, J. V. Newman, N. B. Olivier, L. G. Otterson, B. Andrews, P. A. Boriack-Sjodin, J. N. Breen, P. Doig, J. Dumas, E. Gangl, O. M. Green, S. Y. Guler, M. F. Hentemann, D. Joseph-mccarthy, S. Kawatkar, A. Kutschke, J. T. Loch, A. R. Mckenzie, S. Pradeepan, S. Prasad, G. Martínez-Botella and G. Mart, *ACS Chem. Biol.*, 2012, **7**, 1866–1872.
139. N. Srinivas, P. Jetter, B. J. Ueberbacher, M. Werneburg, K. Zerbe, J. Steinmann, B. Van der Meijden, F. Bernardini, A. Lederer, R. L. A. Dias, P. E. Misson, H. Henze, J. Zumbrunn, F. O. Gombert, D. Obrecht, P. Hunziker, S. Schauer, U. Ziegler, A. Käch, L. Eberl, K. Riedel, S. J. Demarco and J. A. Robinson, *Science*, 2010, **327**, 1010–1013.
140. J. Wallace, N. O. Bowlin, D. M. Mills, P. Saenkham, S. M. Kwasny, T. J. Opperman, J. D. Williams, C. O. Rock, T. L. Bowlin and D. T. Moir, *Antimicrob. Agents Chemother.*, 2015, **59**, 5775–5787.
141. M. A. Farha and E. D. Brown, *Ann. N. Y. Acad. Sci.*, 2015, **1354**, 54–66.
142. K. H. Turner, A. K. Wessel, G. C. Palmer, J. L. Murray and M. Whiteley, *Proc. Natl. Acad. Sci.*, 2015, **112**, 4110–4115.
143. K. H. Turner, J. Everett, U. Trivedi, K. P. Rumbaugh and M. Whiteley, *PLoS Genet.*, 2014, **10**, e1004518.
144. S. A. Lee, L. A. Gallagher, M. Thongdee, B. J. Staudinger, S. Lippman, P. K. Singh and C. Manoil, *Proc. Natl. Acad. Sci.*, 2015, **112**, 5189–5194.
145. S. Zlitni, L. F. Ferruccio and E. D. Brown, *Nat. Chem. Biol.*, 2013, **9**, 796–804.
146. D. Romero, E. Sanabria-Valentín, H. Vlamakis and R. Kolter, *Chem. Biol.*, 2013, **20**, 102–110.
147. V. M. Isabella, A. J. Campbell, J. Manchester, M. Sylvester, A. S. Nayar, K. E. Ferguson, R. Tommasi and A. A. Miller, *Chem. Biol.*, 2015, **22**, 535–547.

CHAPTER 5

The LPS Transport Pathway: A New Target for the Development of Gram-Negative Antibiotics

ALISON M. BEREZUK AND CEZAR M. KHURSIGARA*

Department of Molecular and Cellular Biology, University of Guelph, Guelph, Ontario, Canada
*E-mail: ckhursig@uoguelph.ca

5.1 Introduction

Antibiotics are a cornerstone of modern medicine and the ever-growing resistance of pathogens to antibiotics is accepted as a global threat to public health. This is due to the emergence of newly identified pathogens with multidrug-resistance profiles such as *Acinetobacter baumannii*, and the re-emergence of pathogens such as *Pseudomonas aeruginosa* and *Neisseria gonorrhoeae* in forms that are now resistant to frontline antibiotics. Couple this with the evolution and transmission of resistance genes that target clinically important drugs, such as NDM-1 (New Delhi metallo-β-lactamase 1), and the crisis of antimicrobial resistant infections is a global concern. It has recently been estimated that over 50 000 lives are lost each year to infections by antibiotic resistant bacteria in the US and Europe, with European annual costs in the €1.5M.[1] It has also been speculated that antibiotic infections caused by drug-resistant bacteria will kill more people

Drug Discovery Series No. 58
Antibiotic Drug Discovery: New Targets and Molecular Entities
Edited by Steven M. Firestine and Troy Lister

Published by the Royal Society of Chemistry, www.rsc.org

than cancer by 2050, costing the world up to $100 billion (U.S.). The recent emergence and dissemination of plasmid-encoded (*i.e.* readily transferable) colistin resistance highlights the severity of the problem.[2] This event compromises last resort therapies for combatting many antibiotic resistant bacteria.

While resistance to virtually all classes of current antibiotics seems inevitable, the health crisis is exacerbated by reliance in the clinic on a small number of compound classes and the general withdrawal of 'big pharma' from antibiotic discovery and development. The critical state of this global health care issue has been described in high-profile reports from the Infectious Diseases Society of America,[3] the World Health Organization (Fact Sheet 194, 2014), The US Centers for Disease Control and Prevention (Threat Report, 2013) and the Review on Antimicrobial Resistance commissioned in 2014 by the UK Government in collaboration with the Wellcome Trust. These reports all acknowledge the need for enhanced public awareness and conservation of precious antibiotic resources in medicine and livestock production, but there is also unanimity concerning the dire need for (new) classes of drugs to combat multidrug-resistant bacterial pathogens. It would be unrealistic to claim that efforts in the academic sector can deliver a new antibiotic to the clinic, but it is well within reason to expect that this sector can leverage its deep understanding of the structure and function of pathogens to identify new approaches and viable compound leads that lay an essential foundation to reinvigorate drug development in pharma. The Infectious Diseases Society of America[3] identified the 'ESKAPE' pathogens as priority targets. The 'K' now includes Extended-Spectrum β-Lactamases (ESBL)-containing and carbapenem resistant *Klebsiella pneumoniae* and *Esherichia coli* isolates, while 'P' stands for *Pseudomonas aeruginosa*. In the CDC Threat Report, these groups are ranked as 'urgent' and 'serious', respectively. Therefore, antibiotics with new mechanisms of action are urgently required to combat the growing health threat posed by resistant pathogenic microorganisms.

Gram-negative bacteria, such as *Klebsiella pneumoniae*, *Escherichia coli* and *Pseudomonas aeruginosa*, represent significant treatment challenges due in part to the nature of their cell envelopes. The outer membrane (OM) of Gram-negative bacteria acts as a permeability barrier to many compounds that would otherwise be effective antibacterial agents, including those effective against Gram-positive pathogens. A unique component of the OM is lipopolysaccharide (LPS), which greatly contributes to the structural stability of the cell envelope. Several classes of antimicrobials exist that directly target LPS in the OM. These include the antibiotics polymyxin B and polymyxin E (colistin) that are produced by Gram-positive bacteria such as *Paenibacillus polymyxa*. However, these antibiotics are severely cytotoxic to the host, and as such, have limited clinical use as last resort treatments. In this chapter, we discuss recent advances in understanding the mechanism of LPS biosynthesis and transport, and highlight progress towards targeting these essential processes for antibiotic development.

5.2 Lipopolysaccharide (LPS) and the Gram-Negative Cell Envelope

The cell envelope of a Gram-negative bacterium is a highly dynamic and unique structure. Unlike the single plasma membrane and thick cell wall characteristic of Gram-positive species, the Gram-negative cell envelope contains a thin peptidoglycan layer enclosed within a periplasmic space by two structurally distinct cell membranes (Figure 5.1).[4] In most Gram-negative organisms, it is the presence and composition of the outer membrane (OM) that is the essential defining feature of these bacterial species.

The OM is an asymmetric lipid bilayer, where the inner leaflet is composed of phospholipids and the outer leaflet is predominantly composed of the amphipathic molecule lipopolysaccharide (LPS).[4] Although LPS is not produced by some Gram-negative bacteria (*e.g. Borrelia burgdorferi*),[5,6] its presence is essential for the viability of the vast majority of organisms that do (with few exceptions[7–9]).

LPS typically contains three main structural elements. These include a hydrophobic fatty acyl chain lipid anchor (called lipid A), a well-conserved polysaccharide core region and a highly variable extended polysaccharide chain known as the O-antigen (Figure 5.1, inset).[10,11] The tight packing and hydrophobic properties of the LPS molecule contribute to the excellent permeability barrier formed by the OM.[4] This barrier function adds to the relative ineffectiveness of most antibiotics in treating Gram-negative infections, as many drugs are relatively hydrophobic and are unable to pass through the OM.

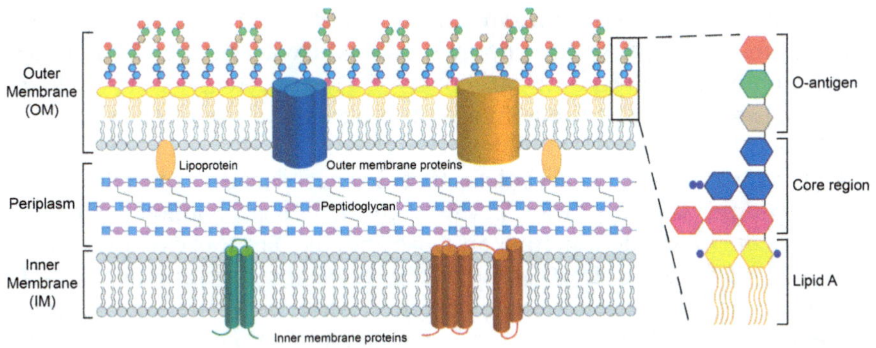

Figure 5.1 *The Gram-negative cell envelope.* The cell envelope of Gram-negative bacteria consists of two membranes, the inner membrane (IM) and the outer membrane (OM). The IM is composed of a phospholipid bilayer, whereas the OM comprises an interior leaflet of phospholipids and an exterior leaflet of lipopolysaccharide (LPS). LPS is composed of three distinct regions: lipid A, the core oligosaccharide and O-antigen (*right inset*). Both the IM and OM contain various transmembrane and membrane-associated proteins. Between the IM and OM is the periplasmic space, which contains the peptidoglycan layer (PG). PG comprises long polymers of repeating disaccharides that form a major structural feature of the cell envelope.

In addition, the enzymes involved in both the biosynthesis and transport of LPS from the inner to outer membranes are highly conserved among Gram-negative organisms and most are essential for cell viability,[6,7] making both pathways promising targets for the development of novel antimicrobials.

5.3 The LPS Biosynthesis Pathway

5.3.1 Kdo_2-Lipid A

In model Gram-negative organisms, such as *E. coli* and *P. aeruginosa*, the biosynthesis of LPS begins with the production of the lipid A moiety through a nine-step process commonly known as the Raetz pathway (Figure 5.2).[11–15] Lipid A anchors the LPS molecule into the OM and is composed of a β-1-6-linked glucosamine disaccharide backbone attached to several acyl chains.[12] The end product of this pathway, 3-deoxy-D-*manno*-oct-2-ulosonic acid $(Kdo)_2$-lipid A (Figure 5.2, inset), is the minimal LPS structure that can support a functional OM necessary for cell viability.[11]

All nine steps in the production of Kdo_2-lipid A occur at the cytosolic face of the inner membrane (IM). The Raetz pathway begins with the addition of an acyl chain to uridine diphosphate *N*-acetylglucosamine (UDP-GlcNAc) by the enzyme LpxA.[16] LpxA is an essential trimeric *O*-acyltransferase that reversibly attaches the acyl chain to the 3-OH group of UDP-GlcNAc using a fatty acyl-linked acyl carrier protein (ACP) as a donor.[16,17] LpxA uses a 'hydrocarbon ruler' within its active site to discriminate between the lengths of acyl chain incorporated into the lipid A structure. For example, LpxA in *E. coli* specifically transfers a β-hydroxymyristate (3-OH-C14:0) acyl chain, whereas LpxA from *P. aeruginosa* incorporates a β-hydroxydecanoate (3-OH-C10:0) residue onto the UDP-GlcNAc backbone.[18–20] This selectivity has been linked to a single amino acid residue in the proposed active site of LpxA in *E. coli* (G173) and *P. aeruginosa* (M169).[19]

Following reversible *O*-acylation, UDP-3-*O*-acyl-GlcNAc is committed to the Kdo_2-lipid A biosynthesis pathway through deacetylation by LpxC.[21] LpxC is a zinc-dependent metalloamidase and, like LpxA, is essential for cell growth.[12,22] Deletion of LpxC is thought to cause toxic accumulation of LPS precursors within the cell.[7] In addition, LpxC does not share high sequence homology with other deacetylases or eukaryotic proteins, and as such, a number of LpxC inhibitors have been developed in an attempt to target Gram-negative pathogens.[23–32] Early inhibitors of LpxC were typically substrate-analogues (*e.g.* TU-514, benzoic acid derivatives) which bound to the hydrophobic tunnel of the LpxC active site (Figure 5.3);[31,33] however, these inhibitors showed limited antimicrobial activity against *E. coli*, *P. aeruginosa* and other clinically important Gram-negative pathogens *in vivo*.[34–36]

More potent LpxC inhibitors have since been developed that include an aromatic ring containing a hydroxamic acid moiety to chelate the catalytic zinc ion and a peptide linker to block the hydrophobic tunnel in a 'warhead and tail' configuration (*e.g.* CHIR-090).[29,31,34]

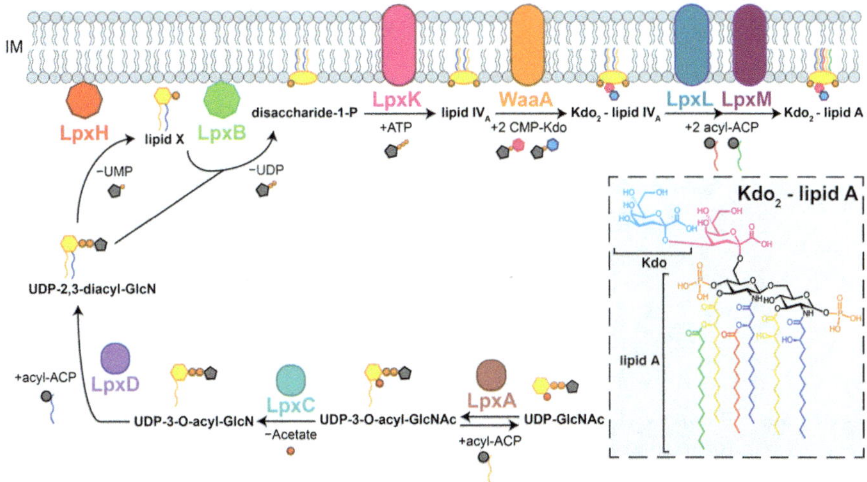

Figure 5.2 *Production of Kdo$_2$ – lipid A via the Raetz biosynthesis pathway*. In *E. coli*, the production of 3-deoxy-D-*manno*-oct-2-ulosonic acid (Kdo)$_2$-lipid A (*inset*) occurs *via* the nine-step Raetz biosynthesis pathway at the cytosolic face of the inner membrane (IM). The process begins with the addition of an acyl chain to uridine diphosphate *N*-acetylglucosamine (UDP-GlcNAc), donated from a fatty acyl-linked acyl carrier protein (acyl-ACP), by the *O*-acyltransferase LpxA. UDP-3-*O*-acyl-GlcNAc is then committed to the Kdo$_2$-lipid A biosynthesis pathway by the removal of an acetate group by the metalloamidase LpxC. LpxD then adds an *N*-linked acyl chain (donated from an ACP) to the free amino group created by deacetylation to form UDP-2,3-diacyl-GlcN. Subsequent condensation and insertion of a phosphorylated, tetra-acylated disaccharide-1-phosphate moiety into the inner leaflet of the IM is catalyzed by the peripheral membrane proteins LpxH and LpxB. Using cytosolic factors, the integral membrane proteins LpxK, WaaA, LpxL and LpxM then sequentially work to build the remainder of the Kdo$_2$-lipid A molecule. This includes phosphorylation by the ATP-dependent kinase LpxK, addition of two Kdo sugars by WaaA, and secondary acylation by the acyltransferases LpxL and LpxM using acyl-ACPs as donors. In *E. coli*, the biosynthesis of Kdo$_2$-lipid A is essential and is the minimal LPS structure that can support a functional OM necessary for cell viability.

The third step of the Raetz pathway is catalyzed by the enzyme LpxD and adds an *N*-linked acyl chain (donated from an ACP) to the free amino group created by deacetylation to form UDP-2,3-diacyl-GlcN.[37,38] Similar in both structure and function to LpxA, LpxD is an essential homotrimeric enzyme that utilizes a hydrocarbon ruler for species-specific substrate preference (β-hydroxymyristate in *E. coli*, β-hydroxylaurate [3-OH-C12:0] in *P. aeruginosa*).[18,39] Deletion of LpxD is lethal in *E. coli* and other Gram-negative bacteria, likely due to enhanced toxicity from the build-up of the detergent-like precursor UDP-3-*O*-acyl-GlcN.[37] Given their similarities and impact on cell viability, competitive peptide inhibitors that target both LpxA and LpxD have been developed.[40–43] These inhibitors prevent access to the ACP binding site

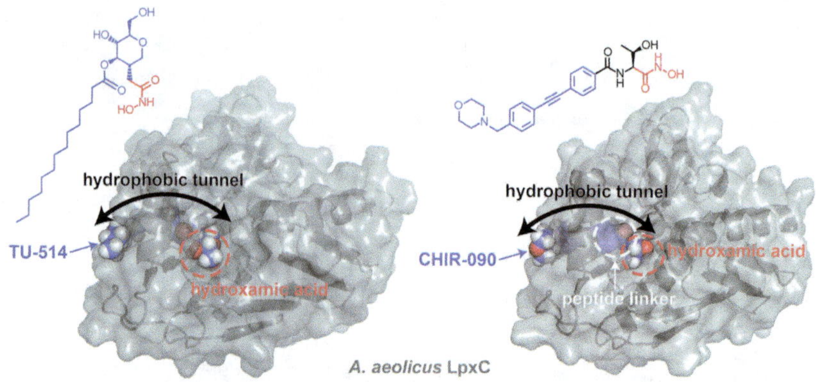

Figure 5.3 *Binding of LpxC inhibitors TU-514 and CHIR-090.* Structures of the LpxC inhibitors TU-514 and CHIR-090 bound to *Aquifex aeolicus* LpxC, as determined by NMR (PDB accession 1XXE [ref. 33]) and X-ray crystallography (PDB accession 2JT2 [ref. 31]), respectively. Typical LpxC inhibitors, such as TU-514 and CHIR-090, are substrate analogs that block access to the hydrophobic active site tunnel and bind the catalytic zinc ion *via* a hydroxamic acid moiety. Increases in *in vivo* antimicrobial activity of these inhibitors were attained upon the addition of a peptide linker, as seen with CHIR-090, resulting in a 'warhead and tail' configuration that more efficiently blocks the active site and chelates the catalytic zinc residue compared to early inhibitor designs.

(Figure 5.4) and make excellent candidates for further development of peptide-based antibiotics against these proteins. In some organisms, it has been shown that inhibition of LpxD has a greater impact on cell growth than the loss of LpxA or LpxC,[44] making it an even more promising target for the development of novel antimicrobials.

In *E. coli*, the next two steps in the biosynthesis of Kdo_2-lipid A are catalyzed by peripheral membrane proteins and result in the insertion of a phosphorylated, tetra-acylated disaccharide-1-phosphate moiety into the inner leaflet of the IM.[13,45] In most organisms, a proportion of the UDP-2,3-diacyl-GlcN population is hydrolyzed by LpxH to produce uridine monophosphate (UMP) and 2,3-diacylglucosamine 1-phosphate (also known as lipid X).[46,47] LpxB then condenses one molecule of UDP-2,3-diacyl-GlcN with one molecule of lipid X to form disaccharide-1-phosphate.[48] This reaction produces the β-1-6-glycosidic linkage characteristic of lipid A molecules. Although most Gram-negative species encode a single copy of each enzyme involved in lipid A biosynthesis, not all Gram-negative organisms contain LpxH.[11,49,50] Rather, α-proteobacteria such as *Caulobacter crescentus*, many δ-proteobacteria and various environmental strains encode a distinct and unrelated enzyme, LpxI.[49] LpxI catalyzes the formation of lipid X *via* hydrolysis of the β-phosphate of UDP-2,3-diacyl-GlcN, rather than the α-phosphate like LpxH.[49,50]

The final four steps of the Raetz pathway occur at the cytosolic face of the IM through the action of four integral membrane proteins: LpxK, WaaA, LpxL and LpxM.[12] Using cytosolic factors, these enzymes sequentially work

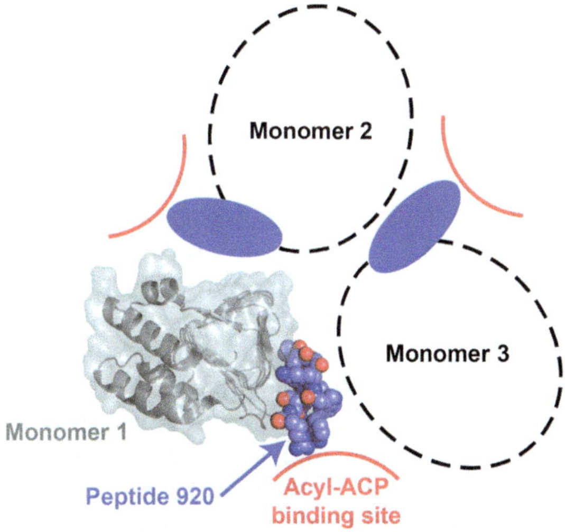

Figure 5.4 *Binding of LpxA inhibitor peptide 920.* Structure of the LpxA inhibitor peptide 920 bound to *E. coli* LpxA, as determined by X-ray crystallography (PDB accession 2AQ9). LpxA peptide-based inhibitors bind between LpxA monomers and prevent access to the fatty acyl-linked acyl carrier protein (ACP) binding sites on the surface of the LpxA homotrimer necessary for the essential acylation of Kdo_2-lipid A.[42]

to build the remainder of the Kdo_2-lipid A molecule. This includes phosphorylation of disaccharide-1-phosphate at the 4′-hydroxyl position to produce lipid IV_A by the ATP-dependent kinase LpxK,[51,52] addition of two Kdo sugars by WaaA[53] and secondary acylation by the acyltransferases LpxL and LpxM, again using acyl-ACPs as donors similar to LpxA and LpxD.[54–56] In *E. coli*, the transfer of the two Kdo sugars is an essential step in the biosynthesis of lipid A, as LpxL and LpxM catalyze the addition of a lauroyl group (C12:0) to the 2′ position and a myristoyl group (C14:0) to the 3′ position only when lipid A contains the Kdo sugars.[54] However, LpxL and LpxM in *P. aeruginosa* do not share this requirement and can produce fully acylated lipid A even if Kdo biosynthesis is inhibited.[57]

5.3.2 Core Polysaccharide

In most Gram-negative organisms, the biosynthesis of LPS continues with the extension of Kdo_2-lipid A by a core oligosaccharide (commonly divided into inner and outer core regions). The two Kdo sugars added during the production of lipid A constitute the first sugars of the LPS polysaccharide inner core. Using nucleotide sugars as donors, membrane-associated glycosyltransferases sequentially add two residues of L-*glycero*-D-*manno*-heptose (L,D-Hep), or in some instances, D,D-Hep.[15,45,58] High conservation of both lipid A and the inner core oligosaccharide is critical for maintaining OM integrity.[15] Compared to the inner core, the outer core polysaccharide is far less conserved among Gram-negative organisms. For example, as many as five different core oligosaccharide types exist among *E. coli* species; however, all are composed of basic linear backbones of six oligosaccharides made of Kdo, Hep, D-glucose and/or D-galactose.[45] Additional units may be added to the basic backbone to create a more complex branched structure. This same variation is seen in other clinically relevant Gram-negative pathogens such as *P. aeruginosa*, whose outer core is typically composed of a combination of D-glucose, L-rhamnose and 2-amino-2-deoxy-D-galactose.[14]

Upon completion of the lipid A-core structure, the LPS precursor is flipped from the inner leaflet to the outer leaflet of the IM where the remainder of LPS biosynthesis can occur (Figure 5.5). The ATP binding cassette (ABC) transporter MsbA mediates this essential process and acts as a quality control checkpoint during the biosynthesis of LPS.[59,60] MsbA most efficiently transports only a completed lipid A-core moiety and has reduced affinity for other incomplete LPS precursors such as the lipid IV_A intermediate or those missing the 1 and 4′ phosphates.[61,62] Depletion of MsbA results in the accumulation of lipid A and invagination of the IM in *E. coli*, ultimately resulting in cell death.[63] However, despite being an essential protein in most Gram-negative

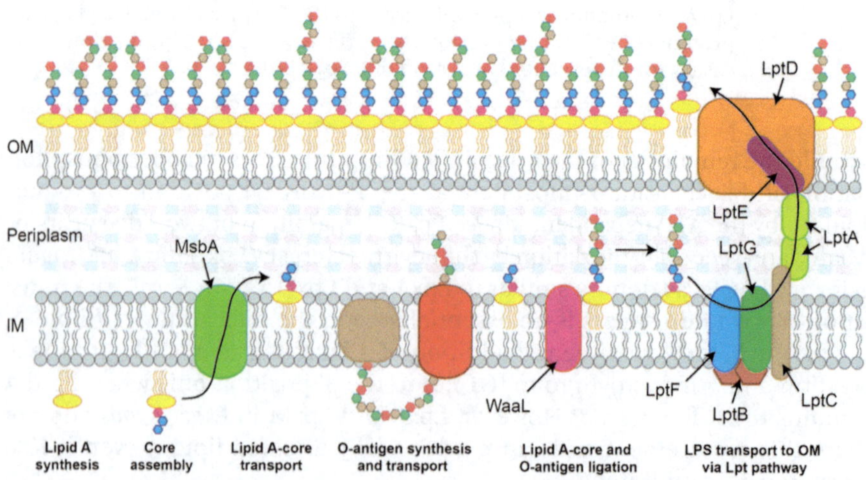

Figure 5.5 *Biosynthesis and transport of lipopolysaccharide.* Kdo_2-lipid A is synthesized and the core region is attached on the cytoplasmic surface of the IM. MsbA transports the lipid A-core complex to the periplasmic side of the IM. O-antigen is synthesized by IM-associated enzyme complexes and transported to the periplasmic face of the IM by one of three pathways: (1) Wzy-dependent, (2) ABC transporter-dependent or (3) synthase-dependent (reviewed in ref. 11 and 64). O-antigen is coupled to the lipid A-core on the periplasmic face of the IM by the WaaL ligase and the complete LPS molecule is transported across the periplasm and inserted into the outer leaflet of the OM by the Lpt (*LPS transport*) system. This protein complex is comprised of $LptB_2FGC$ in the IM, LptA in the periplasm and LptDE in the OM. The functions of each component of the Lpt system are described in the text.

organisms, MsbA may not be as desirable of a therapeutic target as other LPS biosynthetic enzymes as MsbA shares homology with eukaryotic multi-drug resistance (MDR) proteins.[59]

5.3.3 O-Antigen

The final stage in the production of fully mature LPS is the ligation of a highly variable extended polysaccharide chain to the lipid A-core. This polysaccharide, known as the O-polysaccharide or O-antigen, provides the extensive variability seen between LPS structures of different Gram-negative organisms, and even between different strains of the same species.[11] The variation of the O-antigen arises from differences in the composition, order, linkage and chain length of the different repeating sugar units within the polysaccharide structure, and it is this part of the LPS molecule that contributes to the serotype of Gram-negative organisms.[11,14,64,65]

The O-antigen is synthesized at the IM by one of three common mechanisms (Wzy-, ABC transporter- or synthase-dependent; Reviewed in ref. 11 and 64) and is ligated as a completed structure to the outer core region of lipid A (Figure 5.5). Regardless of which pathway is used in the production of the O-antigen, each process requires an undecaprenol phosphate carrier as an acceptor for chain assembly and ends with the addition of the O-antigen to the core at the periplasmic face of the IM by the ligase WaaL.[11,66] Capping of the lipid A-core domain with the O-polysaccharide is not essential for the production of functional LPS and a complex mixture of LPS types (with or without the O-antigen) can be found in the OM of Gram-negative species.[64] However, the presence of the O-polysaccharide, as well as additional modifications to the overall LPS structure, aid in the resistance of Gram-negative organisms to antimicrobial treatment, the evasion of host defences and survival during other harsh environmental pressures.

5.4 LPS Modification and Its Role in Gram-Negative Bacterial Persistence

The presence of LPS on the surface of Gram-negative organisms not only contributes to the effective permeability barrier formed by the OM, but is also a microorganism-associated molecular pattern (MAMP) recognized by the innate immune system.[15,67] The lipid A portion of LPS, also known as endotoxin, triggers the release of proinflammatory mediators through Toll-like receptor 4 (TLR4)-dependent signalling at picomolar concentrations, and, at high concentrations indicative of an advanced bacterial infection, can result in septic shock.[68] To trigger an inflammatory response, the innate immune system recognizes characteristic structural features of the LPS molecule, and as such, Gram-negative bacteria have evolved various mechanisms to fine-tune their LPS structures to promote survival within the host and other hostile environments. These structural modifications, which include changes in lipid A acylation, methylation, phosphate group modification and alterations to the sugar composition and/or linkages within the core or O-antigen regions to name a few, easily increase resistance to antibiotic treatment and the innate immune response. For example, the human gastric pathogen *Helicobacter pylori* produces a highly modified version of LPS compared to *E. coli* that includes dephosphorylation, deacylation and the addition of phosphoethanolamine to its lipid A region.[67,69] These modifications lead to an approximately 1000-fold increased resistance to antimicrobial peptides and reduced recognition by components of the innate immune system compared to mutant strains of *H. pylori* unable to modify their surface.[69] Other species in which LPS modification is advantageous for bacterial virulence include *Salmonella typhimurium*, *Yersinia pestis*, *Francisella tularensis* and *Vibrio cholerae*, among others.[67] How Gram-negative bacteria modify their LPS is highly variable across species. An extensive evaluation of these mechanisms is beyond the scope of this chapter; however, many excellent reviews on this topic are available which highlight the enzymes and regulatory mechanisms responsible for these processes.[10,12,13,45,67]

5.5 LPS Transport: The Lpt Pathway

Once the assembly of LPS is completed and all modifications have been made, the LPS molecule must be transported from the outer leaflet of the IM to the outer leaflet of the OM. Until recently, it was not well understood how this process was carried out given that the amphipathic nature of LPS dictates that it would not be able to cross the cell envelope unassisted, or without an energy source to extract it from the IM. However, recent genetic, biochemical and bioinformatics studies have revealed a *l*ipo*p*olysaccharide *t*ransport (Lpt) complex that spans from the cytoplasm through to the OM that is involved in the complete transport of LPS to the cell surface.[70–72]

The Lpt complex has been predominantly studied in *E. coli*, although the crystal structures of various components of the Lpt complex in other Gram-negative organisms have been determined.[73–79] In *E. coli*, the Lpt complex is composed of seven proteins (LptA-G) that link the inner and outer membranes by a *trans*-envelope protein bridge (Figure 5.5).[15,70,72] All seven of these proteins are essential in *E. coli* and disruption of LPS transport at any stage results in the same cellular phenotype; LPS accumulates in the outer leaflet of the IM and, coupled with compromised OM integrity, leads to cell death.[80] This suggests that the Lpt transport complex operates as a concerted machine, where disruption of any one component results in accumulation of LPS at the beginning of the transport process.

5.5.1 Extraction of LPS from the IM

The first step in LPS transport is extraction of LPS from the outer leaflet of the IM. This is accomplished through the coordinated action of the ABC transporter $LptB_2FG$ and the periplasmic domain of the IM protein LptC.[80–82]

Energy required for LPS transport is supplied by ATP hydrolysis *via* the cytoplasmic ATPase LptB.[77,83] LptB is the soluble nucleotide-binding domain of the ABC transporter and exists as a homodimer at the cytosolic end of the complex.[77,78,82] Upon ATP hydrolysis, the LptB dimer undergoes extensive movement within the cytoplasm that elicits subsequent changes in the two transmembrane domains of the transporter formed by LptF and LptG.[77] LptF and LptG each contain six transmembrane segments and together form the twelve transmembrane domains typically seen in ABC transporters.[81] The periplasmic domains of these two proteins are thought to contribute to the formation of a periplasmic Lpt bridge through their interaction with the bitopic LPS binding protein LptC. LptC forms a stable association with the $LptB_2FG$ complex at the amino-terminal edge of its soluble periplasmic domain.[84] This domain has a β-jellyroll fold (belonging to the organic solvent tolerance protein A [OstA] superfamily) that has been shown to bind LPS.[74,85] LptF and LptG are also thought to have a similar fold within their large periplasmic domains,[81,84] allowing for the unidirectional transfer of LPS from the ABC transporter to LptC using energy generated by ATP hydrolysis.

5.5.2 Traversing the Periplasm: The LptA Protein Bridge

Once LPS has been extracted from the IM, it must pass through the periplasm to the OM. As mentioned previously, this step of the process poses a biochemical challenge as the hydrophobic lipid A portion of the LPS molecule cannot pass through the aqueous environment of the periplasm unassisted. This challenge is overcome by the action of the soluble periplasmic protein LptA.[86] Until recently, it was unclear how LptA mediates the transport of LPS across the periplasm. Like LptC, LptA binds LPS on the inside of its OstA domain.[83,85] Through the use of *in vivo* UV cross-linking and size-exclusion chromatography, the N-terminal domain was shown to interact with the C-terminal periplasmic domain of LptC along the edges of the individual β-jellyrolls.[87,88] This suggests LptC and LptA form a 'head-to-tail' protein bridge with a continuous LPS binding groove. LPS is pushed through the hydrophobic groove of LptC to LptA following a second ATP hydrolysis event by the $LptB_2FG$ transporter.[83]

For a long time, it was unknown how LPS, bound to LptA at the IM, is transported across the periplasm. Two predominant mechanisms were proposed. The first posits that LptA acts as a soluble chaperone that diffuses through the periplasm and docks with the OM complex LptDE, which in turn inserts the LPS molecule into the outer leaflet. This scenario is modeled after the transport of OM lipoproteins by the soluble chaperone LolA.[89] However, recent studies have provided considerable evidence in favour of a second scenario. LptA from *E. coli* crystallizes as a head-to-tail filamentous oligomer[73] and has been shown to readily form stable rod-shaped oligomers *in vitro.*[90] This would suggest that multiple copies of LptA might form a protein bridge spanning the periplasm. In this case, LptA physically connects both the IM and OM transport complexes at the same time and a continuous stream of LPS is pushed through the hydrophobic groove down the middle of the bridge powered by ATP hydrolysis, akin to a PEZ candy dispenser.[70,83] This model, appropriately termed the 'PEZ' model of LPS transport,[70] is further supported by evidence that LPS is still transported to the OM in the absence of soluble periplasmic components,[91] implying that LptA does not freely traverse the periplasm like LolA does. Furthermore, numerous experiments using both sucrose density gradient centrifugation and co-purification assays show that LptA co-fractionates and co-purifies with all inner and outer membrane components of the Lpt transport complex,[92] suggesting that all seven Lpt proteins form a single complex that completely spans the periplasm as a cohesive unit. At this time it is still unknown how many LptA molecules assemble to form the proteinaceous bridge, although estimates suggest that an LptA dimer would be sufficient to span the periplasmic space between LptC and LptD in the OM.[88]

5.5.3 LPS Insertion into the OM

The heterodimeric OM complex LptDE completes the final stage of LPS transport, insertion of LPS into the outer leaflet of the OM.[93] LptD is a large β-barrel OM protein that contains a sizeable periplasmic domain at its N-terminus.[75,76] This domain is homologous to the OstA domains of LptC and LptA, and similar to the interaction between LptC and LptA, binds the C-terminus of LptA along the edges of its β-jellyroll.[88] This interaction completes the head-to-tail LptC-LptA-LptD protein bridge from the IM to the OM. LptE is an OM lipoprotein whose periplasmic domain plugs the lumen of LptD in an extremely stable 1:1 complex.[93,94] This 'plug-and-barrel' conformation is crucial in promoting proper folding of LptD and subsequent insertion of LPS into the OM. LptD is initially inserted into the OM in a non-functional conformation, whereby two cysteine residues (C31 and C173) within the N-terminal OstA domain form a non-native disulfide bond.[95] Interaction of LptE within the lumen of LptD triggers the rearrangement of this disulfide bond by the periplasmic oxidase DsbA.[95] At least one of two new interdomain disulfide bonds (C31-C724 or C173-C725) is formed, connecting the N- and C-terminal regions and aligning the N-terminal domain with the LptA protein bridge.[95,96] This rearrangement requires partial refolding of the β-barrel, which again relies on the interaction of LptD with LptE.[97] Overall, assembly of the completed Lpt complex only occurs after a functional LptDE heterodimer is formed,[88,95] ensuring that LPS is only exported when final insertion into the OM can take place. Lateral insertion of LPS into the outer leaflet is proposed to occur *via* an intramembrane hole in the side of the LptD β-barrel.[75,76,94,98,99] Here, conformational changes in the LptDE complex caused by binding of LPS to the N-terminal OstA domain of LptD would cause opening of two adjacent strands of the β-barrel. The acyl chains of lipid A would then be inserted into the outer leaflet of the OM through this opening while the polysaccharide core and O-antigen move through the lumen of LptD.[98,99] The exact biochemical mechanism by which this process occurs and how LptE interacts with the LPS molecule is still unclear.

Based on our current understanding of the LPS transport system in *E. coli*, there is no stringency placed on which LPS structures are shuttled to the OM through the Lpt complex. For example, LptA has been shown to bind a variety of LPS forms,[85] allowing for the efficient transport of diverse LPS serotypes to the OM. This highlights the importance of the Lpt pathway as an intriguing target for the development of novel antimicrobials since all LPS molecules are transported to the OM using the same system, regardless of what LPS modifications have been made by the cell. Therefore, antibiotics active against this system could potentially target a broader spectrum of Gram-negative pathogens than other more stringent pathways such as the LPS biosynthesis machinery. As such, recent studies involving the development of novel antibiotics against the LPS transport pathway have shown great promise as therapeutics against Gram-negative organisms and are discussed in more detail for the remainder of this chapter.

5.6 LPS Transport as a New Target for the Development of Gram-Negative Antibiotics

The prominent use of antibiotic therapy over the last century has had an overwhelming impact on society through the improved treatment of a multitude of human diseases. For example, the introduction of penicillin for general use in the late 1940s has led to the efficient treatment of many Gram-positive bacterial infections, the most prevalent being *Streptococcal pharyngitis* or strep throat, and has since become the most widely used antibiotic today.[100] Currently, the majority of existing drugs typically target protein, nucleic acid and cell wall synthesis.[101,102] These processes have been thoroughly exploited due to the wealth of biochemical knowledge about these essential pathways. However, despite the positive impact that antibiotics have had, the limited number of novel drug targets combined with clinical overuse of these drugs has contributed to the rising phenomenon of multidrug-resistance in a vast number of pathogenic bacteria.

The treatment of Gram-negative infections becomes even more problematic when you consider the added obstacle of drugs having to pass through the extensive permeability barrier formed by the OM to exert their effects. As mentioned previously, the presence of LPS on the surface of Gram-negative bacteria significantly contributes to this effective barrier. Tight packing of LPS is attributed to a higher number of fatty acyl chains per molecule than regular phospholipids, saturation of the fatty acyl chains and stable lateral interactions facilitated by hydrogen bonds and bridging cations between phosphate groups.[72] In addition, LPS, and in particular the lipid A portion (also known as endotoxin), is a potent activator of the innate immune system.[68,103] Therefore, care must be taken in the treatment of Gram-negative infections as the release of high levels of LPS can lead to septic shock.[68] As such, antimicrobials active against the LPS biosynthesis and transport pathways become intriguing bacterial drug targets given the premise that inhibition of these essential pathways could potentially limit the amount of LPS produced by the cell.

For over two decades, the optimization of LPS biosynthesis inhibitors has made great strides in the development of new potent antimicrobials active against Gram-negative organisms. By far the most extensively targeted component of this pathway is the essential metalloamidase LpxC, likely given that LpxC is a well conserved metallo-enzyme across Gram-negative species and catalyzes the committed step in LPS biosynthesis.[21,22,104] LpxC is inhibited by several different types of compounds (Figure 5.6(a)); these include those that bind the hydrophobic passage leading to the active site (*e.g.* L-161,240),[27] uridine-based inhibitors that target the UDP binding pocket (*e.g.* 1-68A)[23] and

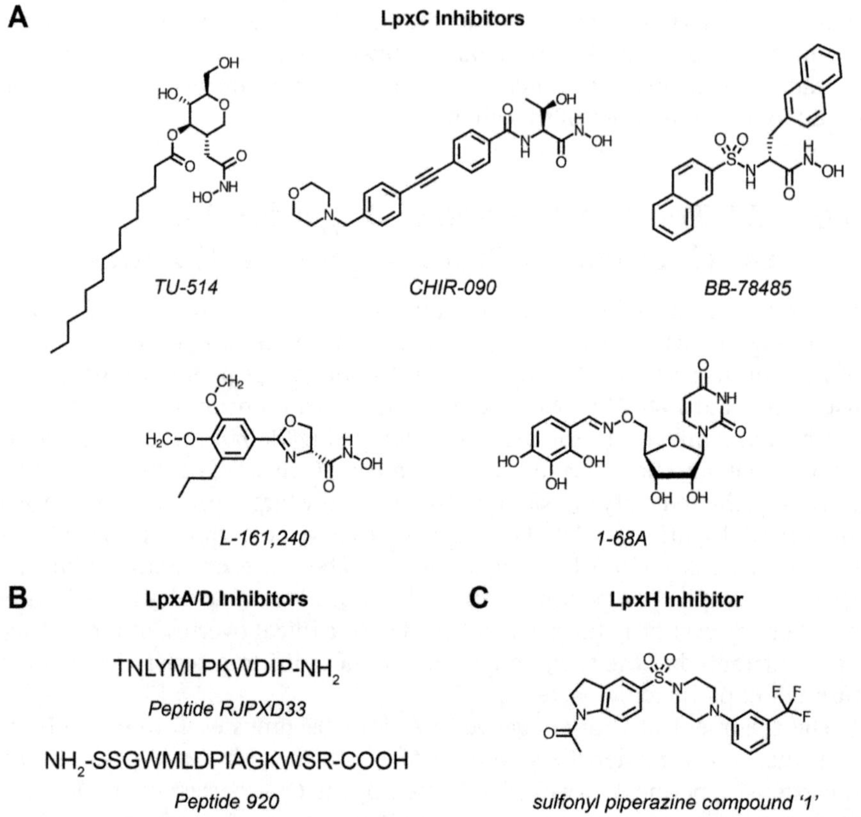

Figure 5.6 *LPS biosynthesis inhibitors.* Chemical structures of current inhibitors of the LPS biosynthesis pathway. (A) LpxC inhibitors typically include hydroxamate-containing compounds, such as TU-514, L-161,240, BB-78485 and CHIR-090, which target the hydrophobic active site tunnel and catalytic zinc ion of LpxC. Additional LpxC inhibitors have also been developed, including the uridine-based inhibitor 1-68A that targets the LpxC UDP binding pocket. (B) LpxA/D dual inhibitors include the peptide-based antibiotics 'peptide RJPXD33' and 'peptide 920', and prevent access to the acyl carrier protein (ACP) binding sites of these enzymes that is required for the essential acylation of Kdo_2-lipid A. (C) Inhibitors of LpxH include the sulfonyl piperazine compound '1'. This compound has been shown to inhibit LpxH mutants *in vivo*, however the precise mode of action of this inhibitor is still unknown (ref. 23, 25, 41, 42 and 105).

most commonly by hydroxamate-containing compounds, which chelate the catalytic zinc residue and are the most potent LpxC inhibitors to date (*e.g.* CHIR-090, BB-78485).[28–32] Few compounds active against other LPS biosynthesis enzymes (such as LpxA/LpxD[40–43] and LpxH[105]) have also been developed (Figure 5.6(b) and (c)). Although current LPS biosynthesis inhibitors demonstrate potency against a broad-spectrum of Gram-negative species that is on par with current antibiotics,[29,32] mechanisms decreasing the susceptibility of organisms to these inhibitors has already been demonstrated in harmful Gram-negative pathogens such as *P. aeruginosa*[106] and highlights the pressing need to identify other novel drug targets.

In comparison to LPS biosynthesis, only a limited number of Lpt inhibitors have currently been developed (Figure 5.7, Table 5.1). The first compounds

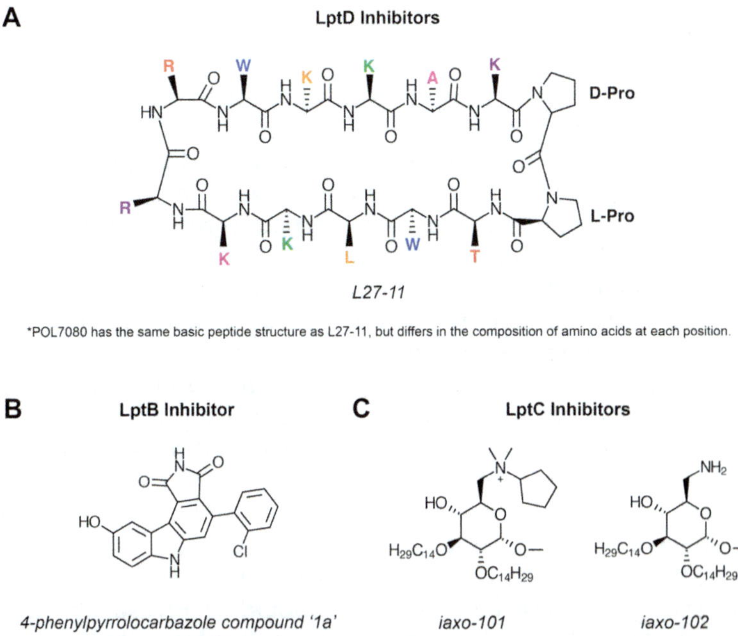

Figure 5.7 *LPS transport inhibitors.* Chemical structures of current inhibitors of the LPS transport pathway. (A) LptD inhibitors, such as L27-11 and POL7080, are peptide mimics that contain a β-hairpin arrangement necessary for function and a varying mixture of hydrophobic, aromatic and cationic amino acids (denoted by their single letter code). These compounds display minimal inhibitory concentrations (MICs) in the nanomolar range strictly against *Pseudomonas* species, but not other Gram-positive or Gram-negative bacteria. (B) LptB inhibitors are based on small molecule kinase inhibitors and include the 4-phenylpyrrolocarbazole derivative compound **'1a'**. (C) Inhibitors of LptC (iaxo-101 and iaxo-102) are composed of a cyclic sugar core, a positively charged nitrogen group and two C_{14} linear aliphatic ether chains. These compounds inhibit LptC *in vitro*, but currently show no *in vivo* antimicrobial activity (ref. 109, 117 and 118).

Table 5.1 Summary of Lpt inhibitors.

		Antibacterial Activity		
Lpt Inhibitor	Target	Organism	*In vivo* activity?	Reference
L27-11	LptD	*P. aeruginosa* (multiple strains)	Yes	107,109,110
		E. coli ATCC 25922	No	107,109,110
		A. baumannii DSM3008	No	107,109,110
		K. pneumonia ATCC 13883	No	107,109,110
POL7080	LptD	*P. aeruginosa* (multiple strains)	Yes	107,108
Compound '**1a**'	LptB	*E. coli* MC4100	No	117
		OM compromised *E. coli*[a]	Yes	117
iaxo-102	LptC	*E. coli* MG1655	No	118
		OM compromised *E. coli*[b]	No	118

[a]OM compromised *E. coli* strain NR698 (ref. 127).
[b]OM compromised *E. coli* strain AS19 (ref. 128).

shown to target the Lpt transport pathway were discovered in 2010 by Srinivas *et al.*, who screened a family of peptidomimetic (protein epitope mimetic or PEM) antibiotics based on the structure of the cationic antibacterial peptide protegrin I.[107,108] These compounds contain a β-hairpin arrangement that is necessary for function and a mix of hydrophobic, aromatic and cationic residues (Figure 5.7(a)).[109]

Two promising candidates, compounds L27-11 and POL7080, displayed minimal inhibitory concentrations (MICs) in the nanomolar range strictly against *Pseudomonas* species and no other Gram-positive or Gram-negative bacteria.[107] A genetic suppressor screen of *P. aeruginosa* isolates found that mutations conferring resistance map to the periplasmic N-terminal domain of LptD.[107] This domain is highly variable between Gram-negative species (*e.g.* 300 amino acids in length in *P. aeruginosa*, but only 100 amino acids in *E. coli*[107]), which could explain the species specificity of these antimicrobials. It was initially unknown whether L27-11 or POL7080 were transported into the cell *via* the lumen of LptD to act on another target, or if LptD was indeed the cellular target of these antibiotics.[110] It would be a reasonable assumption that L27-11 or POL7080 could enter the cell through LptD, especially since the pore formed by LptD is larger than a typical *E. coli* porin (1.8 nm[111] *vs.* 1–1.5 nm[112]). However, conditional depletion of LptD mimics the impact of L27-11 on *P. aeruginosa*, which includes perturbation of the IM and OM blebbing as seen by both transmission and scanning electron microscopy.[110] Also, L27-11 was found to physically bind LptD using a photoaffinity labeled version of the antibiotic,[107] further supporting the hypothesis that LptD is the cellular target of these peptidomimetic compounds. Compared to L27-11, POL7080 displays increased potential as a novel antibiotic given that it has an increased half-life in human plasma compared to L27-11, as well as MICs as low as 0.25 μg mL^{-1} against over 100 clinical isolates of *P. aeruginosa*.[107] POL7080 also provided protection against *P. aeruginosa* in a mouse infection model and exhibited smaller effective doses than gentamicin controls.[107]

This is an extremely promising finding as POL7080 could be developed for the treatment of hospital acquired or life-threatening chronic infections of *P. aeruginosa* such as those typically found in patients with cystic fibrosis.[113] In fact, POL7080 (marketed as Murepavadin) is currently being developed as a novel antibiotic by the Swiss Pharma company Polyphor, in conjunction with Roche, and has entered Phase II clinical trials.[108,110,114]

Given the energy required for transport of LPS from the inner to outer membranes completely relies on ATP hydrolysis, LptB has become another intriguing target for the development of novel therapeutics. In a high-throughput screen of 244 commercially available small molecule kinase inhibitors, Gronenberg *et al.* tested the ability of these compounds to competitively inhibit the ATPase activity of LptB.[115] This screen assessed changes in the activity of purified LptB, both alone and in complex with LptFGC, by coupling ATP hydrolysis with NADH oxidation and measuring decreases in fluorescence intensity.[115] From this screen, two LptB inhibitors were identified. The lead compound (compound '**1a**') is a derivative of 4-phenylpyrrolocarbazole, a potent inhibitor of the eukaryotic kinase Wee1 (Figure 5.7(b)).[116,117] Compound '**1a**' and its optimized derivatives were shown to inhibit LptB activity *in vitro*; however, reduced inhibition was observed when LptB was stabilized in complex with LptFGC.[117] In addition, these compounds exhibited antimicrobial activity *in vivo* and had MIC values as low as 12.5 μM against a strain of *E. coli* harbouring a gene mutation that compromises OM integrity.[117] When tested against a non-leaky wild-type strain of *E. coli*, these compounds showed no antimicrobial activity and had MIC values over 100 μM.[115,117] Given the relatively small size and hydrophobic nature of these compounds, this highlights the impenetrable barrier formed by the OM and the impact it has on the treatment of Gram-negative infections.

The most recent attempt at developing antibiotics that target LPS transport in Gram-negative bacteria draws on the observation that the small molecule compounds iaxo-101 and iaxo-102[118] prevent binding of LPS to CD14, a component of the immune response to LPS.[68] This leads to inhibition of Toll-like receptor 4 (TLR4) signalling and dampening of the inflammatory response that is characteristic of a Gram-negative infection.[119–122] Given components of the Lpt transport pathway also bind LPS, perhaps Lpt inhibitors could be designed based on these high-affinity ligands. The iaxo-101 and iaxo-102 compounds are composed of a cyclic sugar core, a positively charged nitrogen group and two C_{14} linear aliphatic ether chains (Figure 5.7(c)).[118] Using a fluorescent derivative of iaxo-102, Sestito *et al.* assessed the binding of this synthetic ligand with purified LptC through fluorescence intensity measurements and tryptophan fluorescence quenching.[118] Fluorescent iaxo-102 rapidly, specifically and irreversibly bound to LptC *in vitro* and likely competes for the same binding site as LPS.[118] Unfortunately, this compound does not show antimicrobial activity *in vivo* against wild-type or OM compromised *E. coli* strains;[118] however, optimization of these compounds to increase cell penetration and affinity for LptC could lead to more potent competitive inhibitors of the Lpt pathway.

In addition to the development of standalone drugs, it stands to reason that disruption of the cell envelope through the inhibition of LPS transport could also increase the efficacy of current antibiotic treatments. In *Acinetobacter baumannii*, deletion of *lptD* shows a similar growth defect to that of an *lpxC* deletion, however these cells are more sensitive to hydrophobic antibiotics.[123] This would suggest that there is a more severe impact on cell permeability from disruption of LPS transport than LPS biosynthesis. This is an intriguing avenue of study, as it would allow for the development of dual treatments that couple Lpt inhibitors with other current antibiotics typically unable to pass through the Gram-negative OM. Mutations in the essential genes *lptD*[124,125] and *lptE*[97,126] in other clinically relevant pathogens such as *P. aeruginosa* also result in increased OM permeability and susceptibility to an extensive suite of antibiotics, including ciprofloxacin, rifampicin, and erythromycin, further supporting this idea.

5.7 Conclusions

Much remains to be determined about LPS biosynthesis and transport before these systems can be fully exploited as platforms for drug development. Our current understanding of residues crucial for the function of Lpt proteins, although limited, can help guide inhibitor design. For example, residues of LptA and LptC found to cross-link with LPS represent sites at which LPS is bound for longer periods of time[70,88] and could be exploited in the development of novel inhibitors. Additionally, residues essential for nucleotide binding or interaction with the LptFG transmembrane components are candidate sites for LptB inhibition.[77] Of all Lpt transport components, the conserved domain organisation, its exposed OM location and current progress of drugs targeting LptD in clinical trial makes it the most viable target in antimicrobial development to date. Although high-resolution structural information exists for some Lpt components, there are considerable gaps that will be crucial for detailed mechanistic studies of drug–target interactions. The development of inhibitory activity assays and high-throughput screens for small molecules of natural and synthetic origin will also undoubtedly speed the discovery of LPS inhibitors. However, with the pressing need to identify new antimicrobial targets, the LPS biosynthesis and transport pathways present viable and promising targets for the development of new antibiotics against Gram-negative pathogens.

References

1. J. M. A. Blair, M. A. Webber, A. J. Baylay, D. O. Ogbolu and L. J. V. Piddock, *Nat. Rev. Microbiol.*, 2015, **13**, 42–51.
2. C. Zhi, L. Lv, L. F. Yu, Y. Doi and J. H. Liu, *Lancet Infect. Dis.*, 2016, **16**, 292–293.

3. H. W. Boucher, G. H. Talbot, J. S. Bradley, J. E. Edwards, D. Gilbert, L. B. Rice, M. Scheld, B. Spellberg and J. Bartlett, *Clin. Infect. Dis.*, 2009, **48**, 1–12.
4. T. J. Silhavy, D. Kahne and S. Walker, *Cold Spring Harbor Perspect. Biol.*, 2010, **2**, 1–16.
5. K. Takayama, R. J. Rothenberg and A. G. Barbour, *Infect. Immun.*, 1987, **55**, 2311–2313.
6. I. C. Sutcliffe, *Trends Microbiol.*, 2010, **18**, 464–470.
7. G. Zhang, T. C. Meredith and D. Kahne, *Curr. Opin. Microbiol.*, 2013, **16**, 779–785.
8. L. Steeghs, R. den Hartog, A. den Boer, B. Zomer, P. Roholl and P. van der Ley, *Nature*, 1998, **392**, 449–450.
9. J. H. Moffatt, M. Harper, P. Harrison, J. D. F. Hale, E. Vinogradov, T. Seemann, R. Henry, B. Crane, F. St. Michael, A. D. Cox, B. Adler, R. L. Nation, J. Li and J. D. Boyce, *Antimicrob. Agents Chemother.*, 2010, **54**, 4971–4977.
10. R. F. Maldonado, I. Sá-Correia and M. A. Valvano, *FEMS Microbiol. Rev.*, 2016, 480–493.
11. C. R. H. Raetz and C. Whitfield, *Annu. Rev. Biochem.*, 2002, **71**, 635–700.
12. M. S. Trent, C. M. Stead, A. X. Tran and J. V. Hankins, *J. Endotoxin Res.*, 2006, **12**, 205–223.
13. C. R. Raetz, C. M. Reynolds, M. S. Trent and R. E. Bishop, *Annu. Rev. Biochem.*, 2007, **76**, 295–329.
14. J. D. King, D. Kocíncová, E. L. Westman and J. S. Lam, *Innate Immun.*, 2009, **15**, 261–312.
15. C. Whitfield and M. S. Trent, *Annu. Rev. Biochem.*, 2014, **83**, 99–128.
16. M. S. Anderson and C. R. Raetz, *J. Biol. Chem.*, 1987, **262**, 5159–5169.
17. C. R. Raetz and S. L. Roderick, *Science*, 1995, **270**, 997–1000.
18. J. M. Williamson, M. S. Anderson and C. R. H. Raetz, *J. Bacteriol.*, 1991, **173**, 3591–3596.
19. T. J. O. Wyckoff, S. Lin, R. J. Cotter, G. D. Dotson and C. R. H. Raetz, *J. Biol. Chem.*, 1998, **273**, 32369–32372.
20. A. H. Williams and C. R. H. Raetz, *Proc. Natl. Acad. Sci. U. S. A.*, 2007, **104**, 13543–13550.
21. K. Young, L. L. Silver, D. Bramhill, P. Cameron, S. S. Eveland, C. R. H. Raetz, S. A. Hyland and M. S. Anderson, *J. Biol. Chem.*, 1995, **270**, 30384–30391.
22. J. E. Jackman, C. R. H. Raetz and C. A. Fierke, *Biochemistry*, 1999, **38**, 1902–1911.
23. A. W. Barb, T. M. Leavy, L. I. Robins, Z. Guan, D. A. Six, P. Zhou, C. R. Bertozzi and C. R. H. Raetz, *Biochemistry*, 2009, **48**, 3068–3077.
24. M. F. Brown, U. Reilly, J. A. Abramite, J. T. Arcari, R. Oliver, R. A. Barham, Y. Che, J. M. Chen, E. M. Collantes, S. W. Chung, C. Desbonnet, J. Doty, M. Doroski, J. J. Engtrakul, T. M. Harris, M. Huband, J. D. Knafels, K. L. Leach, S. Liu, A. Marfat, A. Marra, E. McElroy, M. Melnick, C. A. Menard,

J. I. Montgomery, L. Mullins, M. C. Noe, J. O'Donnell, J. Penzien, M. S. Plummer, L. M. Price, V. Shanmugasundaram, C. Thoma, D. P. Uccello, J. S. Warmus and D. G. Wishka, *J. Med. Chem.*, 2012, **55**, 914–923.

25. C. J. Lee, X. Liang, X. Chen, D. Zeng, S. H. Joo, H. S. Chung, A. W. Barb, S. M. Swanson, R. A. Nicholas, Y. Li, E. J. Toone, C. R. H. Raetz and P. Zhou, *Chem. Biol.*, 2011, **18**, 38–47.
26. X. Liang, C. J. Lee, X. Chen, H. S. Chung, D. Zeng, C. R. H. Raetz, Y. Li, P. Zhou and E. J. Toone, *Bioorg. Med. Chem.*, 2011, **19**, 852–860.
27. H. R. Onishi, B. A. Pelak, L. S. Gerckens, L. L. Silver, F. M. Kahan, M. H. Chen, A. A. Patchett, S. M. Galloway, S. A. Hyland, M. S. Anderson and C. R. Raetz, *Science*, 1996, **274**, 980–982.
28. K. Bodewits, C. R. H. Raetz, J. R. Govan and D. J. Campopiano, *Antimicrob. Agents Chemother.*, 2010, **54**, 3531–3533.
29. A. L. McClerren, S. Endsley, J. L. Bowman, N. H. Andersen, Z. Guan, J. Rudolph and C. R. H. Raetz, *Biochemistry*, 2005, **44**, 16574–16583.
30. J. M. Clements, F. Coignard, I. Johnson, S. Palan, A. Waller, J. Wijkmans, M. G. Hunter and S. Chandler, *Antimicrob. Agents Chemother.*, 2002, **46**, 1793–1799.
31. A. W. Barb, L. Jiang, C. R. H. Raetz and P. Zhou, *Proc. Natl. Acad. Sci. U. S. A.*, 2007, **104**, 18433–18438.
32. A. W. Barb, A. L. McClerren, K. Snehelatha, C. M. Reynolds, P. Zhou and C. R. H. Raetz, *Biochemistry*, 2007, **46**, 3793–3802.
33. B. E. Coggins, A. L. McClerren, L. Jiang, X. Li, J. Rudolph, O. Hindsgaul, C. R. H. Raetz and P. Zhou, *Biochemistry*, 2005, **44**, 1114–1126.
34. U. Faruk Mansoor, D. Vitharana, P. A. Reddy, D. L. Daubaras, P. McNicholas, P. Orth, T. Black and M. Arshad Siddiqui, *Bioorg. Med. Chem. Lett.*, 2011, **21**, 1155–1161.
35. J. E. Jackman, C. A. Fierke, L. N. Tumey, M. Pirrung, T. Uchiyama, S. H. Tahir, O. Hindsgaul and C. R. H. Raetz, *J. Biol. Chem.*, 2000, **275**, 11002–11009.
36. H. Shin, H. A. Gennadios, D. A. Whittington and D. W. Christianson, *Bioorg. Med. Chem.*, 2007, **15**, 2617–2623.
37. C. M. Bartling and C. R. H. Raetz, *Biochemistry*, 2008, 5290–5302.
38. T. M. Kelly, S. A. Stachula, C. R. H. Raetz and M. S. Anderson, *J. Biol. Chem.*, 1993, **268**, 19866–19874.
39. C. M. Bartling and C. R. H. Raetz, *Biochemistry*, 2009, **48**, 8672–8683.
40. R. J. Jenkins and G. D. Dotson, *ACS Chem. Biol.*, 2012, **7**, 1170–1177.
41. R. J. Jenkins, K. A. Heslip, J. L. Meagher, J. A. Stuckey and G. D. Dotson, *J. Biol. Chem.*, 2014, **289**, 15527–15535.
42. A. H. Williams, R. M. Immormino, D. T. Gewirth and C. R. H. Raetz, *Proc. Natl. Acad. Sci. U. S. A.*, 2006, **103**, 10877–10882.
43. R. E. Benson, E. B. Gottlin, D. J. Christensen and P. T. Hamilton, *Antimicrob. Agents Chemother.*, 2003, **47**, 2875–2881.
44. A. Beceiro, A. Moreno, N. Fernández, J. A. Vallejo, J. Aranda, B. Adler, M. Harper, J. D. Boyce and G. Bou, *Antimicrob. Agents Chemother.*, 2014, **58**, 518–526.

45. X. Wang and P. J. Quinn, *Prog. Lipid Res.*, 2010, **49**, 97–107.
46. K. J. Babinski, A. A. Ribeiro and C. R. H. Raetz, *J. Biol. Chem.*, 2002, **277**, 25937–25946.
47. K. J. Babinski, S. J. Kanjilal and C. R. H. Raetz, *J. Biol. Chem.*, 2002, **277**, 25947–25956.
48. D. N. Crowell, M. S. Anderson and C. R. H. Raetz, *J. Bacteriol.*, 1986, **168**, 152–159.
49. L. E. Metzger and C. R. H. Raetz, *Biochemistry*, 2010, **49**, 6715–6726.
50. L. E. Metzger, J. K. Lee, J. S. Finer-Moore, C. R. H. Raetz and R. M. Stroud, *Nat. Struct. Mol. Biol.*, 2012, **19**, 1132–1138.
51. T. A. Garrett, J. L. Kadrmas and C. R. H. Raetz, *J. Biol. Chem.*, 1997, **272**, 21855–21864.
52. R. P. Emptage, K. D. Daughtry, C. W. Pemble and C. R. H. Raetz, *Proc. Natl. Acad. Sci.*, 2012, **109**, 12956–12961.
53. T. Clementz and C. R. H. Raetz, *J. Biol. Chem.*, 1991, **266**, 9687–9696.
54. K. A. Brozek and C. R. H. Raetz, *J. Biol. Chem.*, 1990, **265**, 15410–15417.
55. T. Clementz, J. J. Bednarski and C. R. H. Raetz, *J. Biol. Chem.*, 1996, **271**, 12095–12102.
56. T. Clementz, Z. Zhou and C. R. H. Raetz, *J. Biol. Chem.*, 1997, **272**, 10353–10360.
57. R. C. Goldman, C. C. Doran, S. K. Kadam and J. O. Capobianco, *J. Biol. Chem.*, 1988, **263**, 5217–5223.
58. O. Holst, in *Bacterial Lipopolysaccharides: Structure, Chemical Synthesis, Biogenesis and Interaction with Host Cells*, ed. Y. Knirel and M. Valvano, Springer, Vienna, 2011, pp. 21–40.
59. M. Karow and C. Georgopoulos, *Mol. Microbiol.*, 1993, **7**, 69–79.
60. W. T. Doerrler, H. S. Gibbons, R. Christian and H. Raetz, *J. Biol. Chem.*, 2004, **279**, 45102–45109.
61. W. T. Doerrler and C. R. H. Raetz, *J. Biol. Chem.*, 2002, **277**, 36697–36705.
62. C. M. Reynolds and C. R. H. Raetz, *Biochemistry*, 2009, **48**, 9627–9640.
63. W. T. Doerrler, M. C. Reedy and C. R. H. Raetz, *J. Biol. Chem.*, 2001, **276**, 11461–11464.
64. L. K. Greenfield and C. Whitfield, *Carbohydr. Res.*, 2012, **356**, 12–24.
65. G. Samuel and P. Reeves, *Carbohydr. Res.*, 2003, **338**, 2503–2519.
66. W. Han, B. Wu, L. Li, G. Zhao, R. Woodward, N. Pettit, L. Cai, V. Thon and P. G. Wang, *J. Biol. Chem.*, 2012, **287**, 5357–5365.
67. B. D. Needham and M. S. Trent, *Nat. Rev. Microbiol.*, 2013, **11**, 467–481.
68. C. E. Bryant, D. R. Spring, M. Gangloff and N. J. Gay, *Nat. Rev. Microbiol.*, 2010, **8**, 8–14.
69. T. W. Cullen, D. K. Giles, L. N. Wolf, C. Ecobichon, I. G. Boneca and M. S. Trent, *PLoS Pathog.*, 2011, **7**, e1002454.
70. S. Okuda, D. J. Sherman, T. J. Silhavy, N. Ruiz and D. Kahne, *Nat. Rev. Microbiol.*, 2016, **14**, 337–345.
71. A. Polissi and P. Sperandeo, *Mar. Drugs*, 2014, **12**, 1023–1042.
72. N. Ruiz, D. Kahne and T. J. Silhavy, *Nat. Rev. Microbiol.*, 2009, **7**, 677–683.

73. M. D. L. Suits, P. Sperandeo, G. Dehò, A. Polissi and Z. Jia, *J. Mol. Biol.*, 2008, **380**, 476–488.
74. A. X. Tran, C. Dong and C. Whitfield, *J. Biol. Chem.*, 2010, **285**, 33529–33539.
75. H. Dong, Q. Xiang, Y. Gu, Z. Wang, N. G. Paterson, P. J. Stansfeld, C. He, Y. Zhang, W. Wang and C. Dong, *Nature*, 2014, **511**, 52–56.
76. S. Qiao, Q. Luo, Y. Zhao, X. C. Zhang and Y. Huang, *Nature*, 2014, **511**, 108–111.
77. D. J. Sherman, M. B. Lazarus, L. Murphy, C. Liu, S. Walker, N. Ruiz and D. Kahne, *Proc. Natl. Acad. Sci. U. S. A.*, 2014, **111**, 4982–4987.
78. Z. Wang, Q. Xiang, X. Zhu, H. Dong, C. He, H. Wang, Y. Zhang, W. Wang and C. Dong, *Biochem. Biophys. Res. Commun.*, 2014, **452**, 443–449.
79. M. Bollati, R. Villa, L. J. Gourlay, M. Benedet, G. Dehò, A. Polissi, A. Barbiroli, A. M. Martorana, P. Sperandeo, M. Bolognesi and M. Nardini, *FEBS J.*, 2015, **282**, 1980–1997.
80. P. Sperandeo, F. K. Lau, A. Carpentieri, C. De Castro, A. Molinaro, G. Dehò, T. J. Silhavy and A. Polissi, *J. Bacteriol.*, 2008, **190**, 4460–4469.
81. N. Ruiz, L. S. Gronenberg, D. Kahne and T. J. Silhavy, *Proc. Natl. Acad. Sci. U. S. A.*, 2008, **105**, 5537–5542.
82. S. Narita and H. Tokuda, *FEBS Lett.*, 2009, **583**, 2160–2164.
83. S. Okuda, E. Freinkman and D. Kahne, *Science*, 2012, **338**, 1214–1217.
84. R. Villa, A. M. Martorana, S. Okuda, L. J. Gourlay, M. Nardini, P. Sperandeo, G. Dehò, M. Bolognesi, D. Kahne and A. Polissia, *J. Bacteriol.*, 2013, **195**, 1100–1108.
85. A. X. Tran, M. S. Trent and C. Whitfield, *J. Biol. Chem.*, 2008, **283**, 20342–20349.
86. P. Sperandeo, R. Cescutti, R. Villa, C. Di Benedetto, D. Candia, G. Dehò and A. Polissi, *J. Bacteriol.*, 2007, **189**, 244–253.
87. A. Bowyer, J. Baardsnes, E. Ajamian, L. Zhang and M. Cygler, *Biochem. Biophys. Res. Commun.*, 2011, **404**, 1093–1098.
88. E. Freinkman, S. Okuda, N. Ruiz and D. Kahne, *Biochemistry*, 2012, **51**, 4800–4806.
89. S. Okuda and H. Tokuda, *Annu. Rev. Microbiol.*, 2011, **65**, 239–259.
90. J. A. Merten, K. M. Schultz and C. S. Klug, *Protein Sci.*, 2012, **21**, 211–218.
91. B. Tefsen, J. Geurtsen, F. Beckers, J. Tommassen and H. De Cock, *J. Biol. Chem.*, 2005, **280**, 4504–4509.
92. S.-S. Chng, L. S. Gronenberg and D. Kahne, *Biochemistry*, 2010, **49**, 4565–4567.
93. S.-S. Chng, N. Ruiz, G. Chimalakonda, T. J. Silhavy and D. Kahne, *Proc. Natl. Acad. Sci. U. S. A.*, 2010, **107**, 5363–5368.
94. E. Freinkman, S.-S. Chng and D. Kahne, *Proc. Natl. Acad. Sci. U. S. A.*, 2011, **108**, 2486–2491.
95. S.-S. Chng, M. Xue, R. A. Garner, H. Kadokura, D. Boyd, J. Beckwich and D. Kahne, *Science*, 2012, **337**, 1665–1668.
96. N. Ruiz, S.-S. Chng, A. Hiniker, D. Kahne and T. J. Silhavy, *Proc. Natl. Acad. Sci. U. S. A.*, 2010, **107**, 12245–12250.

97. G. Chimalakonda, N. Ruiz, S.-S. Chng, R. A. Garner, D. Kahne and T. J. Silhavy, *Proc. Natl. Acad. Sci. U. S. A.*, 2011, **108**, 2492–2497.
98. Y. Gu, P. J. Stansfeld, Y. Zeng, H. Dong, W. Wang and C. Dong, *Structure*, 2015, **23**, 496–504.
99. X. Li, Y. Gu, H. Dong, W. Wang and C. Dong, *Sci. Rep.*, 2015, **5**, 11883.
100. J. Bass, D. Person and D. Chan, *Pediatrics*, 2000, **105**, 423–424.
101. W. Vollmer, *Appl. Microbiol. Biotechnol.*, 2006, **73**, 37–47.
102. R. L. Lock and E. J. Harry, *Nat. Rev. Drug Discovery*, 2008, **7**, 324–338.
103. B. Beutler and E. T. Rietschel, *Nat. Rev. Immunol.*, 2003, **3**, 169–176.
104. S. Langklotz, M. Schäkermann and F. Narberhaus, *J. Bacteriol.*, 2011, **193**, 1090–1097.
105. A. S. Nayar, T. J. Dougherty, K. E. Ferguson, B. A. Granger, L. McWilliams, C. Stacey, L. J. Leach, S. ichiro Narita, H. Tokuda, A. A. Miller, D. G. Brown and S. M. McLeod, *J. Bacteriol.*, 2015, **197**, 1726–1734.
106. R. E. Caughlan, A. K. Jones, A. M. DeLucia, A. L. Woods, L. Xie, B. Ma, S. W. Barnes, J. R. Walker, E. R. Sprague, X. Yang and C. R. Dean, *Antimicrob. Agents Chemother.*, 2012, **56**, 17–27.
107. N. Srinivas, P. Jetter, B. J. Ueberbacher, M. Werneburg, K. Zerbe, J. Steinmann, B. Van der Meijden, F. Bernardini, A. Lederer, R. L. A. Dias, P. E. Misson, H. Henze, J. Zumbrunn, F. O. Gombert, D. Obrecht, P. Hunziker, S. Schauer, U. Ziegler, A. Kach, L. Eberl, K. Riedel, S. J. DeMarco and J. A. Robinson, *Science*, 2010, **327**, 1010–1013.
108. D. T. Moir, T. J. Opperman, M. M. Butler and T. L. Bowlin, *Curr. Opin. Pharmacol.*, 2012, **12**, 535–544.
109. J. Schmidt, K. Patora-Komisarska, K. Moehle, D. Obrecht and J. A. Robinson, *Bioorg. Med. Chem.*, 2013, **21**, 5806–5810.
110. M. Werneburg, K. Zerbe, M. Juhas, L. Bigler, U. Stalder, A. Kaech, U. Ziegler, D. Obrecht, L. Eberl and J. A. Robinson, *ChemBioChem*, 2012, **13**, 1767–1775.
111. R. Haarmann, M. Ibrahim, M. Stevanovic, R. Bredemeier and E. Schleiff, *J. Phys.: Condens. Matter*, 2010, **22**, 454124.
112. K. Zeth and M. Thein, *Biochem. J.*, 2010, **431**, 13–22.
113. J. B. Lyczak, C. L. Cannon and G. B. Pier, *Clin. Microbiol. Rev.*, 2002, **15**, 194–222.
114. D. Wilbraham, S. DeMarco, A. Ograbek, J. Agren, A. Wach, D. Obrecht and K. Dembowsky, *21st European Congress of Clinical Microbiology and Infections Diseases - 27th International Congress of Chemotherapy*, 2011.
115. L. S. Gronenberg and D. Kahne, *J. Am. Chem. Soc.*, 2010, **132**, 2518–2519.
116. J. B. Smaill, H. H. Lee, B. D. Palmer, A. M. Thompson, C. J. Squire, E. N. Baker, R. J. Booth, A. Kraker, K. Hook and W. A. Denny, *Bioorg. Med. Chem. Lett.*, 2008, **18**, 929–933.
117. D. J. Sherman, S. Okuda, W. A. Denny and D. Kahne, *Bioorg. Med. Chem.*, 2013, **21**, 4846–4851.
118. S. E. Sestito, P. Sperandeo, C. Santambrogio, C. Ciaramelli, V. Calabrese, G. E. Rovati, L. Zambelloni, R. Grandori, A. Polissi and F. Peri, *ChemBioChem*, 2014, **15**, 734–742.

119. F. Peri, F. Granucci, B. Costa, I. Zanoni, C. Marinzi and F. Nicotra, *Angew. Chem., Int. Ed.*, 2007, **46**, 3308–3312.
120. M. Piazza, C. Rossini, S. D. Fiorentina, C. Pozzi, F. Comelli, I. Bettoni, P. Fusi, B. Costa and F. Peri, *J. Med. Chem.*, 2009, **52**, 1209–1213.
121. I. Bettoni, F. Comelli, C. Rossini, F. Granucci, G. Giagnoni, F. Peri and B. Costa, *Glia*, 2008, **56**, 1312–1319.
122. M. Piazza, V. Calabrese, C. Baruffa, T. Gioannini, J. Weiss and F. Peri, *Biochem. Pharmacol.*, 2010, **80**, 2050–2056.
123. J. Bojkovic, D. L. Richie, D. A. Six, C. M. Rath, W. S. Sawyer, Q. Hu and C. R. Dean, *J. Bacteriol.*, 2016, **198**, 731–741.
124. B. A. Sampson, R. Misra and S. A. Benson, *Genetics*, 1989, **122**, 491–501.
125. C. J. Balibar and M. Grabowicz, *Antimicrob. Agents Chemother.*, 2015, **60**, 845–854.
126. M. Grabowicz, J. Yeh and T. J. Silhavy, *J. Bacteriol.*, 2013, **195**, 1327–1334.
127. N. Ruiz, B. Falcone, D. Kahne and T. J. Silhavy, *Cell*, 2005, **121**, 307–317.
128. M. Sekiguchi and S. Iida, *Proc. Natl. Acad. Sci.*, 1967, **58**, 2315–2320.

CHAPTER 6

The Discovery of Teixobactin

L. LAZARIDES*[a], J. Y. C. CHIVA[a], M. JONES[a] AND V. A. STEADMAN[a]

[a]Selcia, Ongar, Essex, CM5 0GS UK
*E-mail: linos.x.lazarides@selcia.com

6.1 Introduction

In 1943 Selman Waksman and Albert Schatz co-discovered the aminoglycoside antibiotic Streptomycin.[1] This was the first effective antibiotic for the treatment of tuberculosis and is still in use to this day. Following on from that landmark discovery Waksman received the Nobel Prize in 1952 for "your ingenious, systematic and successful studies of the soil microbes that have led to the discovery of streptomycin." Unlike the earlier chance discovery of penicillin by Alexander Fleming, who observed that a mold contaminant on a Petri dish culture had inhibited the growth of a bacterial pathogen, Waksman and his team had developed and established a successful discovery platform for identifying new antibiotic agents. This platform allowed the systematic screening of soil-derived actinomycetes for antimicrobial activity against susceptible test microorganisms on agar plates. Employing this platform during the 1940s, Waksman and his students isolated more than fifteen antibiotics including streptomycin and neomycin.

This successful discovery platform was widely adopted by the pharmaceutical industry and yielded many major classes of antibiotics over the following decades such as the macrolides, cephalosporins, tetracyclines and

Drug Discovery Series No. 58
Antibiotic Drug Discovery: New Targets and Molecular Entities
Edited by Steven M. Firestine and Troy Lister

Published by the Royal Society of Chemistry, www.rsc.org

rifamycins. Unfortunately, the screening of soil microorganisms stopped producing fruit in the mid-sixties, the reason being that this source of cultivable bacteria had been heavily mined. In fact, this resource of cultivable bacteria only represented 1% of microbes in the external environment, the remaining 99% had until recent years been uncultivable and inaccessible for drug discovery endeavours. Such uncultured bacteria provide a previously untapped source of natural products which in combination with the discovery platform developed by Waksman could provide several novel antibiotics.

6.2 Cultivating the Unculturable – The iChip

Exploiting the rich source of potential therapeutic natural products produced by uncultivable organisms has proved until recent years an insurmountable task. A breakthrough technology was achieved by researchers at Northeastern University and Novobiotic Pharmaceuticals through a process of *in situ* cultivation. This technology uses a multichannel device, named the iChip,[2] to isolate and grow uncultured bacteria in their natural environments (Figure 6.1).[3]

The process involves diluting a soil sample to such an extent that approximately one bacterial cell is delivered to each of the chambers of the iChip. The device is then covered with two semi-permeable membranes, to allow the diffusion of nutrients and growth factors, and placed back in the soil. This enables the uncultured bacteria to grow in their natural environment. Once a colony is produced it can then be transferred to the laboratory and grown *in vitro*. Employing this technique afforded extracts from 10 000 isolates which were subsequently screened for antibacterial activity on plates overlaid with *S. aureus*. This resulted in the identification of a new species of Gram-negative organism provisionally named *Eleftheria terrae* (Figure 6.2), which produced a compound that displayed good antibacterial activity. Following purification and analysis of the active extract, a novel antibiotic was discovered, it was named teixobactin.[4,5]

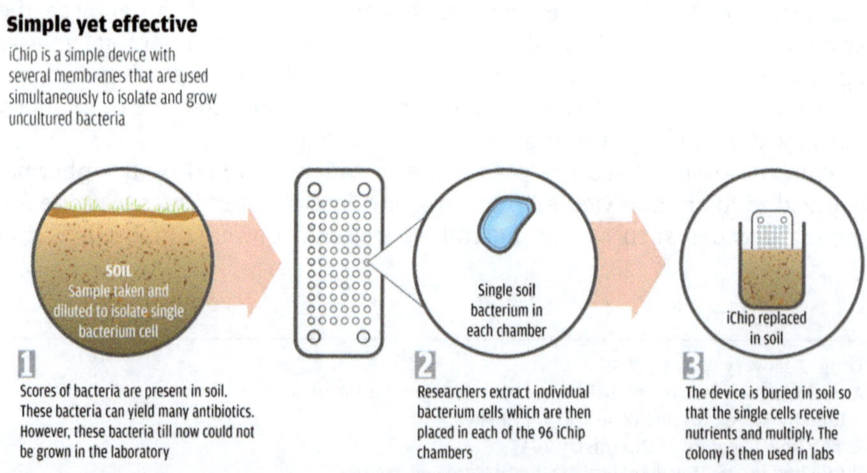

Figure 6.1 Cultivation of Bacteria Using the iChip, reprinted with permission from Down to Earth, 30th January 2015.

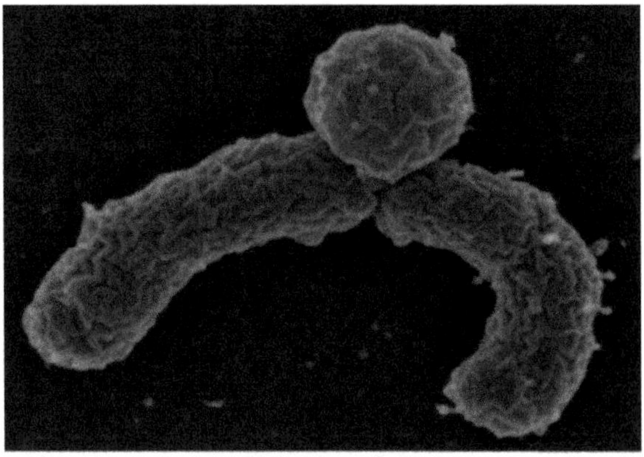

Eleftheria terrae

Figure 6.2 Teixobactin Producing Microorganism (Credit: William Fowle, Northeastern University).

6.3 Teixobactin – A Novel Antibiotic

Teixobactin displays excellent activity against Gram-positive pathogens this includes vancomycin resistant enterococci (VRE) and methicillin resistant *Staphylococcus aureus* (MRSA), both of which have been designated as serious threats to public health by the Centers for Disease Control and Prevention (CDC). Furthermore teixobactin shows exceptional potency against *Clostridium difficile* (designated as an urgent threat to public health by CDC) and *Bacillus anthracis*. This activity is translated *in vivo* where teixobactin displays efficacy in various mouse models of infection, specifically a septicemia model (MRSA), thigh model (MRSA) and a lung model (*S. pneumoniae*) of infection.

Another remarkable facet of the antibacterial properties of teixobactin is the non-development of resistance during various experiments. For example, no mutants of *S.aureus* or *M.tuberculosis* resistant to teixobactin were obtained during plating on media at low doses and no resistant mutants were observed during serial passaging experiments. This is a major property when considering that one of the biggest global issues facing humanity is the development of antimicrobial resistance.

The mechanism of action of teixobactin has been shown to involve binding to two important bacterial cell wall lipids; lipid II which is a key precursor of peptidoglycan, and lipid III which is a precursor of wall teichoic acid (WTA). The later acids play an important role in preventing uncontrolled hydrolysis of peptidoglycan, so potential inhibition of WTA synthesis by teixobactin may help to liberate autolysins which kill the bacteria. Lipid II is also a known target of the glycopeptide antibiotic vancomycin, but teixobactin binds to a different region, as it is active against vancomycin resistant enterococci (VRE) which have modified lipid II.

Currently teixobactin is in preclinical development and is displaying a good ADMET profile. Possible market opportunities will include treatment of acute bacterial skin and skin structure infections, particularly where MRSA is the causative pathogen, enterococcal endocarditis (VRE), and infections which require long-term treatment such as bone and joint and various indwelling infections.

6.4 Structural Determination of Teixobactin

The structure of teixobactin was determined through extensive analysis by liquid chromatography mass spectrometry (LC-MS) and nuclear magnetic resonance (NMR) techniques. In addition, the biosynthetic gene cluster was elucidated and was consistent with the spectroscopically determined amino acid order of teixobactin. These analyzes revealed an unusual depsipeptide containing 11 amino acids, which included a non-proteinogenic amino acid known as enduracididine (Figure 6.3). Moreover, four of the amino acids were contained within a macrolactone ring.

Central to any further development work and medicinal chemistry endeavors was the stereochemical assignment of all the constituent amino acids. One could not assume that all were of the natural *L* configuration. A peer into the literature reveals that cyclic peptides of a similar nature derived from various microorganisms have differing L- and D- amino acid residues (*e.g.* mannopeptimycins,[6] A-3302-B[7] and Hypeptin,[8] Figure 6.4). Additionally, it was known that two forms of enduracididine (L- and D-allo-) have been found in the cyclic peptide enduracidin (Figure 6.4).[9]

To elucidate the absolute stereochemistry of teixobactin we employed an advanced Marfey's analysis.[10,11] This process attaches a UV-active derivatising agent of known configuration (*e.g.* L-FDLA: 1-fluoro-2,4-dinitrophenol-5-L-leucinamide) to the unknown hydrolysis products of a peptide. This allows the stereoisomers of the constituent amino acids to be chromatographically resolved using simple reverse phase chromatographic methodology, with the added advantages that they have both a strong chromophore and are readily ionizable by electrospray ionization.

For all but one of the amino acids (Enduracididine) reference markers were available to allow determination of the configuration based on chromatographic retention time. All other possible amino acids contained within teixobactin were commercially available (alanine, isoleucine, threonine,

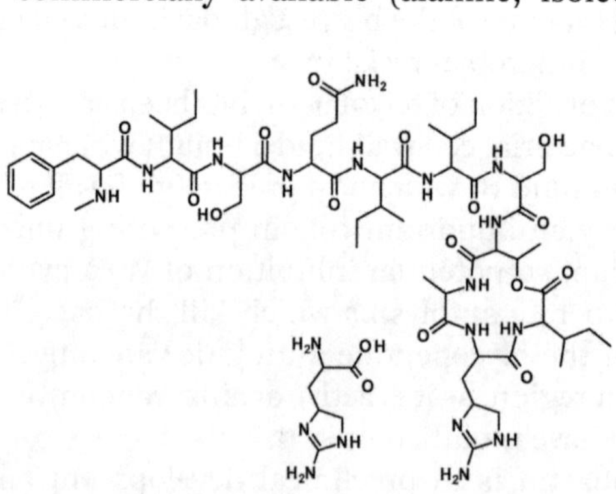

Figure 6.3 Teixobactin and enduracididine.

serine, *N*-Me-phenylalanine and glutamic acid) and their corresponding L-FDLA derivatives synthesized (Figure 6.5). For glutamine, glutamic acid was used as the reference as glutamine is converted to its acid counterpart under the hydrolytic conditions used in preparation of samples for analysis. The L-FDLA standard derivatives were, initially, analyzed using full scan Liquid Chromatography-Mass Spectrometry (LC-MS) techniques to determine the elution order of the amino acids derivatives and their enantiomers.

The only amino acid not available as a commercial standard was enduracididine. Hence, to determine the absolute configuration, a synthesis of all four potential diastereomers of enduracididine was necessary. A scan of the literature revealed a route to the diastereomers but, unfortunately, this was unfruitful in our hands. We therefore, devised an alternative synthesis,

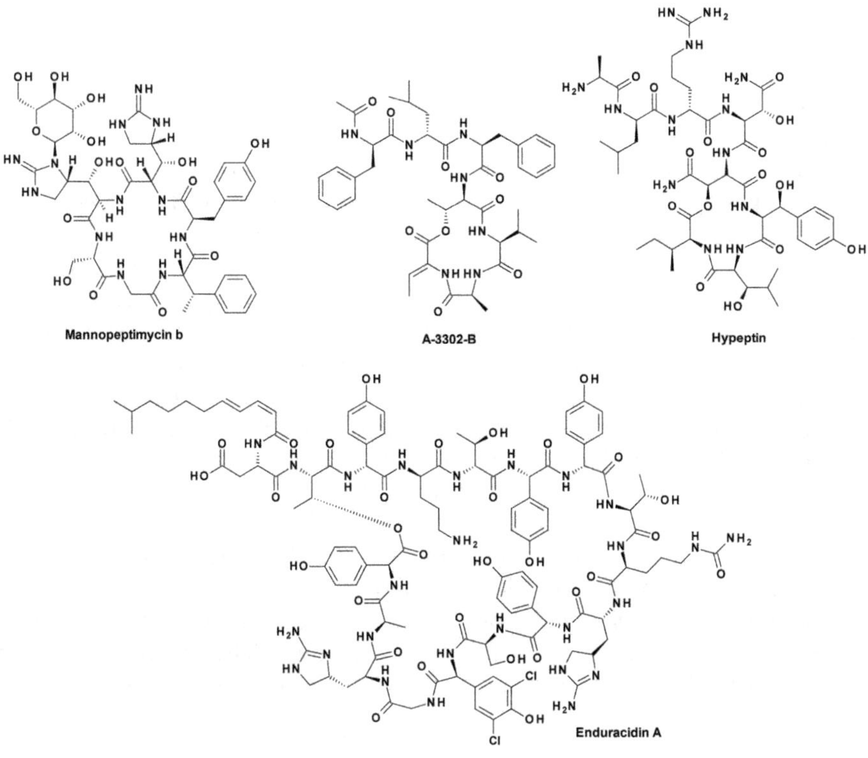

Figure 6.4 Cyclic depsipeptides derived from nature.

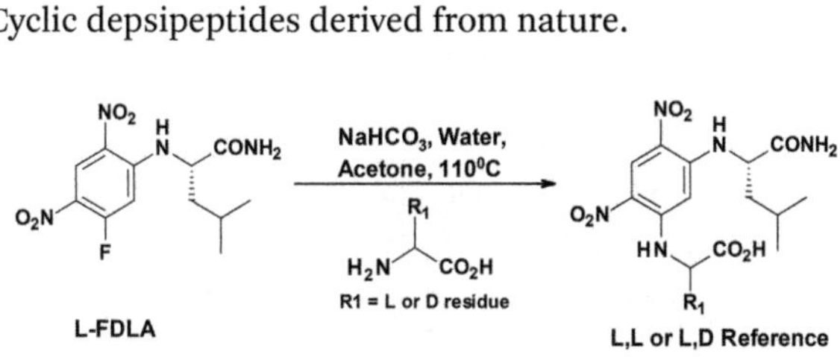

Figure 6.5 Standards for Marfey's analysis.

adapting various literature procedures (Figure 6.6) to provide the four diastereomers required for the synthesis of the Marfey's standards.

The synthesis commenced with the commercially available α-*tert*-Butyl-*N*-Boc-L-aspartate, which was converted to the nitroketone by reaction of the intermediary imidazoyl ester with nitromethane. The ensuing ketone then underwent a 1, 3 stereoselective reduction, at low temperature, utilizing L-selectride. This reduction provided the alcohol in an 85 : 15 mixture in favor of the (2S, 4R) configuration over the (2S, 4S).[12] This mixture could easily be separated by normal phase silica gel chromatography but, for the purpose of the Marfeys analysis, this was not necessary as, provided the ratio was known, the mixture could be processed to the end. The nitro functionality was reduced *via* a transfer hydrogenation using ammonium formate under palladium catalysis. The resulting amine was then converted immediately following simple workup to the Boc-protected guanidine using *N*,*N*′-Di-Boc-1*H*-pyrazole-1-carboxamidine. The linear guanidine was converted to the cyclic guanidine by forming the triflic ester with *in situ* cyclization. The final step to reveal the amino acid enduracididine involved simple deprotection of all acid-labile protecting groups with trifluoroacetic acid. Enduracididne was isolated as a 6 : 1 mix in favor of L-enduracididine over L-allo-enduracididine. Likewise, starting from α-*tert*-Butyl-*N*-Boc-D-aspartate we were able to access the remaining two diastereomers, D-enduracididine and D-allo-enduracididine. At this stage no effort was expended to separate the mixture as both were required to make Marfey's standards to compare with the authentic material. For this exercise it was sufficient to know which diastereomer corresponded to the major and which to the minor LC-MS peaks.

With the reference L-FDLA amino acids in hand we proceeded to subject a sample of teixobactin to acid hydrolysis conditions (6 N HCl, 110 °C over 22 h). This resulted in complete hydrolysis to its individual amino acid components which were then subject to derivatization using L-FDLA. The resultant mixture of hydrolyzed teixobactin L-FDLA derivatives was analyzed by LC-MS, alongside the reference standards, to determine the stereochemistry of the amino acids contained within the natural product.

A sample LC-MS chromatogram, used to elucidate the stereochemistry of the threonine residue, is shown in Figure 6.7. The molecular ions of the respective derivatives were targeted using selective ion recording (SIR)

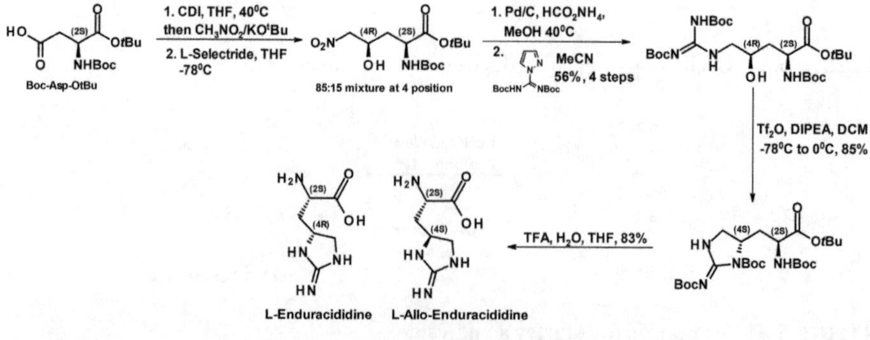

Figure 6.6 Synthesis of enduracididine diastereomers.

analyzes. This provided both improved sensitivity and signal to noise as it only targets the ions of interest. Comparison of the data generated for the amino acid L-FDLA standards and the L-FDLA derivatives of the native, hydrolyzed, material enabled us to unequivocally determine the stereochemistry of all eleven amino acids of teixobactin; D-*N*-methyl-phenylalanine, two L-serines, D-glutamine, D-threonine, L-alanine and L-allo-enduracididine, three L-isoleucines and D-allo-isoleucine.

Following Marfey's analysis it was evident that teixobactin contained a 3 : 1 mixture of L-isoleucine and D-allo-isoleucine. This raised the problem; which amino acid position of teixobactin corresponded to D-allo-isoleucine? To address this, teixobactin had to be partially hydrolyzed to yield fragments

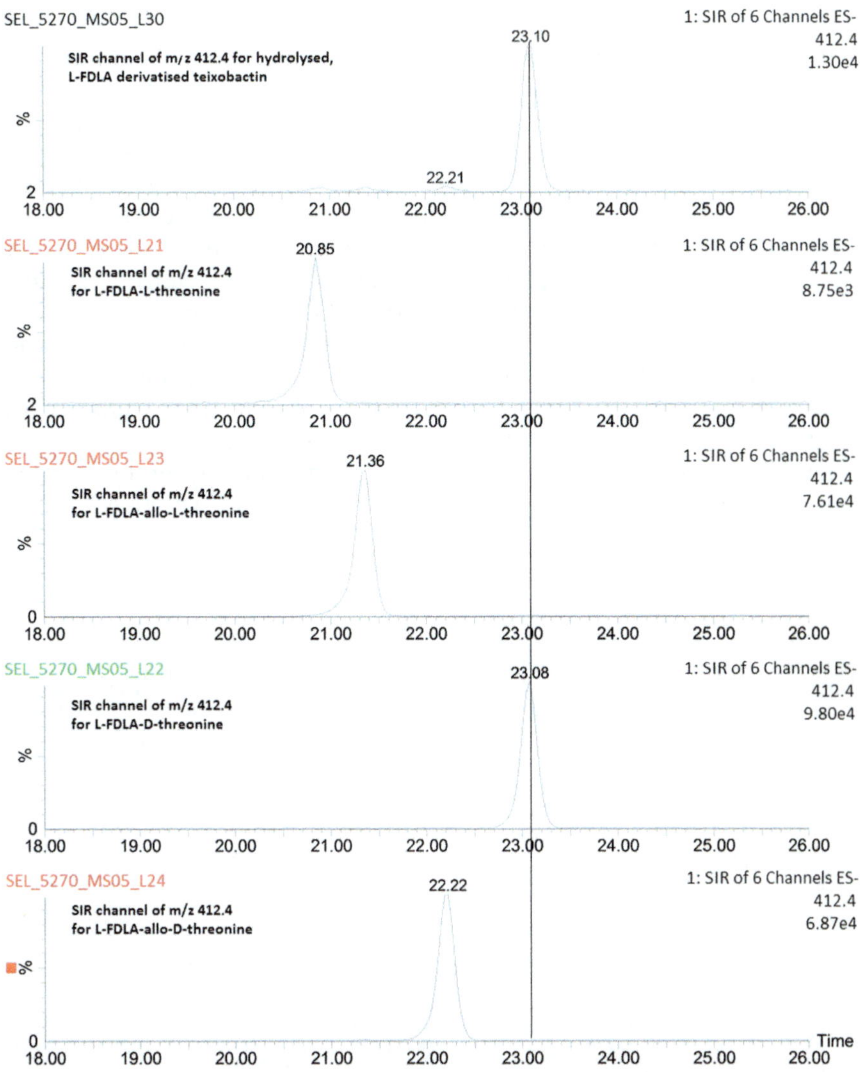

Figure 6.7 Sample SIR chromatograms (m/z 412.4) of threonine L-FDLA derivatives used to elucidate threonine stereochemistry within teixobactin.

rather than single amino acids; identify those which contained isoleucine *via* mass spectrometry and then analyze such fragments *via* Marfeys analysis to identify the isoleucine diastereomer.

Following optimization, it was found that subjecting teixobactin to hydrolytic conditions (6 N HCl, 110 °C) for one hour generated several workable fragments. Figure 6.8 shows the High Performance Liquid Chromatography (HPLC) UV trace of the partial hydrolysis of teixobactin and the fragments that were isolated.

Following isolation, the fragments were identified by their molecular weight and then subjected to full hydrolysis and stereochemical identification by Marfey's analysis. We first focused on the teixobactin N-terminal fragment (MWt = 292) containing the *N*-methylphenylalanine and one of the isoleucines. Marfey's analysis determined that the isoleucine in this fragment was of the natural L configuration. Analysis of the hexapeptide fragment of the linear arm of teixobactin (MWt = 734) revealed a mixture in the ratio of 2 : 1 L-isoleucine to D-allo-isoleucine. Hence, by the process of elimination, the ring isoleucine was deemed to be of the L configuration. It is worth noting that the gentle hydrolysis failed to identify any fragments from the cyclic tetrapeptide portion of teixobactin. From the above observations, it was concluded that the single D-allo-isoleucine was one of the two adjacent isoleucines within the linear arm of teixobactin, but which one? After assessing several chemical methods to try and differentiate the C-terminal isoleucine from its adjacent partner it was decided to synthesize both possible '734' fragments and then compare them by their retention times on HPLC to the corresponding fragment derived from teixobactin. The results (Figure 6.9) indicated the correct placement of the allo-D-isoleucine.

Hence a combination of Marfey's analysis, and some further investigative chemistry, afforded the absolute stereochemistry of teixobactin (Figure 6.10).

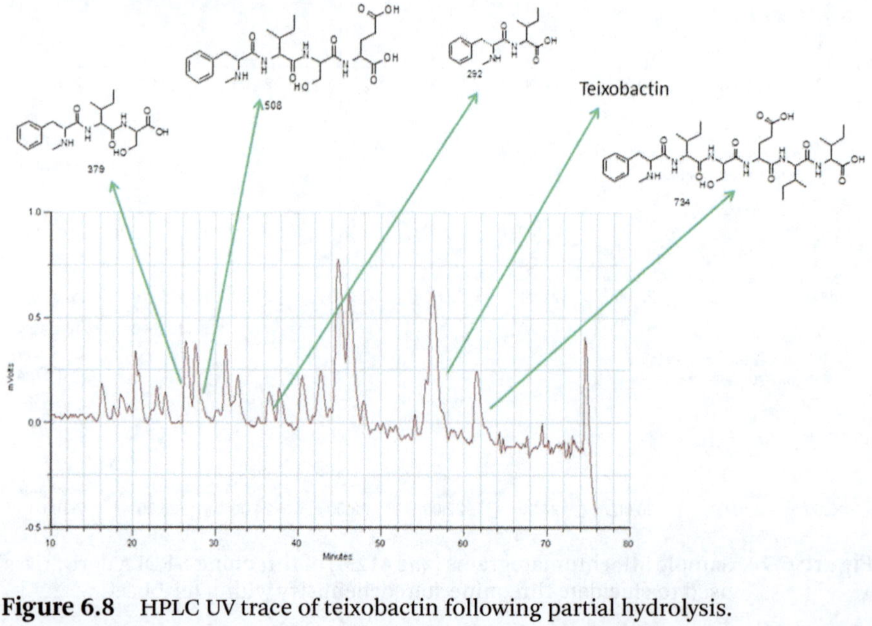

Figure 6.8 HPLC UV trace of teixobactin following partial hydrolysis.

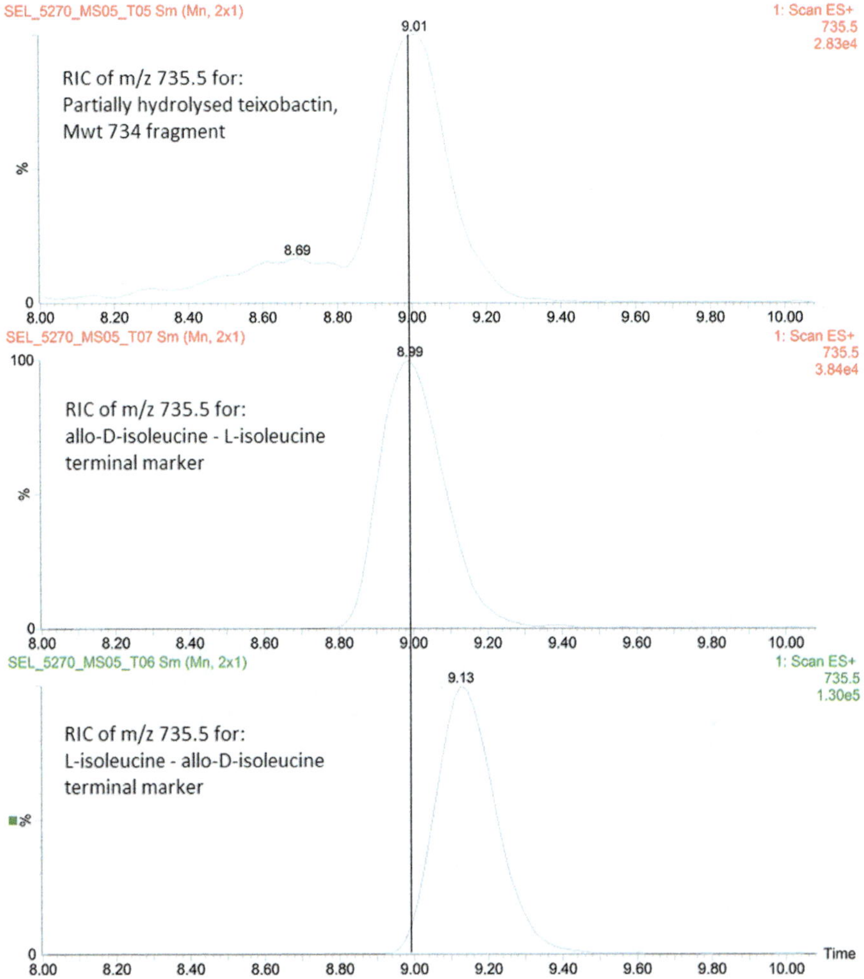

Figure 6.9 Determination of teixobactin fragment stereochemistry by comparison of LCMS retention times.

6.5 Synthesis of Teixobactin and Analogues

Following the elucidation of the teixobactin structure, a successful total synthesis of a teixobactin analogue has been made by at least two groups.[13,14] This analogue contains an L-arginine in place of the L-*allo*-enduracididine. Both syntheses reported were conducted by solid phase peptide synthesis employing the acid labile 2-Chlorotritylchloride resin, with the L-alanine of the teixobactin cyclic tetradepsipeptide providing the point of attachment (Figure 6.11). Following construction of peptide **1** using primarily an Fmoc-coupling strategy, the resin was removed in dilute acid, preserving the protecting groups of the amino acid side chains. A different cyclization protocol was used by the two groups for successfully closing the tetradepsipeptide between the L-alanine and L-arginine residues; PyAOP/Oxyma Pure/DIEA/DCM and HATU/DIPEA/DMF. (Different conditions to cyclize *via* a

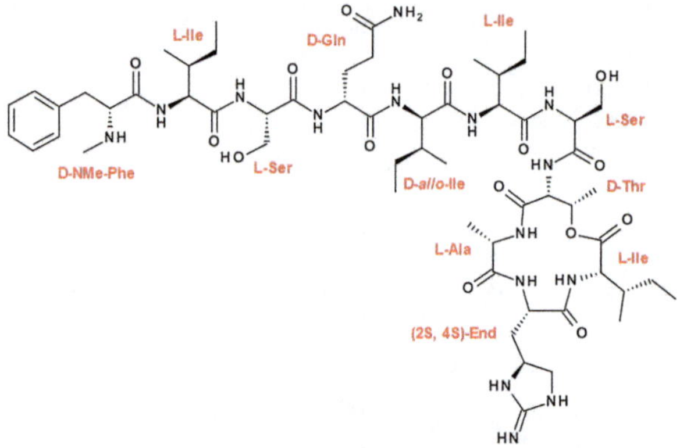

Figure 6.10 Absolute stereochemistry of teixobactin.

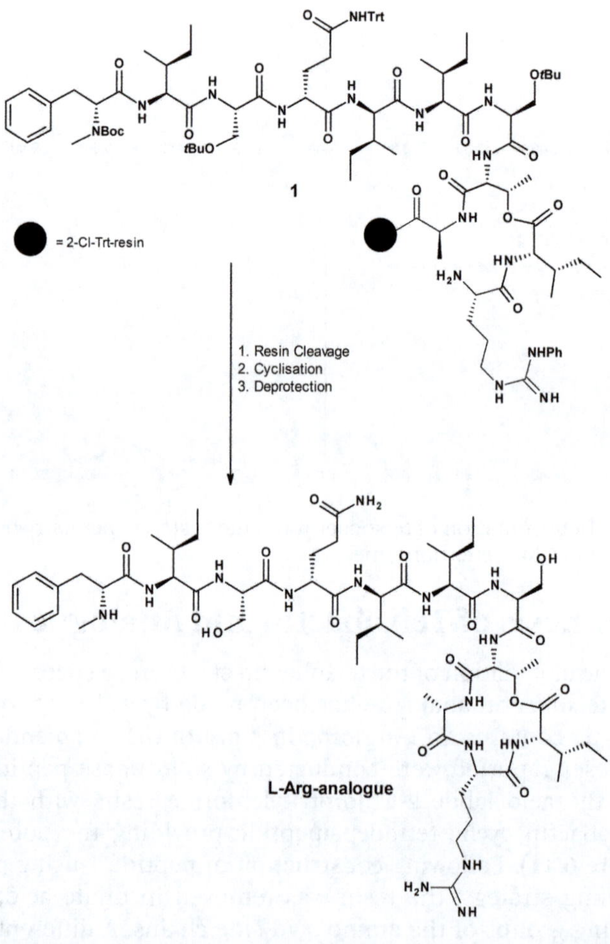

Figure 6.11 Solid phase synthesis of an L-arginine teixobactin analogue.

lactonization have been unsuccessful despite this being the way *Eleftheria Terrae* synthesizes teixobactin). Final global deprotection yielded the L-arginine teixobactin analogue.

The analogue maintains potency against Gram-positive pathogens such as *Staphylococcus aureus* although this was slightly inferior to teixobactin. Interestingly, an analogue which replaces the D-amino acids of the linear portion of teixobactin (D-*N*-Me-Phe, D-Gln and D-allo-Ile) with the corresponding L-residues eliminates the antibacterial activity of teixobactin, underlining the importance of these residues and their configuration.

At the time of completing this chapter the total synthesis of teixobactin was reported by two separate research groups, namely that of Richard Payne at The University of Sydney[15] and of Xuechen Li at the University of Hong Kong.[16] Solid phase synthesis was employed for both respective total syntheses but using different approaches. The Payne synthesis, unlike the syntheses of the analogues, entailed performing the key macrolactamization step between the D-threonine and L-alanine residues (Figure 6.12). This cyclization

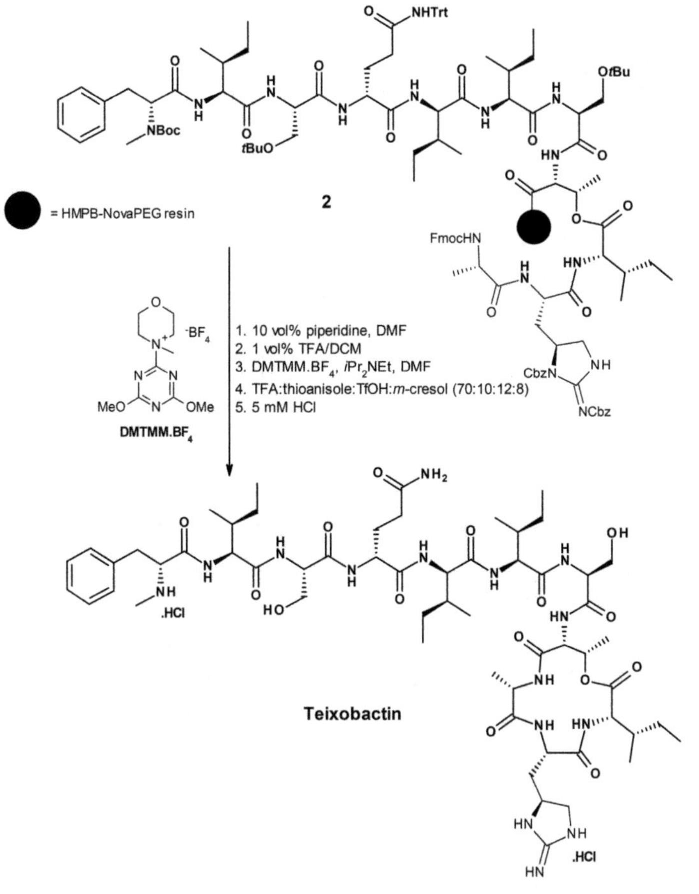

Figure 6.12 Payne's solid phase total synthesis of teixobactin.

was conducted successfully using 4-(4,6-dimethoxy-1,3,5-triazin-2-yl)-4-methylmorpholinium tetrafluoroborate (DMTMM.BF_4) as the activating agent, following cleavage of linear peptide **2** from the resin. The resin chosen was (4-(hydroxymethyl)-3-methoxyphenoxy) acetic acid (HMPB) functionalized polyethylene glycol-based NovaPEG as difficulties were encountered when attempting to esterify the resin linked D-threonine with the ring L-isoleucine when utilizing the 2-Cl-Trt resin. Teixobactin was synthesized in 24 steps in an overall yield of 3.3%. The synthetic teixobactin thus prepared displayed the same spectroscopic and antibacterial profile as the natural product. Thus, also confirming the integrity of the analytical and chemistry work employed to determine the absolute stereochemistry of the natural product.

Dissimilar to the previous syntheses of teixobactin and analogues, Li's strategy involved a convergent approach utilising a neat serine ligation to join the linear and cyclic components of teixobactin (Figure 6.13). Thus

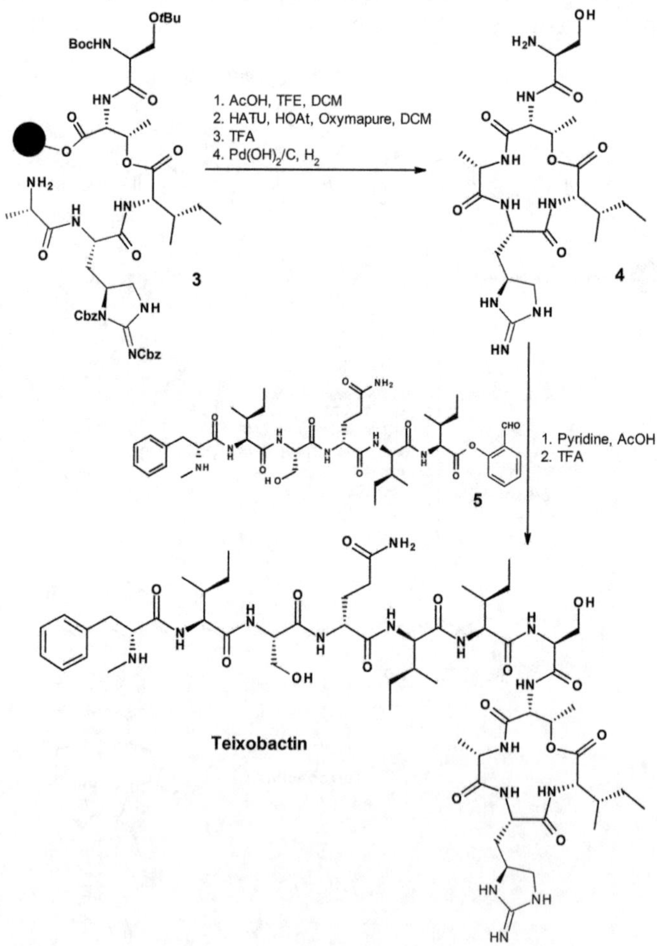

Figure 6.13 Li's solid phase total synthesis of teixobactin.

solid phase synthesis was used to form ring precursor **3**, which was cyclized using 1-[Bis(dimethylamino)methylene]-1H-1,2,3-triazolo[4,5-*b*]pyridinium 3-oxide hexafluorophosphate (HATU), following cleavage from the resin. The resulting macrolactone **4** was linked to the linear hexapeptide **5** containing a C-terminal salicylaldehyde ester (also synthesized by solid phase synthesis) using serine ligation. This step afforded teixobactin in 37% yield following HPLC purification. Worthy of note is that the final coupling involved no protection of the amino acid residues.

6.6 Conclusion

In conclusion, a novel antibiotic teixobactin has been discovered and its absolute configuration elucidated *via* various analytical techniques. This structural elucidation has been verified by the total synthesis of an active teixobactin analogue possessing a single point of difference from the parent and the discovery that replacement of the linear chain D-amino acids with the equivalent L-amino acids abolishes the antibacterial activity of teixobactin.

Despite the current shift in focus towards combating Gram-negative pathogens there is still a medical requirement for new compound classes with novel mechanisms of action and low resistance profiles against Gram-positive bacteria. It is our desire that teixobactin or a suitably derived analogue will satisfy this need and find its use in the clinic against various ailments.

References

1. American Chemical Society National Historic Chemical Landmarks, *Selman Waksman and Antibiotics*, http://www.acs.org/content/acs/en/education/whatischemistry/landmarks/selmanwaksman.html, accessed April, 2016.
2. T. Kaeberlein, K. Lewis and S. S. Epstein, *Science*, 2002, **296**, 1127.
3. *Down To Earth*, 2015, http://www.downtoearth.org.in/news.
4. L. L. Ling, T. Schneider, A. J. Peoples, A. L. Spoering, I. Engels, B. P. Conlon, A. Mueller, T. F. Schäberle, D. E. Hughes, S. Epstein, M. Jones, L. Lazarides, V. A. Steadman, D. R. Cohen, C. R. Felix, K. A. Fetterman, W. P. Millett, A. G. Nitti, A. M. Zullo, C. Chen and K. Lewis, *Nature*, 2015, **517**, 455.
5. A. J. Peoples, D. Hughes, L. L. Ling, W. P. Millet, A. G. Nitti, A. L. Spoering, M. Jones, L. Lazarides, V. A. Steadman, J. Y. C. Chiva, K. G. Poullenec and K. Lewis, WO2014089053 A1, June 12th 2014.
6. H. He, R. T. Williamson, B. Shen, E. I. Graziani, H. Y. Yang, S. M. Sakya, P. J. Petersen and G. T. J. Carter, *J. Am. Chem. Soc.*, 2002, **124**, 9729.
7. Y. Kitajima, M. Waki, J. Shoji, T. Ueno and N. Izumiya, *FEBS Lett.*, 1990, **270**(1–2), 139.
8. J. Shoji, H. Hinoo, T. Hattori, K. Hirooka, Y. Kimura and T. Yoshida, *J. Antibiot.*, 1989, 1462.

9. S. Horii and Y. Kameda, *J. Antibiot.*, 1968, **21**, 665.
10. P. Marfey, *Carlsberg Res. Commun.*, 1984, **49**, 591.
11. R. Bhusan and H. Brückner, *Amino Acids*, 2004, **27**, 247.
12. J. Rudolph, F. Hannig, H. Theis and R. Wischnat, *Org. Lett.*, 2001, **3**(20), 3153.
13. Y. E. Jad, G. A. Acosta, T. Naicker, M. Ramtahal, A. El-Faham, T. Govender, H. G. Kruger, B. G. de la Torre and F. Albericio, *Org. Lett.*, 2015, **17**, 6182.
14. A. Parmar, A. Iyer, C. S. Vincent, D. Van Lysebetten, S. H. Prior, A. Madder, E. j. Taylor and I. Singh, *Chem. Commun.*, 2016, **52**, 6060.
15. A. M. Giltrap, L. J. Dowman, G. Nagaingam, J. L. Ochoa, R. G. Linington, W. J. Britton and R. J. Payne, *Org. Lett.*, 2016, **18**(11), 2788.
16. K. Jin, I. Hou Sam, K. Hiu Laam Po, D. Lin, E. H. Ghazvini Zadeh, S. Chen, Y. Yuan and X. Li, *Nat. Commun.*, 2016, **7**, 12394.

CHAPTER 7

Emerging Targets in Anti-Tubercular Drug Design

KEITH D. GREEN[a], SELINA Y. L. HOLBROOK[a], HUY X. NGO[a] AND SYLVIE GARNEAU-TSODIKOVA*[a]

[a]Department of Pharmaceutical Sciences, College of Pharmacy, University of Kentucky, 789 South Limestone Street, Lexington, KY, 40536-0596, USA
*E-mail: sylviegtsodikova@uky.edu

7.1 Introduction

"Dr Waksman! You have discovered a new and powerful weapon in the deadly battle against one of the oldest foes of mankind, tuberculosis. This battle is as old as medical science and we now have a definite impression that at last the enemy is beginning to yield..."

Harald Cramér, member of the Royal Swedish Academy of Sciences

Professor Cramér congratulated the Ukrainian-American scientist Selman A. Waksman regarding his discovery of streptomycin (STR) at Waksman's Nobel banquet in 1952.[1,2] Professor Cramér's optimism resonated with the rest of the scientific world. STR could penetrate the highly lipophilic cell envelope of *Mycobacterium tuberculosis* (*Mtb*), which was previously considered to be an impossible feat.[3] Tuberculosis (TB), also known as phthisis, consumption, or the white plague, was untreatable for centuries and has devastated humanity since the dawn of time.[4] To cope with the TB epidemic, during the 19th century, many writers, such as John Keats and George Sand,

Drug Discovery Series No. 58
Antibiotic Drug Discovery: New Targets and Molecular Entities
Edited by Steven M. Firestine and Troy Lister

Published by the Royal Society of Chemistry, www.rsc.org

romanticized this disease.[5] In fact, TB was considered to be artistic, poetic, and fashionable, which influenced many upper-class young ladies to pale their skin to mimic the appearance of the diseased.[6] With the discovery of STR, the world finally possessed an agent that could treat TB. Like polio and leprosy, it was thought that TB would eventually be a banal affliction. Unfortunately, the euphoric feeling about STR gradually evaporated, as various clinical studies reported the rapid development of STR resistance in *Mtb* clinical isolates (typically after two months of treatment) as well as the lack of clinical efficacy in the landmark trial by the British Medical Research Council.[7,8] Today, TB still remains a public health threat and is increasingly becoming a global healthcare emergency. According to the latest report from the World Health Organization (WHO), in 2014, about 9.6 million new cases of active TB per year are estimated to occur, leading to 1.5 million casualties.[9] Globally, the prevalence of TB is highest in Southeast Asia with 5.4 million people currently living with TB. Along with Southeast Asia, Africa and the Western Pacific make up more than 85% of the global TB prevalence, whereas the Eastern Mediterranean, Europe, and the Americas account for the remaining ~15% (Figure 7.1).[9]

Furthermore, *Mtb* strains continue to evolve to resist the most clinically useful first-line anti-tubercular agents (isoniazid (INH), rifampicin (RIF), pyrazinamide (PZA), and ethambutol (EMB)) (Figure 7.2). Multidrug-resistant TB (MDR-TB) is caused by isolates that are resistant to at least INH and RIF. The current rate of MDR-TB is estimated to appear in 3.3% of all new TB cases and 20% of previously treated TB cases.[9] Extensively drug-resistant TB (XDR-TB) is a type of MDR-TB, with additional resistance to second-line drugs such as any fluoroquinolone (FQ) and at least one of three injectable drugs (*i.e.* the aminoglycosides (AGs) kanamycin (KAN) and amikacin (AMK), or the cyclic polypeptide capreomycin (CAP)). From the latest estimation, 9.7%

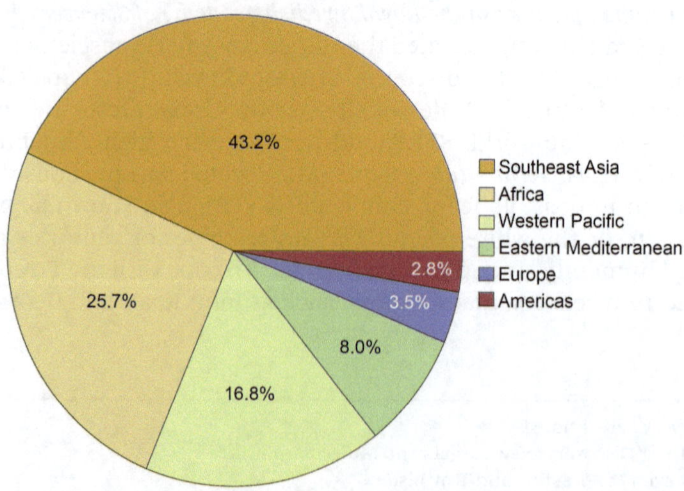

Figure 7.1 Pie chart representing the current global distribution of TB prevalence.

of MDR-TB cases eventually become XDR-TB. Moreover, the number of TB cases that are unresponsive to the most potent drugs continues to rise year after year. As reported by the WHO, the number of RIF-resistant (RR) and MDR-TB cases increased by approximately 10-fold from 2005 to 2014 (Figure 7.3). Hence, there is an immediate clinical necessity for the discovery

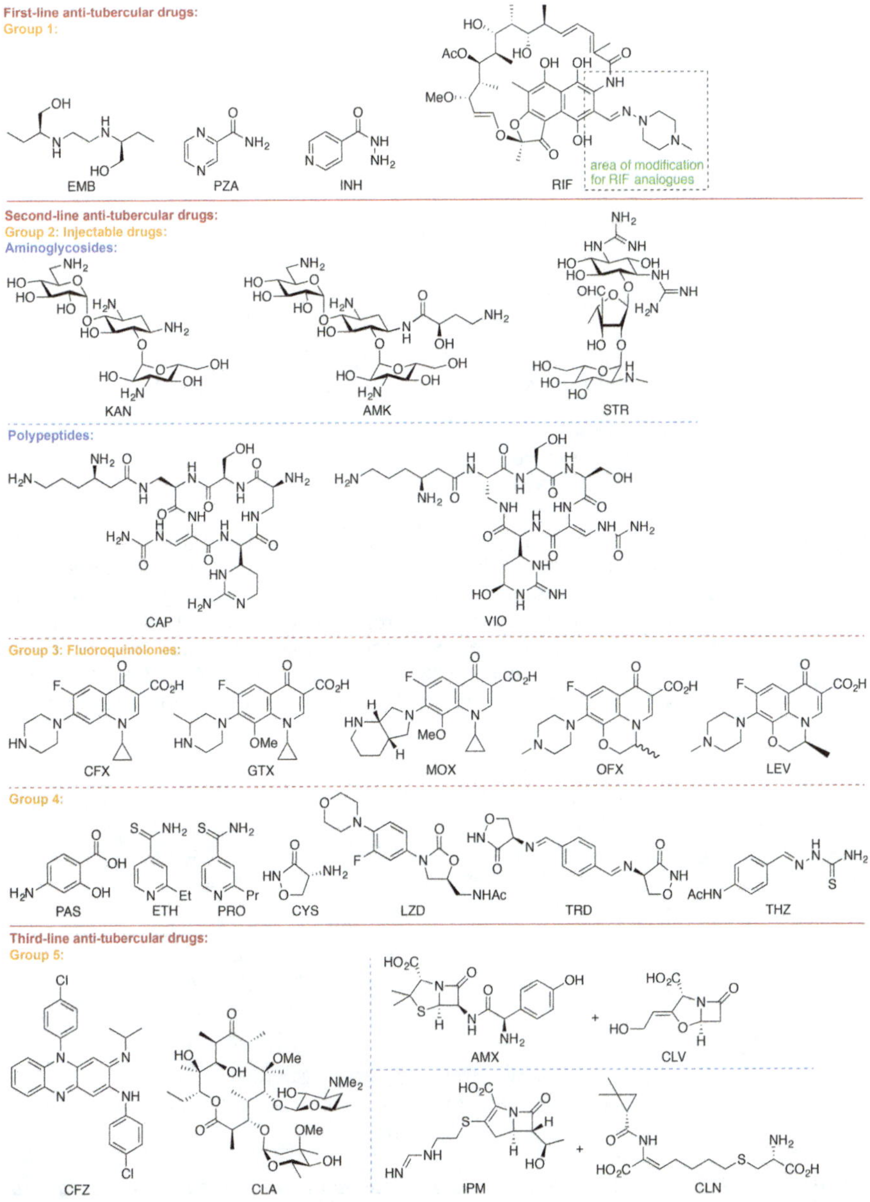

Figure 7.2 Structures of the current anti-tubercular drugs based on their therapeutic classification (groups 1–5).

and development of anti-tubercular compounds with novel drug targets and mechanisms of action.

This story of anti-tubercular drug discovery is mirrored by the PubMed publication log (Figure 7.4). For almost a century after the first published report on TB, very few publications on this respiratory disease (<100 per year from 1853 to 1943) appeared in the literature. Starting from the monumental

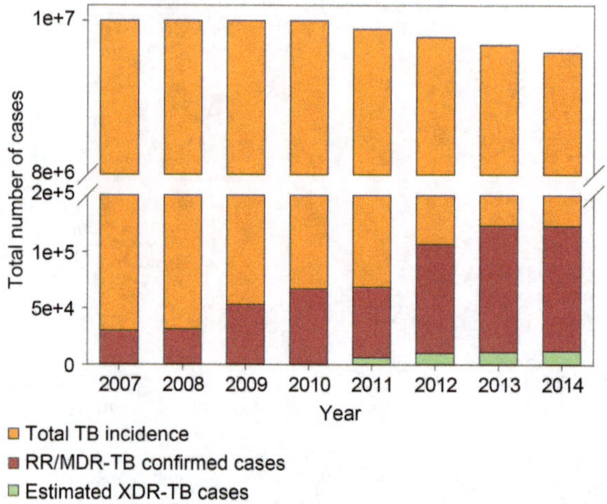

Figure 7.3 Bar graph indicating the incidence of global TB of any variety (orange bars), RR/MDR-TB (red bars), and XDR-TB (green bars). The data points for incidence of global total TB and RR/MDR-TB were extracted from the Global Health Observatory data repository. The data points for XDR-TB were calculated from the percentage of XDR-TB related to the MDR-TB data reported from 2011 to 2014 in the annual WHO global TB reports. Before 2010, the percentage of XDR-TB was minimal and not reported.

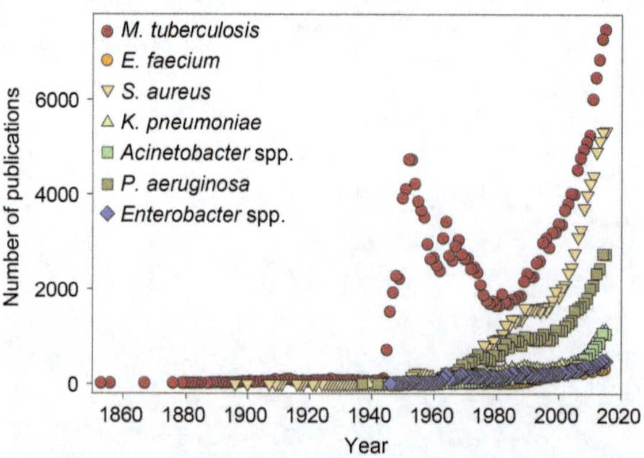

Figure 7.4 PubMed publication record over the years for *Mtb* (red circles), *E. faecium* (orange circles), *S. aureus* (light orange inverted triangles), *K. pneumoniae* (yellow triangles), *Acinetobacter* spp. (green squares), *P. aeruginosa* (brown squares), *Enterobacter* spp. (purple diamonds).

discovery of STR, there was an explosion of TB-related publications ignited by several new and effective anti-tubercular drugs, such as *para*-aminosalicylic acid (PAS), INH, PZA, cycloserine (CYS), and KAN, from 1944 to 1952 (Figure 7.2). The introduction of RIF in the 1960s established the efficacy of the combination therapy regimen, including INH, RIF, PZA, and EMB, and eventually shortened the duration of therapy to as low as six months. Due to the initial effectiveness of the first-line agents, there was a gradual decline in interest in TB research up until 1980 when MDR-*Mtb* strains began to emerge. This observation deeply troubled the public and propelled an immense effort in TB research. In fact, the number of publications on TB has exceeded those related to the ESKAPE pathogens (*Enterococcus faecium*, *Staphylococcus aureus* (*Sau*), *Klebsiella pneumoniae*, *Acinetobacter* spp., *Pseudomonas aeruginosa*, and *Enterobacter* spp.) in recent years.

Inspired by this renaissance in TB drug discovery, herein, we will first briefly survey the pathophysiology of TB, the current targets of clinically approved medications, and the resistance mechanisms associated with these drugs. These fundamental topics lay the groundwork and context for our eventual important discussions on emerging anti-tubercular targets and their novel modulators.

7.1.1 The Biology and Pathology of TB

The genus *Mycobacterium* is so named because of the apparent mold-like growth of the bacteria on the surface of liquid media. Other characteristics of this genus include slow growth, the ability to enter a latent state, a complex cell envelope, and a fairly homogeneous genetic content across isolates.[10] These bacteria are resistant to most staining techniques and are classified as acid-fast bacilli. This resistance to staining is due to the complex and lipocentric composition of the mycobacterial cell envelope, which is also responsible for their high resistance to environmental stresses, including drying and antiseptics.[11] This unique cell envelope also precludes mycobacteria from being concretely classified as either Gram-negative or Gram-positive, since it does not adhere to the strict definition of either class (Figure 7.5).[12] While the bacteria have a peptidoglycan layer, they only become faintly colored by Gram staining.

In *Mtb*, the inner most layer of the *Mtb* cell envelope consists of diacylphosphatidylinositol dimannoside, a fairly unusual lipid.[13] This particular lipid is thought to endow the bacteria with poor membrane fluidity and permeability. The peptidoglycan layer is next and linked to a layer of arabinogalactan, which is in turn attached to a layer of mycolic acids, and finally coated with an outermost layer of phospholipids resulting in a fairly thick overall cell envelope.[14,15] Amid the cell envelope, porin proteins can be found that form channels filled with water allowing the passage of hydrophilic molecules. Only one porin has been reported for *Mtb*, an OmpA-like porin,[16] which plays a role in pH adaptation and does not appear to function as a transport channel.[16,17] Interestingly, when *Mtb* expressed the MspA porin

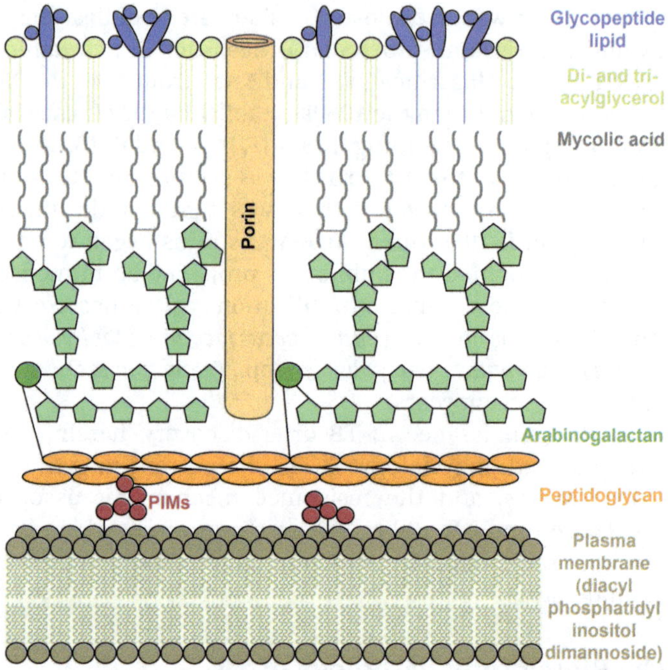

Figure 7.5 Schematic representation of the *Mtb* cell envelope showing the major components discussed herein.

from *Mycobacterium smegmatis* (*Msm*), a non-infectious and fast-growing mycobacterium commonly used as a model for *Mtb*, it becomes sensitized to β-lactams, INH, EMB, and STR.[18] The *Mtb* peptidoglycan layer differs slightly from those of typical Gram-positive and Gram-negative bacteria. In *Mtb*, just like in other bacteria, the peptidoglycan layer consists of alternating units of *N*-acetylglucosamine and *N*-acetylmuramic acid with peptidic side chains of L-alanyl-D-isoglutamyl-*meso*-diaminopimelyl-D-alanine with further amidation of the glutamyl side chain. However, in *Mtb*, the *N*-acetylmuramic acid is further acylated with glycolic acid and cross-linking occurs not only between *meso*-diaminopimelyl groups, but also between *meso*-diaminopimelyl and D-alanyl groups (a detailed structure of the peptidoglycan layer cross-linkages is depicted in Figure 7.17 where a new target is discussed).[19] The arabinoglycan layer is composed of the furanose forms of arabinose and galactose,[19,20] typically arranged in a linear galactan chain bearing several branched arabinose chains, each ending in four arabinose dimers, which hold two points of attachment for the mycolic acids.[21,22] Over 60% of the mycobacterial cell envelope dry-weight consists of lipids.[12] The three major components of the lipid content are mycolic acids, cord factor (better known as trehalose 6,6′-dimycolate, TDM[19,23,24]), and wax-D. Mycolic acids are high molecular weight branched α-alkyl-β-hydroxy fatty acids that form lipophilic tails of glycolipids or are esterified at the end of the arabinogalactans.[25] TDM is responsible for the serpentine cord into which *Mtb* cells develop. This lipid is also toxic to mammals and abundantly produced in virulent strains of *Mtb*. Wax-D is the antigen found in Freund's complete adjuvant, which is used in immunopotentiation.[12] This unique construction of the cell envelope is associated with the bacterium's characteristic impermeability and resistance to antibiotics, acidic and basic compounds, osmotic lysis, and lethal oxidations, and it allows for survival inside the macrophage. Of the ~4000 encoding genes of *Mtb*, 525 are involved in the synthesis of the cell envelope and 200 are related to fatty acid metabolism.

In addition to the highly specialized cell envelope, *Mtb* also utilizes some metabolic pathways to protect itself from its environment. The fact that *Mtb* can enter a non-replicating state indicates a period of metabolic shutdown.[10] During this phase, treatment of TB becomes difficult since the etiologic agent is not replicating and, therefore, not utilizing the enzymes and pathways targeted by the current anti-tubercular agents, discussed in Section 7.1.2. While the molecular basis for the metabolic shutdown of *Mtb* is still not completely understood,[10] several recent studies have shed some light on the mechanisms and pathways involved in the ability of *Mtb* to enter this non-replicating and slowed metabolic state. One particular gene found to be involved in this process is *dosR* (*rv3133c*) that is responsible for mediating the hypoxic response[26] and aids the rebound of the bacteria back to a normal metabolic state.[27] Changes in glucose phosphorylation have also been documented to be required in the non-replicating state.[28] The genome of *Mtb* appears to have an abundance of extra metabolic and biosynthetic proteins, which is suggestive of its ability to adapt to its environment. *Mtb* has the potential to synthesize all the essential amino acids, vitamins, and cofactors needed for its survival.[10] It also has the ability to catabolize a wide variety of carbon sources, including, but not limited to, carbohydrates, hydrocarbons, and alcohols.[10]

TB is most commonly acquired by the inhalation of dried particles containing one to three *Mtb* bacilli. When these particles reach the lungs, they are phagocytized by macrophages in the alveoli whereupon most bacteria are usually destroyed. In certain instances, whether caused by a compromised immune system or another underlying condition, the body's defenses can fail to terminate the inhaled bacilli. In this case, the infection progresses in five distinct steps: (i) the tubercle bacilli reach the alveoli of the lung and are ingested by macrophages, however, some bacilli survive. At this point, the infection is asymptomatic, indicative of a latent infection. (ii) The bacilli replicate in the macrophage, recruiting more macrophages to the area, which then become hosts for the mycobacteria. The macrophages form an early tubercle and excrete cytokines and other enzymes that cause lung-damaging inflammation. (iii) After a few weeks of inflammation, symptoms begin to appear as a result of many macrophages dying and releasing their load of bacilli forming a caseous center in the tubercle. The anaerobic environment is not ideal for the propagation of bacteria and growth is halted. However, many cells survive in dormant or latent states to become activated later. At this point, the lesions can become calcified and the disease progression is stalled. (iv) If the growth of the caseous center continues, the center can become liquefied and bacilli are able to multiply outside of macrophages.

(v) The tubercle eventually ruptures from the growth of the bacilli and they are released into the bronchiole. From this location, the bacteria can be disseminated to the rest of the lung, the blood stream, and the lymphatic system.

Transcriptional analysis of *Mtb* cultivated in macrophages reveals that several genes related to fatty acid degradation and lipid synthesis are attenuated along with transcriptional regulators, proteins involved in the glyoxylate cycle, citrate synthesis, mycobactin synthesis, and α-crystallin, along with many other genes.[29–31] Additionally, several genes with unidentified products are also noted. One of the cell envelope components, lipoarabinomannan is essential for the virulence of *Mtb*. This particular lipoglycan binds the macrophage mannose receptor facilitating entry into the macrophage. It is also responsible for modulating the host immune response and preventing the phagosome-lysosome fusion. All these factors combined lead to intracellular survival and persistence.[32] The virulence of *Mtb* was also found to be reliant on lipoamide dehydrogenase, which is a core protein in three separate multi-enzyme complexes affecting the biosynthesis of amino acids, acetyl coenzyme A (AcCoA), and reduction of reactive nitrogen intermediates.[33]

7.1.2 Current Drug Targets

The current drug targets span three main areas: (i) cell envelope synthesis, (ii) translation, and (iii) transcription and DNA replication (Figure 7.6). The structures of these molecules, which are classified as first-, second-, and third-line agents are drawn in Figure 7.2.

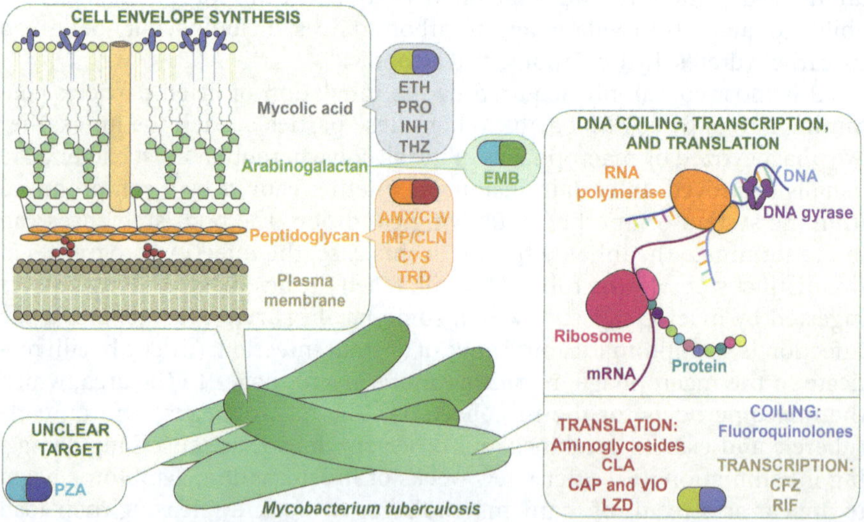

Figure 7.6 A schematic of a *Mtb* cell showing where the current drug targets affect the cell, including the cell envelope, protein/RNA/DNA synthesis, and other areas.

7.1.2.1 Cell Envelope

While not among the core treatment regimen of TB, the amino acid-derived CYS and terizidone (TRD) display broad-spectrum activity and inhibit the synthesis of the peptidoglycan layer.[34] These amino acid derivatives target alanine racemase and D-alanine:D-alanine ligase, both of which are involved in the synthesis of the peptidoglycan layer.[35] β-Lactams (*e.g.* amoxicillin (AMX)) work in a similar fashion inhibiting transpeptidase and preventing the cross-linking of the peptidoglycan layer.[36] The penicillin AMX and the carbapenem imipenem (IPM) are used as third-line TB treatments in combination with the β-lactamase inhibitor clavulanate (CLV) and the carbapenemase inhibitor cilastatin (CLN), respectively, to prevent the degradation of the anti-tubercular compounds. EMB targets an arabinosyltransferase responsible for building the arabinogalactans found in the cell envelope.[37,38] Ethionamide (ETH), prothionamide (PRO), and INH all inhibit the enoyl-acyl carrier protein (ACP) reductase InhA.[39,40] All of these nicotinamide derivatives are prodrugs that need to be activated. INH is activated by the catalase-peroxidase hemoprotein KatG. However, ETH is activated by the monooxygenase EthA to an *S*-oxide metabolite.[41] Thioacetazone (THZ) is also a prodrug that gets transformed by EthA into its active form, 2-ethyl-4-amidopyridine.[42] While THZ is not a nicotinamide analogue, it targets the same pathway as PRO, ETH, and INH. These drugs specifically target the cyclopropane synthase enzymes, disrupting fatty acid synthesis by forming adducts with NADH and preventing InhA from performing its catalytic activity.[43]

7.1.2.2 Translation

The AGs AMK, KAN, and STR function by binding tightly to the 16S rRNA in the 30S ribosomal subunit.[44,45] When bound to the ribosome, they prevent normal protein translation, which inevitably leads to cell death. Clarithromycin (CLA), a macrolide antibiotic, also inhibits the ribosome. However, while AGs bind the 30S ribosomal subunit, CLA binds the 50S ribosomal subunit, which also results in a cessation of protein synthesis.[46] CAP and viomycin (VIO) are cyclic peptides from the tuberactinomycin family, which are known to inhibit protein synthesis by binding the ribosome, hence their usual coupling with AGs. CAP also has significant activity against the persistent form of TB and is thought to have a second target outside of the ribosome.[47] Linezolid (LZD), an oxazolidinone, is responsible for inhibiting protein translation in a completely different way than other drugs.[48] While most compounds bind one of the two large halves (30S and 50S) of the ribosome, preventing translation from proceeding, LZD binds the 23S rRNA, thereby inhibiting translation in the early stages by preventing formyl-methionine tRNA from binding the complex.[49]

7.1.2.3 Transcription and DNA Replication

The FQs used to treat TB, including gatifloxacin (GTX), levofloxacin (LEV), moxifloxacin (MOX), ciprofloxacin (CFX), and ofloxacin (OFX), all trap the ATP-dependent type II topoisomerases (DNA gyrase and topoisomerase IV) in an enzyme-DNA-inhibitor complex. When trapped, the enzymes no longer perform DNA uncoiling, freezing that portion of DNA and shutting down DNA replication, translation of genes in the region, and DNA repair.[50] Clofazimine (CFZ) binds the guanine bases of bacterial DNA, preventing DNA from being used as a template and inhibiting bacterial proliferation.[51,52] It also increases activity of bacterial phospholipase A2, which causes the release and accumulation of toxic lysophospholipids.[53,54] PAS was originally proposed to target dihydropteroate synthase, the usual target of sulfonamide drugs.[55] Several years later, it was discovered that in fact PAS inhibits the thymidylate synthase, preventing thymidine from being synthesized and, therefore, disrupting DNA replication.[56] RIF and its analogues rifabutin, rifalazil, and rifapentine all inhibit the β-subunit of DNA-dependent RNA polymerase, halting transcription.[57,58] It is thought that RIF blocks the transit of the growing RNA chain after the addition of two or three nucleotides. In *Escherichia coli* (*Eco*), the suicide gene *mazEF* is triggered, and the same system is found in *Mtb*.[59]

PZA has a poorly understood mechanism of action. As a prodrug, it is activated by the pyrazinamidase/nicotinamidase PncA to pyrazinoic acid (POA), which has been shown to inhibit many functions in low-pH environments that are often found when *Mtb* is internalized to the macrophage.[60,61] POA is pumped out of the cell, and if the extracellular environment is acidic enough, POA becomes protonated and reenters the cell at a faster rate than it can be egested. This leads to an increase in intracellular acidity, and this change in pH eventually affects numerous cellular cycles.

7.1.3 Resistance Related to TB

Mtb resists the action of many drugs. The mechanisms of resistance in *Mtb* comprise (i) intrinsic cellular properties and (ii) the presence of efflux pumps and several chromosomally encoded resistance genes, as well as mutation of several genes resulting in reduced efficacy of the antibiotics.[62] It is important to note that *Mtb* has no reported instances of acquiring resistance mechanisms from mobile genetic elements. In addition to the modes of resistance employed by *Mtb*, the mycolic acid-rich cell envelope decreases the permeability of many antibiotics and antimicrobial compounds.

7.1.3.1 Efflux Pumps

One mechanism of resistance shared by many bacteria is the presence of efflux pumps. At their basal levels, efflux pumps contribute little to resistance, however, when overexpressed or mutated to have higher affinity for a particular substrate, they can greatly affect resistance, particularly in mycobacteria considering their low rate of influx. Molecularly, efflux pumps are active transporters and use either an energy source (*e.g.* ATP) or an ionic gradient to shuttle xenobiotics and other toxic compounds out of the cell. An example of a *Mtb* efflux pump includes P55 (Rv1410c). P55, a member of the major facilitator superfamily of efflux pumps, acts on tetracycline, STR, gentamicin, and netilmicin. Interestingly, other efflux pumps associated with AG resistance are from the resistance-nodulation-cell division (RND) superfamily of proteins.[63] Additionally, the *p55* gene forms an operon with *p27*, a gene encoding an antigenic lipoprotein. The P55 efflux pump is sensitive to inhibition by verapamil and reserpine, two known efflux pump inhibitors, and is dependent on the cellular proton gradient of the bacteria. In mycobacteria, efflux pumps are also responsible for resistance to FQs (in addition to other methods of resistance (See Section 7.1.3.2(a))).[64,65]

7.1.3.2 Gene Mutations

Genetic mutations occur in two major areas: (i) mutations or modifications of drug targets, including methylation of RNA, and (ii) failure to convert prodrugs into active compounds.

7.1.3.2(a) Mutations or Modifications of Drug Targets. *Mutations*: Anti-tubercular agents interrupt the cellular metabolism preventing requisite biomolecules from being made. Most often the interactions of these agents with their protein targets is dependent on a few key interactions. A simple and common mechanism for bacteria to prevent this from happening is the mutation of the target's binding site. This can happen either naturally during replication error, or through selective pressure. The resulting mutations decrease the ability of anti-tubercular drugs from binding to their respective targets allowing the bacteria to survive the therapy. This mechanism is commonly found in *Mtb* strains resistant to RIF, EMB, and FQs by single residue mutations in the β-subunit of RNA polymerase,[57] a glycosyltransferase,[66] and/or type II topoisomerases.[67] Slight mutations in the 16S rRNA (*e.g.*, A1401G) often confer resistance to the injectable drugs AMK, KAN, and CAP.[68] Two mutations in the *rrs* gene from *Mtb* encoding three rRNA residue alterations (A1400G, C1401A, and G1483T) cause resistance to KAN.[69]

Modifications: An alternative mode of resistance in *Mtb* is the inactivation of rRNA methyltransferase enzymes (*e.g.* TlyA and GibB). TlyA mutations were documented to hinder the methylation activity of rRNA 2′-*O*-methyltransferase at nucleotides C1409 of the 16S rRNA and C1920 of the 23S rRNA, thereby causing resistance to CAP and VIO.[70] Mutations in GidB were found to confer minor resistance to STR.[71,72]

7.1.3.2(b) Inability to Activate Prodrugs. Among the anti-tubercular agents, some prodrugs are found (*e.g.* INH, PZA, and ETH). These prodrugs are activated by KatG (INH), EthA (ETH), and PncA (PZA). Mutations in the *katG*, *ethA*, and *pncA* genes were shown to reduce the ability of the corresponding enzymes to activate INH,[73,74] ETH,[41,75,76] and PZA,[77,78] respectively. Interestingly, mutations in promoter regions were additionally found to result in overexpression of drug targets[41,75,79] and to prevent binding of the active form of INH and ETH to their targets.[64,80]

7.1.3.3 Enzymatic Modifications and Inactivation of Drugs

Perhaps the most prevalent and well-studied mechanism of resistance in non-mycobacteria is drug modification. Modifications can include degradation of the core structures or alteration of the chemical appearance of the antibiotic, to the point where the compound no longer binds or inactivates its target. *Mtb* naturally harbors a chromosomally encoded class A β-lactamase, BlaC, which is constitutively expressed to provide intrinsic resistance to penicillins by hydrolysis of their β-lactam ring.[81] *Mtb* also contains two genes encoding for AG-modifying enzymes in its genome: *aac(2′)-Ic* and *eis* encoding an AG 2′-*N*-acetyltransferase[45,82,83] and the enhanced intracellular survival (Eis) protein, respectively.[84] Both enzymes acetylate AGs, reducing the ability of these drugs to bind the ribosome. In the case of Eis, increased expression due to mutations in the *eis* promoter or the 5′-untranslated region of the transcriptional activator WhiB7 leads to clinically relevant, low-level resistance to KAN.[84] Eis has a unique ability to modify AGs at multiple sites and multiple times *in vitro*, completely inactivating these compounds.[85] Crystallographic studies recently showed that the AG tobramycin (TOB) is able to bind in two distinct modes, allowing for a minimum of two acetylations (Figure 7.7).[86] Eis was also reported to acetylate CAP at its β-lysine side chain *in vitro*, but a clinical link to CAP resistance still needs to be elucidated.[87]

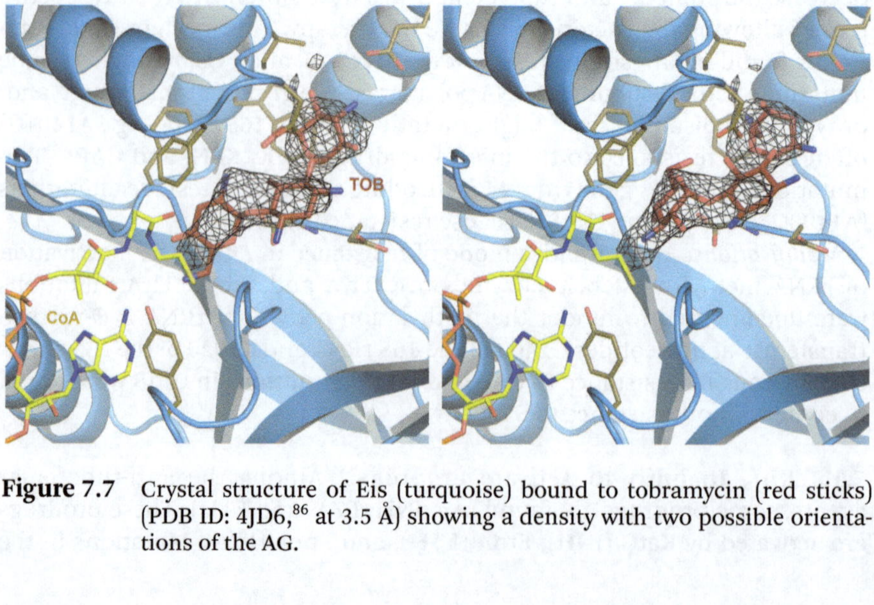

Figure 7.7 Crystal structure of Eis (turquoise) bound to tobramycin (red sticks) (PDB ID: 4JD6,[86] at 3.5 Å) showing a density with two possible orientations of the AG.

This particular resistance enzyme has been extensively studied, both for its AG acetylating properties[88–90] and its other *in vivo* functions.[91–93] Interestingly, Eis enzyme homologues are found in a slew of bacteria, both from the genus *Mycobacterium* as well as other genera.[94–97]

7.2 Discovery and Validation of Novel Drug Targets in TB

The accelerated emergence of resistant TB has spurred extensive research seeking novel anti-tubercular drug targets. Herein, we discuss the novel drug targets from the biosynthesis of the distinct *Mtb* cell envelope (Section 7.2.1) and from general metabolic processes in *Mtb* (Section 7.2.2). We also present representative inhibitors of these enzymatic targets (Figure 7.8).

7.2.1 Targets Involved in *Mycobacterium tuberculosis* Cell Envelope Biosynthesis

The distinct and complex structure of the *Mtb* cell envelope gives the bacteria intrinsic protection. However, it also makes the *Mtb* cell envelope an attractive target for anti-tubercular research, as mammalian cells do not possess a similar cellular structure. In addition to the established and well-recognized targets discussed in Section 7.1.2.1, the biosynthesis of the cell envelope of *Mtb* has recently provided many new additional enzymatic targets involved

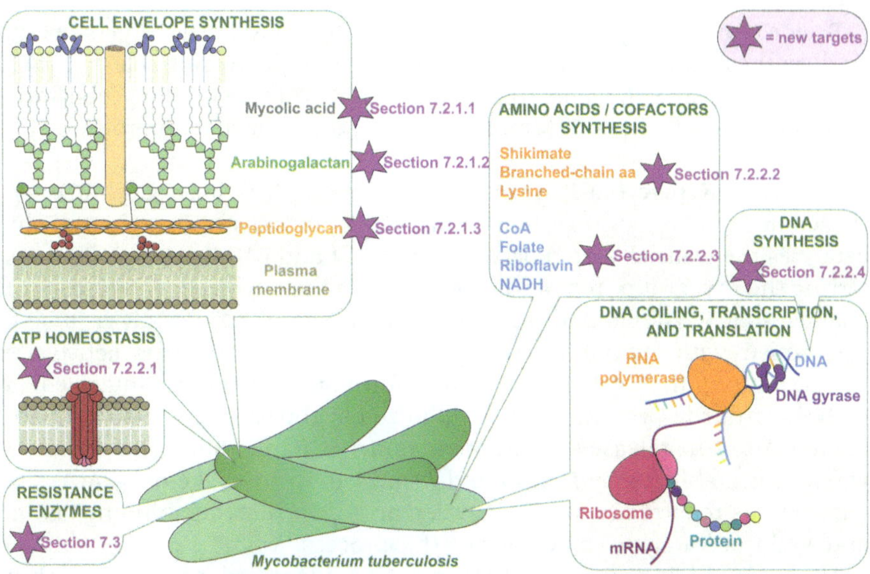

Figure 7.8 The areas of *Mtb* metabolism, protein/DNA/RNA synthesis, and cell wall envelope generation that have novel targets being probed for new anti-tubercular compounds.

in the formation and processing of (i) mycolic acid, (ii) arabinogalactan, and (iii) peptidoglycan.

7.2.1.1 *Mycolic Acid Biosynthesis and Processing*

One prominent feature in the *Mtb* cell envelope is the presence of the covalently-linked mycolate-arabinogalactan-peptidoglycan (mAGP) complex. Mycolic acids are major components of the *Mtb* cell envelope, and are crucial for *Mtb*'s viability under adverse conditions by providing structural integrity.[98,99]

While human fatty acids are biosynthesized by fatty acid synthase I (FAS I), the generation of mycolic acids in *Mtb* is mostly accomplished by fatty acid synthase II (FAS II) (Figure 7.9).[100] FAS I and Fas II co-exist in *Mtb* and cooperatively produce diverse fatty acid products, including linear meromycolic acids and their α-branched counterparts.[100–102] It is proposed that the fatty acid products of FAS I are used by FAS II for the biosynthesis of very-long-chain fatty acids.

Enzymes involved in mycolic acid biosynthesis have been heavily investigated as anti-tubercular targets. As mentioned in Section 7.1.2.1, one of the current first-line treatments, INH, targets the enoyl-ACP reductase InhA.[103] Even though InhA is a known anti-tubercular drug target, new classes of InhA inhibitors with improved efficacy, such as Genz-10850,[104–106] hirsutellone A,[107] moiramide B and andrimide,[108] and their analogues, as well as rhodanine[109] and triazole derivatives have recently been developed (Figure 7.10).[110,111]

Besides InhA, research has also identified other novel drug targets in fatty acid synthesis and mycolic acid biosynthesis and processing that may help us combat these pathogens: (i) β-ketoacyl-ACP synthases, (ii) mycobacterial membrane protein large 3 (MmpL3), and (iii) the antigen 85 complex.

7.2.1.1(a) β-Ketoacyl-ACP Synthases. There are three β-ketoacyl-ACP synthases in the *Mtb* FAS II system: FabH, KasA, and KasB (Figure 7.9). FabH catalyzes the transition step between the FAS I and FAS II systems by transferring the acyl group from acyl-CoA (FAS I) to malonyl-ACP (FAS II) to form β-ketoacyl-ACP. The homodimeric *Mtb*_FabH (Figure 7.11) catalyzes a bi–bi mechanism with the initial binding of acyl-CoA into its narrow active site. The acyl group then gets transferred to form an acylthioester intermediate with the catalytic Cys112, which is stabilized by the backbone of Cys112 and Ala306. After the release of CoA, the malonyl-ACP then binds in this pocket where it is stabilized by His244 and Asn274 to form a β-ketoacyl-ACP. The enzyme was reported to switch between an "open/substrate-binding" mode and a "closed/catalytic" mode during this process.[112,113]

Although thiolactomycin (TLM) was first identified as a weak inhibitor of FabH by binding to the acyl-enzyme intermediate in 1982,[114–118] research aiming to improve this potential anti-tubercular agent is still ongoing. Recently, a series of TLM-inspired compounds containing a

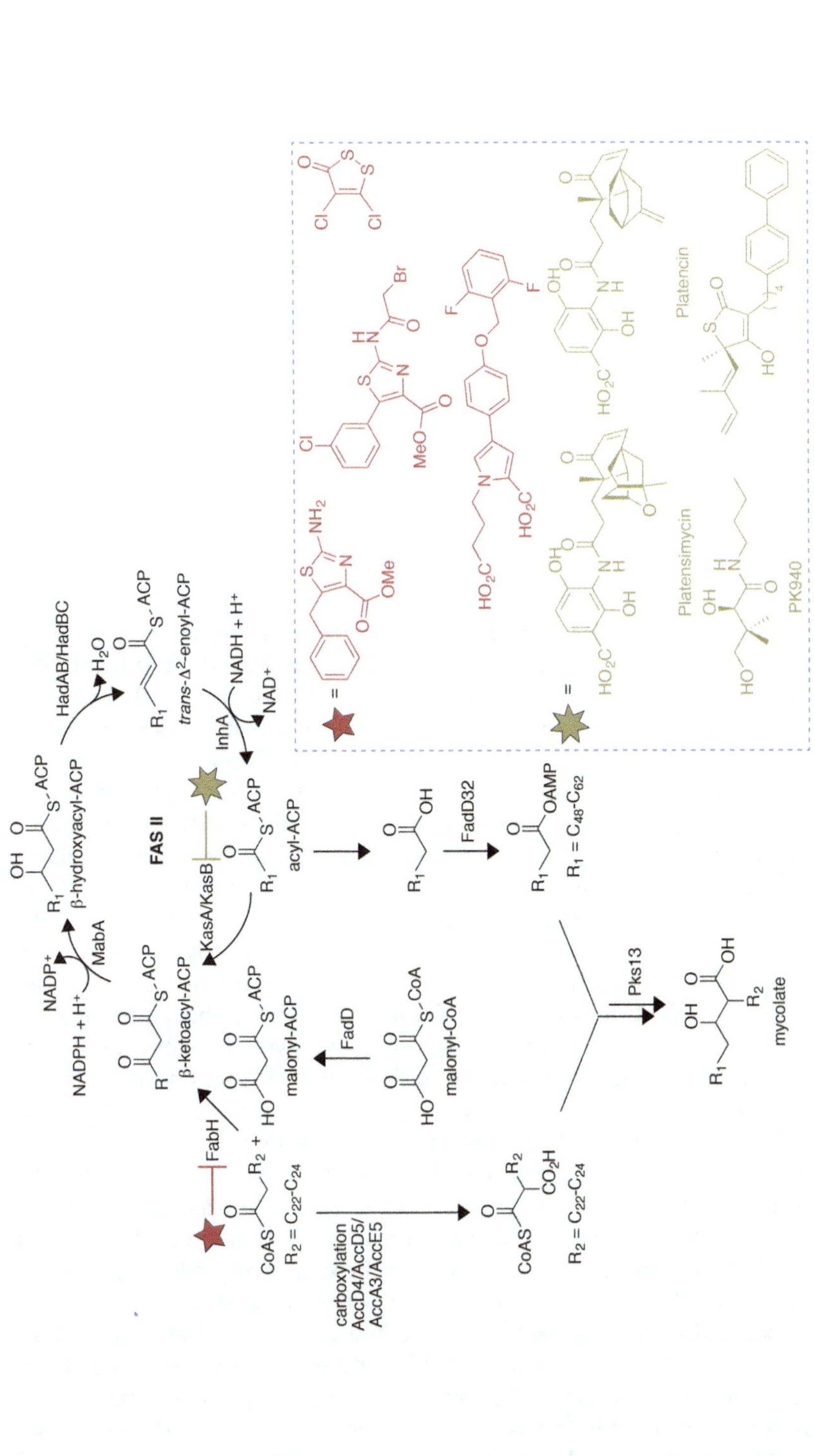

Figure 7.9 The FAS I and II pathways in *Mtb* mycolic acid biosynthesis along with the structures of the inhibitors of the FAS enzymes discussed.

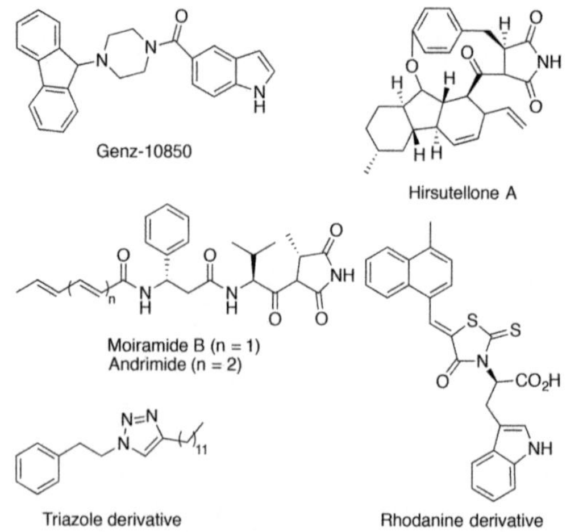

Figure 7.10 Various classes of InhA inhibitors developed in recent years.

2-aminothiazole-4-carboxylate core was generated.[119] The best compounds identified were methyl 2-amino-5-benzylthiazole-4-carboxylate and methyl 2-(2-bromoacetamido)-5-(3-chlorophenyl)thiazone-4-carboxylate (Figure 7.9). Methyl 2-amino-5-benzylthiazole-4-carboxylate was found to inhibit *Mtb* H37Rv with an MIC value of 0.06 μg mL^{-1} (compared to the MIC value of TLM being 13 μg mL^{-1}), but to display no inhibitory activity against the FabH enzyme alone. However, methyl 2-(2-bromoacetamido)-5-(3-chlorophenyl) thiazone-4-carboxylate was shown to inhibit FabH with an IC_{50} value of 0.95 μg mL^{-1} (compared to TLM at 16 μg mL^{-1}), but to not be active against *Mtb* H37Rv. In addition a series of 1,2-dithiole-3-ones were discovered to be inhibitors of FabH in *E. coli* and *S. aureus*, by searching compounds that were structurally similar to TLM in the National Cancer Institute (NCI) database.[120] The most potent inhibitor, 4,5-dichloro-1,2-dithiole-3-one, displayed IC_{50} values of 2 and 0.16 μM against *Eco*_FabH and *Sau*_FabH, respectively. Some of these 1,2-dithiole-3-ones were later found to also be highly active against *Mtb*_FabH.[121,122] Moreover, a series of 1-(4-carboxybutyl)-4-(4-(substituted benzyloxy)phenyl)-1*H*-pyrrole-2-carboxylic acid derivatives were reported.[123] The best inhibitor in this series contained a 2,6-difluorobenzyloxyl moiety and displayed an MIC value of 12.5 μg mL^{-1} against *Mtb* H37Rv.

Besides inhibiting FabH, TLM and its derivatives are also known to halt the action of the KasA and KasB condensing enzymes.[124] KasA and KasB are part of the elongation cycle in FAS II and form the same kind of products as FabH.[125,126] As TLM mimics malonyl-ACP in the active site of KasA (as determined by NMR), PK940 and TLM derivatives were also synthesized and found to be potent inhibitors of KasA.[127] As overexpression of KasA and/or KasB in *Mycobacterium bovis* BCG is associated with increased TLM resistance both *in vitro* and *in vivo*, additional TLM analogues that target these enzymes in *Mtb* H37Rv were evaluated.[128] The *Streptomyces platensis* secondary metabolites platensimycin and platencin were also found to inhibit KasA and

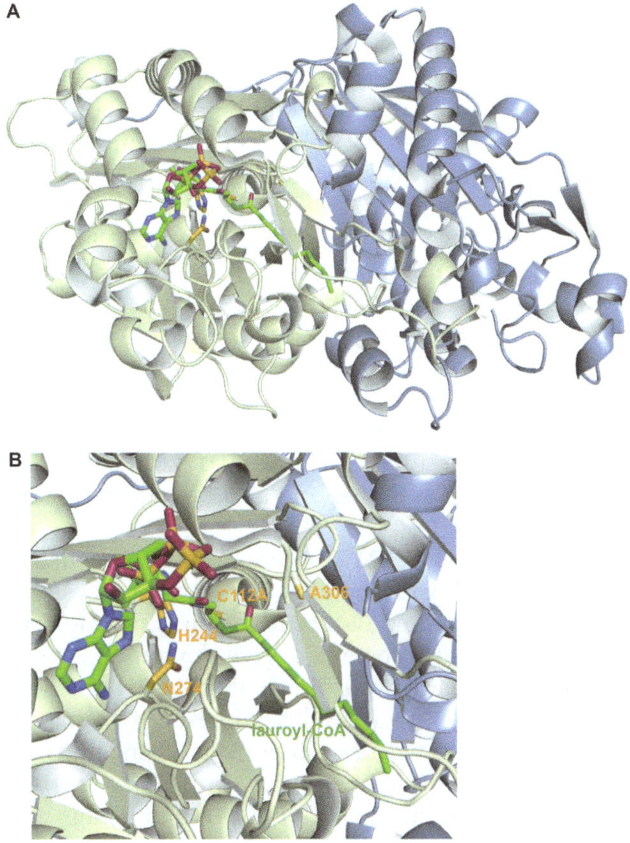

Figure 7.11 (a) Crystal structure of *Mtb*_FabH (pale yellow and blue) (PDB ID: 1U6S,[113] at 2.3 Å) in homodimeric form co-crystallized with lauroyl-CoA (green sticks with oxygen, nitrogen, sulfur, and phosphorous atoms colored red, blue, dark green, and orange, respectively). The residues important for substrate binding, including the catalytic Cys112 as well as Ala306, His244, and Asn274 are depicted as orange sticks in the zoomed-in view presented in panel (b).

KasB, but not FabH, and to stop the growth of *M. smegmatis* and *Mtb* with MIC values of 14 and 12 μg mL^{-1}, respectively.[129] The lack of mammalian toxicity of platensimycin and platencin renders them great candidates as potential anti-tubercular agents.[130] Overall, these results confirm the potential of *Mtb* β-ketoacyl-ACP synthases as anti-tubercular targets.

7.2.1.1(b) Mycobacterial Membrane Protein Large 3 (MmpL3) (Cell Wall Mycolic Acid Transporter). The synthesized mycolic acids from the FAS I and FAS II systems undergo further sequential modifications: (i) transport into the periplasmic space (discussed in this section), (ii) attachment to trehalose to form trehalose monomycolate (TMM) or attachment to TMM to form TDM, and (iii) transfer from TDM to the arabinoglycan layer (discussed in

Section 7.2.1.1(c)).[131–133] The export of mycolic acids from the *Mtb* cytoplasm to the extracellular environment relies on the mycobacterial membrane protein large (MmpL) transporters. Specifically, MmpL3, a RND superfamily transporter, is responsible for the export of TMM.[134,135] Interestingly, research has recently validated the important role of MmpL3 in heme uptake along with its better-known activity of mycolic acid secretion.[136] These findings render MmpL3 an attractive drug target for anti-tubercular drug discovery.

SQ109 (Figure 7.12), which was originally developed as an EMB derivative, was later identified as an inhibitor of MmpL3.[137] Contrary to EMB, SQ109 does not affect the total level of mycolic acids in *Mtb*, but instead results in an accumulation of TMM. It was not until 2014 that SQ109 was reported to also inhibit additional processes in *Mycobacteria*, including disruption of membrane potential and ATP homeostasis.[138] The multi-target nature of SQ109 has made it an attractive drug candidate for TB patients and is currently in phase II clinical trials. The urea derivative AU1235 (Figure 7.12) contains the same adamantyl group as SQ109 and also targets MmpL3 *via* a similar mechanism.[139]

In addition to SQ109 and AU1235, other scaffolds have also been shown to inhibit mycolic acid transport by targeting MmpL3. These newly discovered inhibitors/scaffolds include BM212 (a 1,5-diarylpyrrole derivative), the indoleamide class, tetrahydropyrazolo [1,5-*a*]pyrimidine-3-carboxamides (THPPs), and *N*-benzyl-6′,7′-dihydrospiro[piperidine-4,4′-thieno[3,2-*c*]pyran] analogues (spiros) (Figure 7.12).[134,140–143] Moreover, by using a whole cell-based HTS of 20 000 small molecules against *Mtb*, a benzimidazole molecule (later named C215) was identified to inhibit MmpL3.[144] Surprisingly, a large portion of the *Mtb* growth inhibitors from this HTS campaign were shown to not inhibit the growth of *M. smegmatis* and/or *M. bovis* BCG, suggesting the limitations of using in *M. smegmatis* and/or *M. bovis* BCG as model organisms for anti-tubercular drug discovery.

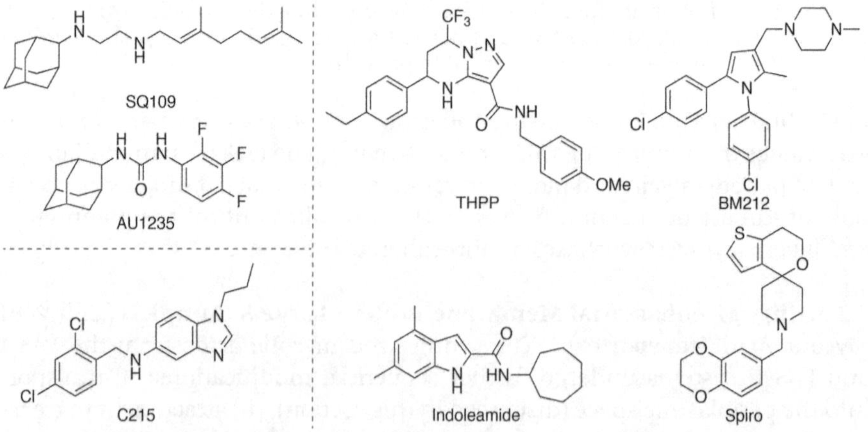

Figure 7.12 Structures of MmpL3 inhibitors discussed in this section.

7.2.1.1(c) Antigen 85 Complex. The Ag85 complex consists of three homologous mycolyltransferase enzymes Ag85A/B/C, which catalyze the transfer of a mycolyl group from TMM to either another TMM molecule to generate TDM or to arabinose to form mycolyl-arabinogalactan.[145,146] Crystal structure analysis revealed that all three isoforms of the *Mtb* Ag85 complex contained highly conserved active sites that could all be inhibited by the same compound.[147–149] Knockout studies revealed that the fibronectin binding protein genes encoding the three enzymes of the Ag85 complex are essential for *Mtb* viability.[150] Additionally, the Ag85 complex stimulates the complement-mediated phagocytosis of *Mtb* by macrophages. It has been proposed that the binding of the Ag85 complex to human fibronectin facilitates uptake by macrophages.[151–153] Traditionally, crossing the cell envelope is a significant challenge in novel anti-tubercular agent development. Unlike other drug targets, the Ag85 complex is excreted. Hence, the accessibility to this complex is high, and intracellular resistant mechanisms such as efflux pump or enzymatic modifications are potentially avoided.[25,154] Collectively, these findings suggest that the Ag85 complex could be a promising novel *Mtb* drug target.

The crystal structures of *Mtb*_Ag85B-trehalose and *Mtb*_Ag85C-octylthioglucoside complexes were determined and found to be instrumental in providing evidence supporting the proposed mechanism of mycolyl transfer (Figure 7.13A).[147–149] Based on the arrangements of the active sites, the mechanism was expected to be similar to that of other related serine hydrolases involving the Ser124, Glu228, and His260 (Ag85C numbering) catalytic triad.[155,156] As Ser124 was demonstrated to be crucial for mycolyl transfer, a popular strategy towards discovery of Ag85 complex inhibitors consists of targeting this amino acid residue.[145] A known serine hydrolase inhibitor, diethyl *p*-nitrophenylphosphate, was first shown to target Ag85C by covalently binding Ser124 in a co-crystallization study (Figure 7.13D).[147] Although diethyl *p*-nitrophenylphosphate is a potent covalent inhibitor for Ag85C, it is also highly toxic.[157] Inspired by the diethyl *p*-nitrophenylphosphate inhibitor, analogous phosphonates designed to mimic the tetrahedral intermediate of TMM bound to Ser124 were found to inhibit Ag85C at low micromolar concentration.[158] However, these compounds were discovered to be either inactive or have high MIC values against *Mycobacterium avium*.

A straightforward approach to generate Ag85 complex inhibitors is to mimic the trehalose portion of TMM, which occupies the carbohydrate-binding pocket. 6-Azido-6-deoxytrehalose was a first-generation trehalose-based analogue that displayed inhibition of the *Mtb* Ag85 complex and antibacterial properties against *Mycobacterium aurum* (Figure 7.14).[145] Additional trehalose analogues modified with alkyl chains displayed moderate growth inhibition of *M. smegmatis*.[159] A library of *N,N′*-dialkylamino- and 6,6′-bis (sulfonamido)-trehalose analogues, which displayed good antibacterial activity against *Mtb* strain H37Ra (as low as 4 μg mL^{-1}) was reported.[160] These first-generation trehalose analogues set the foundation for future optimizations of trehalose-based inhibitors. In 2011, a fluorophosphonate trehalose analogue designed by combining the two popular strategies discussed

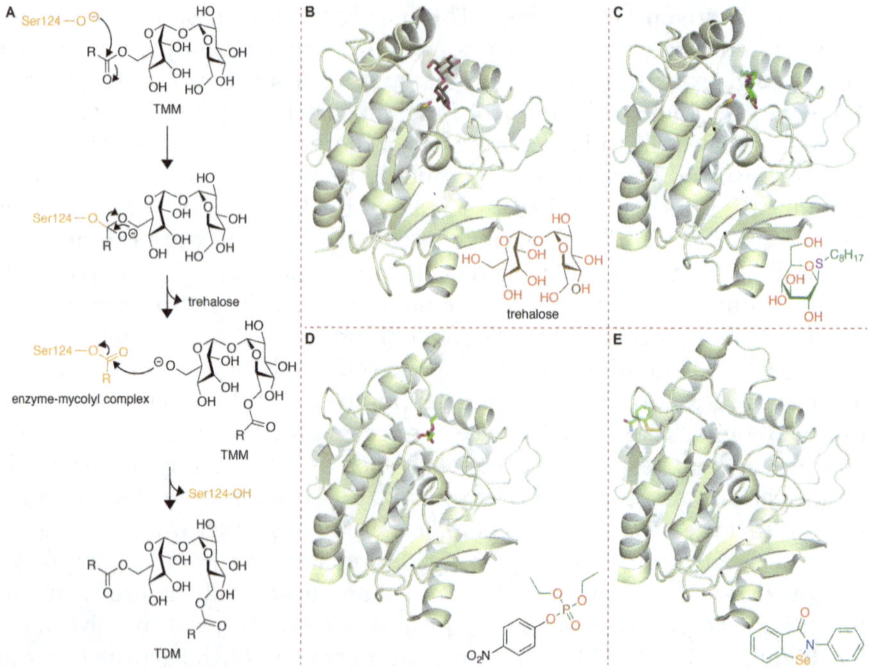

Figure 7.13 (a) Proposed enzyme mechanism of the Ag85 complex. (b) Crystal structure of Ag85B (pale yellow) with the substrate trehalose (sticks) (PDB ID: F10P,[148] at 1.9 Å). (c) Crystal structure of Ag85C (pale yellow) with octylthioglucoside (sticks) (PDB ID: 1VA5,[149] at 2.0 Å). (d) Crystal structure of Ag85C (pale yellow) with the inhibitor diethyl *p*-nitrophenylphosphate (sticks) (PDB ID: 1DQY,[147] at 1.8 Å). (e) Crystal structure of Ag85C (pale yellow) with the allosteric inhibitor ebselen (sticks) (PDB ID: 4QDU,[25] at 1.4 Å).

thus far: (i) targeting Ser124 and (ii) mimicking TMM, was reported (Figure 7.14).[161] Unlike previous generations of phosphonate inhibitors that were promiscuous with many off-target effects, the fluorophosphonate trehalose inhibitor displayed strong selectivity towards the Ag85 complex. Mass spectrometry analysis revealed no detectable serine modifications against other prokaryotic and eukaryotic serine proteases tested.

Mimicking another substrate of the Ag85 complex, arabinogalactan, was hypothesized to be an alternative viable strategy to generate inhibitors of this complex.[162] However, arabinogalactan-based inhibitors were found to display weak inhibitory activity against Ag85C.[163,164] Previously, the 2-amino-4,5,6,7-tetrahydrobenzo[*b*]-thiophene-3-carbonnitrile scaffold was identified to inhibit Ag85C based on NMR spectroscopy-based fragment screening.[165,166] The thiophene-based scaffold was hypothesized to occupy the hydrophobic mycolyl pocket of Ag85C. In light of these reports, an arabinogalactan inhibitor that is conjugated with the thiophene scaffold was created; however, this arabinose-thiophene conjugate inhibitor still did not inhibit the growth

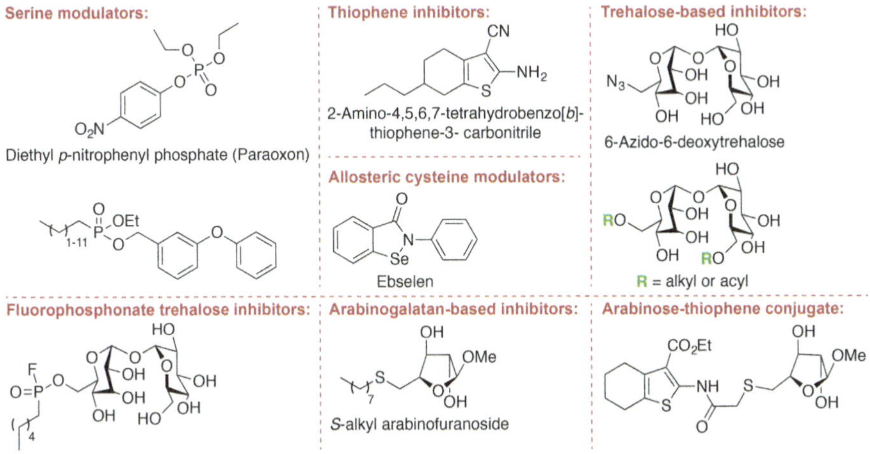

Figure 7.14 Structures of inhibitors of the Ag85 complex categorized by class.

of *M. smegmatis*.[156] Overall, attempts to mimic the arabinogalacatan have not been as fruitful as those of TMM.

Moving away from the active site, researchers began a search for allosteric modulators of the Ag85 complex. Recently, ebselen, also known as PZ-51, was discovered to be a nanomolar inhibitor of Ag85C by covalently binding to the conserved cysteine residue (Cys209) *via* the formation of a seleno-sulfide bond (Figure 7.13E).[25,167] This covalent modification forces a conformational change leading to disruption of the hydrogen-bond network within the catalytic triad and enzyme inactivation. Additionally, ebselen and a library of ebsulfur analogues were shown to display antibacterial properties against *M. smegmatis*, which provides further evidence that inhibition of the Ag85 complex is an effective strategy for anti-tubercular drug development.[168] What truly makes ebselen a remarkable scaffold is that it already surpassed safety evaluations in phase I (USA) and phase III (Japan) clinical trials.[169–171] As the pharmacokinetics of ebselen are also well known, the transition from bench to bedside could be expedited. However, there are concerns that ebselen could be too reactive due to its Se–N bond, which potentially could contribute to off-target binding.[172]

Overall, it can be concluded that mycolic acid biosynthesis and processing present many great potential drug targets. In addition to the enzymatic targets that have been extensively discussed above, other key enzymes have also been evaluated, such as the fatty acid degradation protein D32 (FadD32) and the condensing enzyme polyketide synthase (Pks13) (Figure 7.9).[173–176] The trehalose biosynthetic enzymes have also been postulated as potential targets for anti-tubercular drug development. Trehalose is an α-linked disaccharide comprised of two glucose monomers with many important biological functions in *Mtb* pathophysiology, such as virulence promotion and protection against environmental stress.[177–179] Hence, the trehalose biosynthetic pathway potentially contains valuable drug targets since trehalose biology is absent in humans.[177] There are three pathways for trehalose biosynthesis in *Mtb*: the OtsA-OtsB pathway, the TreS pathway, and the TreY-TreZ pathway.[178] At first glance, targeting the biosynthesis of trehalose seems like a challenging feat because of the existences of three independent biosynthetic routes potentially capable of compensating for each other. However, mutagenesis and biochemical studies suggested that the OtsA-OtsB pathway is the predominant route.[180] Additionally, *otsA* gene deletion resulted in a decrease in *Mtb* virulence in mice.[181] However, inhibitors have yet to be identified for the FadD32, Pks13, and enzymes in the three trehalose biosynthetic pathways.

7.2.1.2 Arabinogalactan Biosynthesis

In addition to the arabinosyltransferase EMB target, recent research has extensively explored the mycobacterial decaprenylphosphoryl-β-D-ribofuranose-2′-epimerase (DprE), encoded by *dprE1* and *dprE2*, as a potential anti-tubercular target.[182] DprE1 and DprE2 catalyze a sequential series of redox reactions. DprE1 converts decaprenylphosphoryl-β-D-ribofuranose (DPR) to decaprenylphosphoryl-2-keto-β-D-*erythro*-pentofuranose (DPX), which is further reduced by DprE2 to decaprenylphosphoryl-β-D-arabinose (DPA) (Figure 7.15).[183] As DPA is the sole precursor for the biosynthesis of arabinan, which is critical for cell envelope integrity, inhibition of DprE1 and/or DprE2 represents a valid method for development of novel anti-tubercular agents.[184,185]

Compounds discovered to target DprE1 include benzothiazinone derivatives, such as BTZ043, which inhibits *Mtb* growth at 1 ng mL^{-1} (Figure 7.15).[186] A crystal structure of *Msm*_DprE1 in complex with BTZ043 revealed that this compound forms a covalent semimercaptal linkage with the active site cysteine (Cys387 in *Mtb*, Cys394 in *M. smegmatis*) of DprE1 (Figure 7.16). Lys425

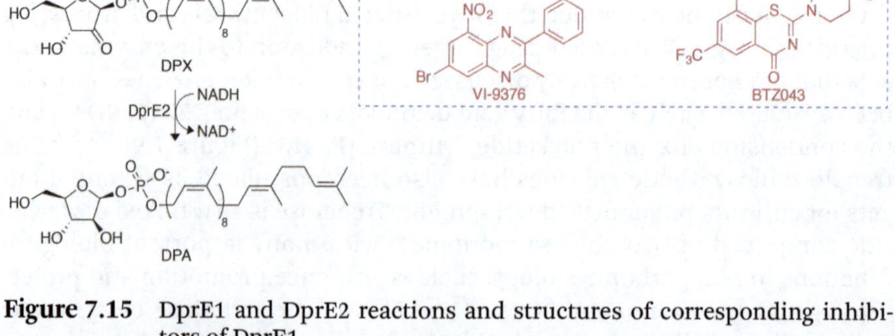

Figure 7.15 DprE1 and DprE2 reactions and structures of corresponding inhibitors of DprE1.

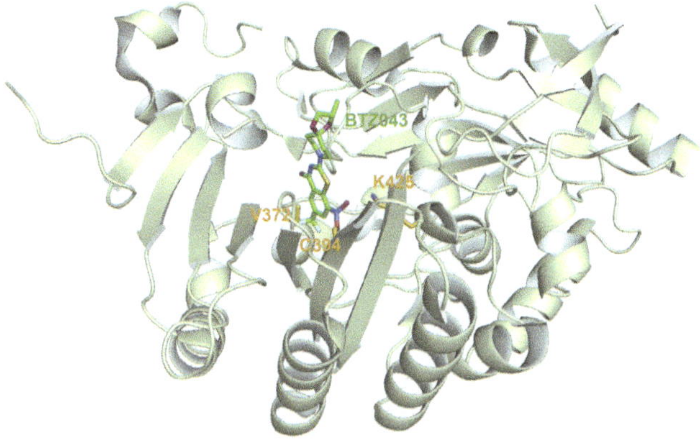

Figure 7.16 Crystal structure of *Msm*_DprE1 (pale yellow) (PDB ID: 4F4Q,[182] at 2.6 Å) co-crystallized with the inhibitor BTZ043 (green sticks with oxygen, nitrogen, sulfur, and fluorine atoms colored red, blue, orange, and turquoise, respectively) in the active site. The covalent semimercaptal linkage between BTZ043 and the catalytic Cys394 and other important residues, including Lys425 and Val372, are highlighted as orange sticks (with oxygens, nitrogen, and sulfur atoms colored).

and Val372 are also important for forming hydrogen bonds and hydrophobic interactions between BTZ043 and the enzyme.[182] Dinitrobenzamides (*e.g.* DNB1), a class of anti-tubercular compounds structurally related to the benzothiazinone family, and nitrobenzoquinoxaline (*e.g.* VI-9376) also inhibit DprE1 by the same mechanism, but with slightly lower efficiencies (Figure 7.15).[187,188] A nitrotriazole (*e.g.* 377790) was identified as a DprE1 inhibitor in a whole cell-based HTS, with an IC_{90} value of 0.5 μM against *Mtb*_DprE1.[140,144]

In addition to DprE1 and DprE2, other enzymes involved in the biosynthesis of arabinogalactan are also essential for the viability of *Mtb.* These targets include the arabinofuranosyltransferase AftA,[185,189] the decaprenylphosphate-5-phosphoribosyltransferase (*Rv3806c*),[190] a UDP-galactofuranosyltransferase,[191] dTDP-deoxyhexulose reductase,[192] and RmlA-D.[193] Although these enzymes do not have any validated inhibitors, they could provide insights in the discovery of novel drug targets in future anti-tubercular research endeavors.

7.2.1.3 Peptidoglycan Biosynthesis and Maintenance

The inner most layer of the mAGP complex is the peptidoglycan layer. When *Mtb* transitions from an aerobic replicating phase to a non-replicating phase, the bacterium undergoes adaptive remodeling of the peptidoglycan network by substituting the predominant 4 → 3 cross-linkages (catalyzed by D,D-transpeptidase) with 3 → 3 cross-linkages (catalyzed by L,D-transpeptidase 2, Ldt_{mt2}, Figure 7.17).[194] This further contributes to the extensive resistance to β-lactam antibiotics in persistent *Mtb*, as the D,D-transpeptidase is the target of most β-lactam antibiotics. The reconstruction of the peptidoglycan layer

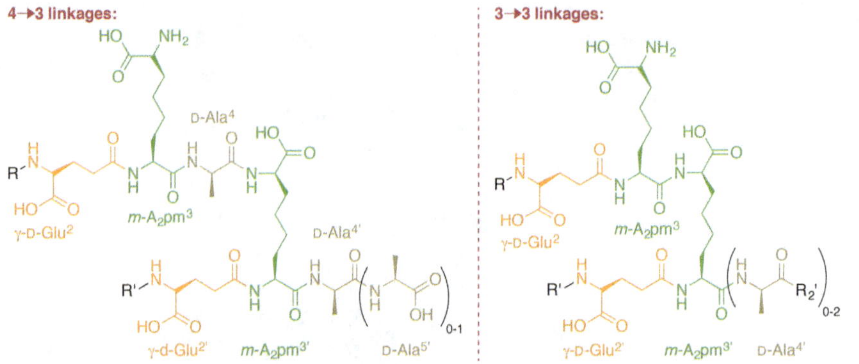

Figure 7.17 The various forms of cross-linkages between the peptide chains in the *Mtb* peptidoglycan layer. Left panel: the 4 → 3 cross-linkages that are predominant in aerobic replicating *Mtb* as well as other non-mycobacterial species. Right panel: the 3 → 3 cross-linkages that are predominant in non-replicating *Mtb*. Abbreviation: *m*-A_2pm = *meso*-diaminopimelic acid.

in *Mtb* depends on Ldt_{mt2}, which is upregulated in persistent *Mtb*.[195] The special 3 → 3 linkages are frequently associated with virulence and β-lactam resistance, not only in *Mtb*, but also in other species, such as *E. faecium*[196] and *Mycobacterium abscessus*.[197] Loss of this protein leads to increased susceptibility to the AMX-CLV combination during chronic infection.[195] A crystal structure of the extramembrane portion of Ldt_{Mt2} has been determined with the tautomerized form of meropenem (MEM) forming a covalent adduct with Cys354 (Figure 7.18(a)).[198–200] In order to identify inhibitors of Ldt_{mt2}, an *in silico* structure-based modeling study of 4.5 million compounds led to the discovery of one sulfonamide/triazole-based molecule with limited H37Ra growth inhibition potential (Figure 7.18(b)).[201]

Along with Ldt_{mt2}, peptidoglycan amidases have been proposed to play important roles in bacterial surface growth, cell division, adhesion of host cells, and bacterial pathogenesis.[202–204] Biochemical and structural analyses suggested that the *N*-acetylmuramyl-L-alanine amidase Rv3717 is important for peptidoglycan fragment recycling.[205] Heterologous expression of another peptidoglycan amidase (*Mtb*_Rv0024) in *M. smegmatis* led to significant increases in biofilm formation, cell wall hydrophobicity, and adherence to lung epithelial cells, which resulted in an increase in resistance against INH and PZN.[206] Although there are no inhibitors reported, these results suggest that these unexplored peptidoglycan amidases could be useful drug targets.

7.2.2 Targets Involved in *Mycobacterium tuberculosis* General Metabolism

While the featured *Mtb* cell envelope is an excellent source of anti-tubercular drug targets, researchers have also identified various weak links in the general metabolism of *Mtb*. Growing evidence suggests that the key to shorten

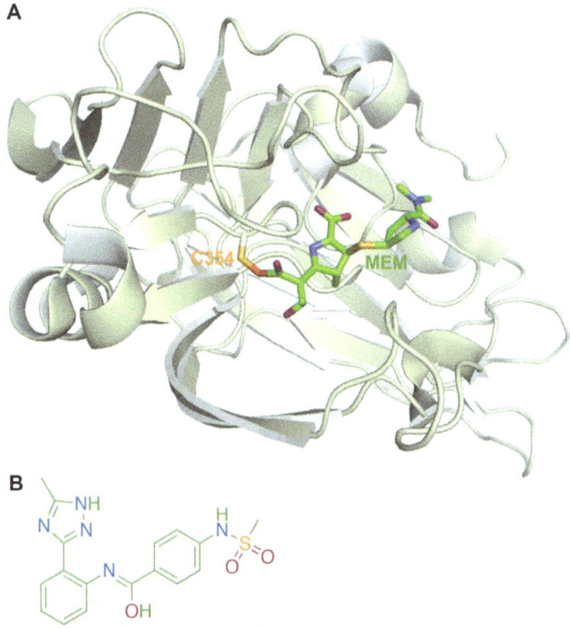

Figure 7.18 (a) Crystal structure of the extramembrane portion of Ldt_{Mt2} (pale yellow) (PDB ID: 3VYP,[305] at 1.4 Å) with the active site Cys354 (orange sticks) forming a covalent adduct with the inhibitor MEM (in tautomerized form, green sticks with oxygen, nitrogen, and sulfur atoms colored red, blue, and orange, respectively). (b) Structure of the Ldt_{Mt2} inhibitor discussed.

the lengthy treatment regimen for TB patients is to effectively eliminate the subpopulation of bacilli that survive after phagocytosis.[29] These determined bacilli endure the oxygen and nutrient-depleted environment inside the phagosome vacuoles by forfeiting/reducing cellular functions, exiting their cell cycle, and transitioning into a non-replicating phase.

In order to study these non-replicating bacilli, *in vitro* models that generate quiescent organisms have been developed to elucidate how *Mtb* adapts to (i) an oxygen-depleted and nutrient-rich environment (Wayne model)[207] or (ii) an oxygen-rich and nutrient-deprived environment (Loebel model).[208] In the Wayne model, *Mtb* is placed in a nutrient-rich, sealed container, which leads to gradual oxygen depletion. When the oxygen level drops to 0.06%, *Mtb* shifts from an aerobic replicating phase to a well-defined, anaerobic, non-replicating phase, and becomes tolerant to many anti-tubercular agents.[209] In fact, comparison showed that upon transitioning into a non-replicating phase, RIF and MOX became 50 times less bactericidal against non-replicating *Mtb* than against exponentially replicating *Mtb*, whereas INH and EMB became totally inactive.[210] These non-replicating bacilli's adaptive survival strategies include, but are not limited to, down-regulating key complexes in the electron transport chain[26,30,31,210,211] and important pathways for essential cofactor biosynthesis and regulation. The bacilli heavily control these sensitive processes in order to conserve energy to endure harsh conditions, which simultaneously provided inspirations for the discovery of new anti-tubercular drug targets.

The potential targets from the general metabolism of *Mtb* include enzymes that affect (i) ATP homeostasis, (ii) amino acid biosynthesis, (iii) cofactor biosynthesis, (iv) DNA synthesis, and (v) other non-centralized pathways.

7.2.2.1 ATP Homeostasis

Studies revealed that under hypoxic conditions the ATP levels in *Mtb* are five to six times lower than in its replicating phase.[210] While this lower level of ATP is carefully maintained in the non-replicating bacilli, it also sensitizes them to further fluctuations in ATP concentrations. Research has verified that the *de novo* ATP synthesis and a fully energized cytoplasmic membrane are essential for the viability of non-replicating mycobacteria under hypoxic conditions. The proton gradient-dependent F_0F_1 ATP synthase is the last complex in the electron transport chain that synthesizes ATP by using ADP and inorganic phosphate (P_i) when additional ATP is in demand, or hydrolyzes ATP, when in excess, to restore the proton gradient across the plasma membrane.

Bedaquiline (also named R207910 or TMC207, Figure 7.19) is a diarylquinoline that targets the F_0F_1 ATP synthase in *Mtb*, and shows bactericidal effect in both drug-sensitive and drug-resistant bacilli.[212] It further reduces the ATP level in the non-replicating bacilli in a dose-dependent manner, which decreases their viability under hypoxic and starvation conditions, leading to cell death.[213,214] With an MIC value of 0.03 μg mL^{-1} against *Mtb* H37Rv and an MIC range of 0.01–0.06 μg mL^{-1} against various drug-resistant strains, bedaquiline exceeded the efficacy of RIF and INH by at least 10 times both *in vitro* and *in vivo*. Studies in murine models suggested that replacing a first-line anti-tubercular agent with bedaquiline resulted in culture conversion in two months of treatment.[212] Single doses of bedaquiline at 50 or 100 mg kg^{-1} resulted in extended bacteriostatic or bactericidal effects, respectively, for up to eight days. The extended half-life and potent anti-tubercular activity suggested the potential for a less frequent anti-tubercular dosing regimen.[212] Studies have shown that adding

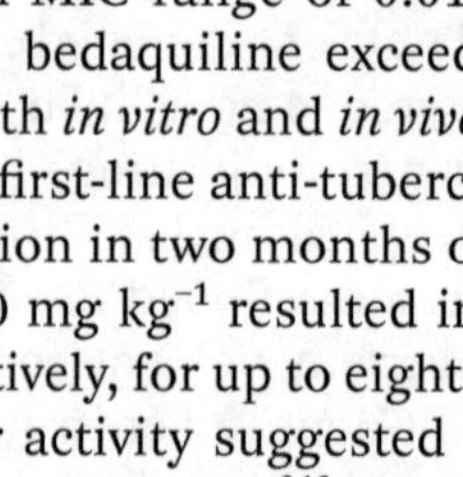

Bedaquiline

Figure 7.19 Structure of bedaquiline, an *Mtb*_F_0F_1 ATP synthase inhibitor.

bedaquiline to a standard anti-tubercular therapy for MDR-TB increased the number of patients with conversion of sputum culture and reduced the time for culture conversion, with most adverse effects observed being mild to moderate reactions such as nausea.[215] A combinatorial approach of bedaquiline and the MmpL3 inhibitor SQ109 was shown to synergistically kill *Mtb*.[216] The ability of SQ109 to weaken the *Mtb* cell envelope was postulated to allow for an increased intracellular bedaquiline concentration in order to act on ATP synthase.

Besides bedaquiline, recent research has also explored other chemical/biochemical entities as potential F_0F_1 ATP synthase inhibitors. Maganins are antimicrobial peptides (AMPs) that have been investigated for anti-tubercular purposes due to their low mammalian cytotoxicity. A maganin-I analogue peptide (MIAP, amino acid sequence GIGKFLKSKGKFGKA; MIC = 0.3 mg mL^{-1}) was found to be three times more potent than maganin-I against *Mtb* H37Ra.[217] Unlike bedaquiline and MIAP that have a clear cellular target, three classes of compounds, including benzimidazoles, thiophenes, and imidazopyridines, were identified in a HTS to impact mycobacterial ATP homeostasis. About 800 out of 600 000 compounds screened were confirmed to reduce intracellular ATP levels in a dose-dependent manner in *M. bovis* BCG, but their exact enzymatic target(s) remain(s) to be validated.[218]

7.2.2.2 Targets in Amino Acid Biosynthetic Pathways

In addition to targeting the bacterial cell wall, an attractive strategy in antimicrobial drug discovery is to target enzymes that are specific and unique to the pathogens and not present in mammals. Since *Mtb* can generate all of its own amino acids, attractive targets include enzymes involved in amino acid biosynthetic pathways that humans do not have. For humans these include lysine, leucine, isoleucine, valine, threonine, tryptophan, and histidine. The pathways responsible for the production of some of these amino acids or their precursors in *Mtb* include the (i) shikimate pathway, (ii) branched-chain amino acid pathway, and (iii) the lysine biosynthetic pathway.

7.2.2.2(a) Targets in the Shikimate Pathway. The shikimate pathway encompasses seven steps leading to chorismate, a precursor of folic acid, ubiquinone, aromatic amino acids, and other aromatic products (Figure 7.20).[219] It is present in microorganisms and plants,[220] but absent in mammals as they need to acquire tryptophan from their diet. The pathway was first elucidated and proved essential in bacteria by mutagenesis of essential enzymes.[221] Given the dire need for novel anti-tubercular agents, many groups are investigating various compounds capable of disrupting the shikimate pathway in *Mtb*.

3-Deoxy-D-arabino-heptulosonate 7-phosphate (DAH7P) synthase catalyzes the first committed step in the shikimate pathway.[222] Based on the understanding of the mechanism of the 3-deoxy-D-manno-octulosonate 8-phosphate synthase, a structurally and evolutionarily related enzyme,

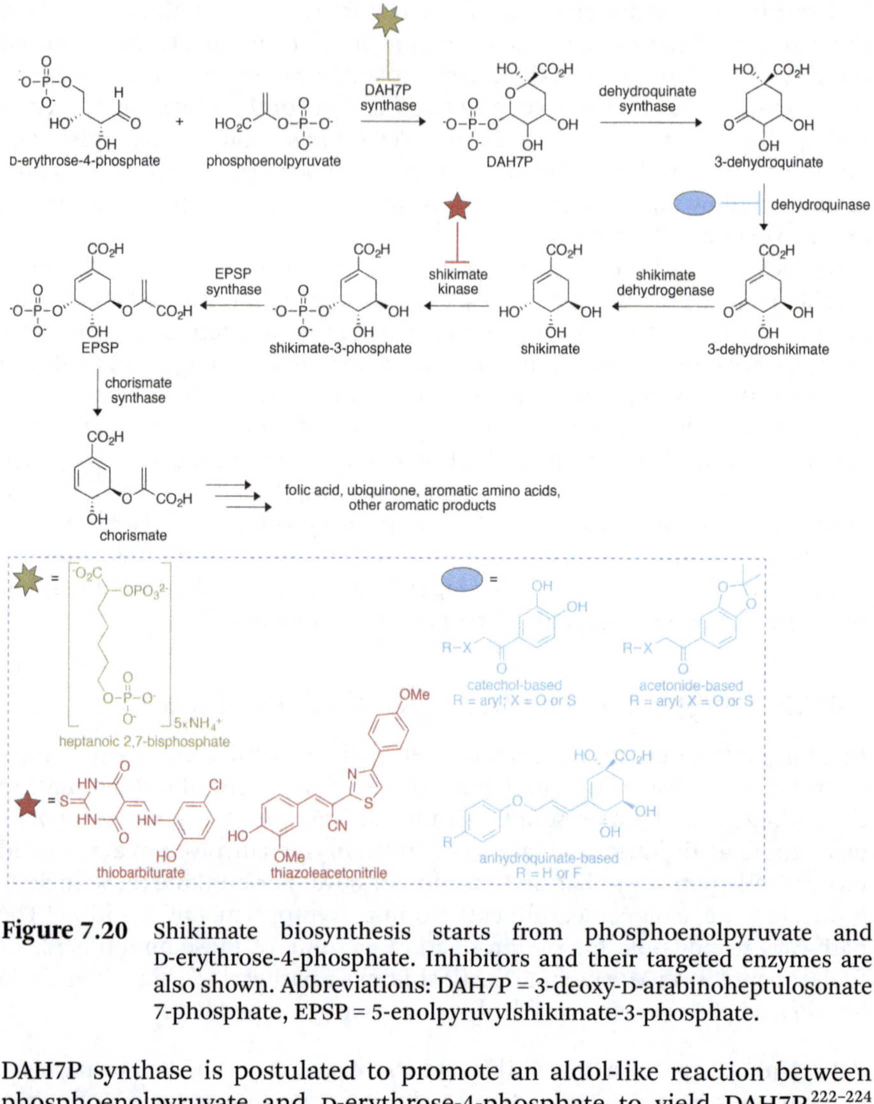

Figure 7.20 Shikimate biosynthesis starts from phosphoenolpyruvate and D-erythrose-4-phosphate. Inhibitors and their targeted enzymes are also shown. Abbreviations: DAH7P = 3-deoxy-D-arabinoheptulosonate 7-phosphate, EPSP = 5-enolpyruvylshikimate-3-phosphate.

DAH7P synthase is postulated to promote an aldol-like reaction between phosphoenolpyruvate and D-erythrose-4-phosphate to yield DAH7P.[222–224] Novel DAH7P synthase inhibitors were designed to mimic the intermediate of this enzymatic step.[222] Computational modeling suggested that heptanoic 2,7-bisphosphate would be a simplified analogue of an intermediate on the way to DAH7P and an efficient inhibitor of the enzyme given the arrangement of the binding pocket. The racemic mixture of heptanoic 2,7-bisphosphate was then synthesized and evaluated *in vitro* with DAH7P synthase from *Mtb*. Co-crystallization of a DAH7P synthase-heptanoic 2,7-bisphosphate complex confirmed binding of the inhibitor to the enzyme and suggested that the (*S*)-enantiomer of the bisphosphate displays slightly lower binding affinity due to possible steric hindrance with an active water molecule. However, the negatively-charged nature of heptanoic 2,7-bisphosphate causes difficulty for membrane penetration of this compound. Nevertheless, these findings offer great opportunities for researchers to optimize the bisphosphate scaffold, maybe its development as prodrugs, in the future.

Dehydroquinase is the enzyme responsible for the essential step of converting 3-dehydroquinate to 3-dehydroshikimate (Figure 7.20).[225] A number of potent dehydroquinase inhibitors containing catechol or acetonide scaffolds were reported with K_i in the low micromolar range.[226] Ten potent anhydroquinate inhibitors with K_i in the nanomolar range were also identified *via* docking studies.[227]

A third enzyme in the shikimate pathway that has received attention is shikimate kinase (SK), which generates shikimate-3-phosphate (Figure 7.20).[228] SK was shown to be essential in *Mtb* by the disruption of *aroK*, which encodes this kinase.[229] From a HTS of 1000 antibacterial compounds, five molecules were found to be ATP-competitive inhibitors of SK with IC_{50} values of 5.1–10 μM.[230] Out of these five SK inhibitors, a thiobarbiturate and a thiazoleacetonitrile were found to be active against a drug-sensitive *Mtb* H37Rv strain (MIC = 4 μg mL^{-1}) and a MDR-*Mtb* clinical isolate (MIC = 1 μg mL^{-1}), respectively (Figure 7.20). Unfortunately, the thiazoleacetonitrile displayed toxicity against HepG2, likely as a result of competitive binding with ATP, indicating that development of future SK inhibitors should focus on the shikimate binding pocket of an allosteric binding site.

7.2.2.2(b) Targets in the Branched-Chain Amino Acid Pathway. A second amino acid biosynthetic pathway to exploit is the branched-chain amino acid (BCAA) biosynthetic pathway. Humans do not have the BCAA biosynthetic pathway and solely acquire these amino acids *via* food consumption. All enzymes in the BCAA pathway were shown to be essential for the survival of *Mtb*.[231] Furthermore, by using penoxsulam, an effective general biocide with low mammalian toxicity, the enzymes of the BCAA biosynthetic pathway was validated as potential antibacterial targets.[232] The ketol-acid reductoisomerase (KARI) is a bifunctional enzyme, which first catalyzes isomerization of 2-acetolactate or 2-aceto-2-hydroxy-butyrate into their corresponding intermediates, then catalyzes the reduction of these intermediates to the 2,3-dihydroxy-3-alkyl-butyrate products (Figure 7.21). *N*-Hydroxy-*N*-isopropyloxamate (IpOHA) was first reported to be an inhibitor of KARI from spinach and later identified to also inhibit KARI from *Mtb*, but displayed virtually no activity when tested against *Mtb* cells.[233] The crystal structure of KARI from *Mtb* was recently determined and will likely facilitate future rational design of inhibitors of this enzyme.[234]

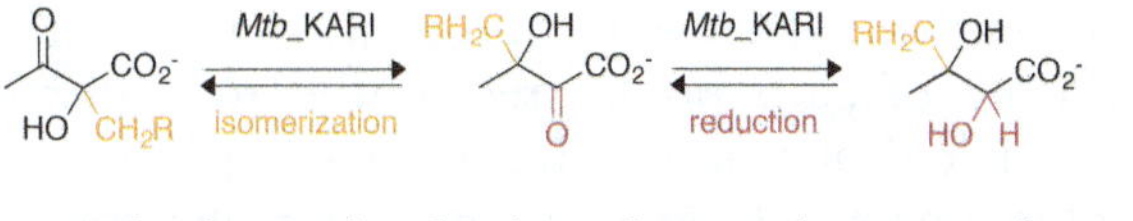

Figure 7.21 Isomerization and reduction of 2-acetolactate to the 2,3-dihydroxy-3-alkyl-butyrate product by the bifunctional keto-acid reductoisomerase (KARI) enzyme.

7.2.2.2(c) Targets in the Biosynthesis of Lysine and Its Precursors. A third amino acid biosynthetic pathway, which has been investigated and validated for development of anti-tubercular agents, results in the production of lysine from aspartic acid through the aspartic acid cycle (Figure 7.22). Aspartic acid is first phosphorylated by asparto kinase (AK) and then dehydrogenated by α-aspartate semialdehyde dehydrogenase (ASADH) to yield aspartate-4-semialdehyde. This metabolite is a branching point for the *de novo* synthesis of several amino acids. Condensation of the aldehyde with pyruvate and loss of two water molecules yields 2,3-dihydrodipicolinate. Using NADPH and a proton, 2,3-dihydrodipicolinate is converted to 2,3,4,5-tetrahydrodipicolinate. This reductive step is common to many bacteria[235] and its inhibition would prevent the synthesis of most lysine's precursors, including some that are used in cell envelope biosynthesis.[236] Through HTS and virtual screening, several compounds that inhibit the reductase were identified (Figure 7.22).[237] Without the formation of the tetrahydrodipicolinate, diaminopimelate, used in the peptidoglycan layer, and lysine, used in general protein synthesis, neither the cell envelope nor proteins can be generated, hitting the bacteria in two essential areas required for its survival. In addition to dihydrodipicolinate reductase, diaminopimelate decarboxylase, which catalyzes the penultimate decarboxylation of *meso*-2,6-diamino-heptanedionate to lysine, has been categorized as an essential protein for *Mtb* viability.[238]

On top of the three amino acid pathways discussed above, the histidine and aspartic acid pathways have also been proposed as new drug targets. Inhibition of ATP phosphoribosyltransferase, in the biosynthesis of histidine, catalyzes the first committed step in the pathway.[239] This enzyme is essential for the survival of *Mtb* and is governed by a feedback loop. When the cell has an adequate amount of histidine available, the amino acid allosterically inhibits the phosphoribosyltransferase. If an inhibitor was designed for this target, the source of histidine for *Mtb* would be essentially non-existent. The lack of histidine biosynthesis in humans also makes this an ideal target and/or pathway to target for new anti-tubercular compounds. As mentioned before, ASADH from the aspartic acid pathway, a key enzyme in the conversion of aspartic acid to other amino acids (*e.g.* threonine, lysine, methionine, and isoleucine), has shown promise in knockdown studies.[240] While no compounds have been found to inhibit this enzyme, yet, its essentiality to *Mtb* makes it a viable drug target.

7.2.2.3 Cofactors Biosynthesis

While inhibition of amino acid biosynthetic pathways severely limits the ability of *Mtb* to reproduce, there is always a chance that the bacilli could absorb these essential molecules from the environment. The inhibition of the biosynthesis of cofactors would prevent the production of many amino acids and other biological processes from occurring. Cofactor pathways that have been explored as new targets for anti-tubercular compounds include: (i) CoA, (ii) folate, and (iii) riboflavin (vitamin B_2) biosyntheses, as well as (iv) the redox of NADH.

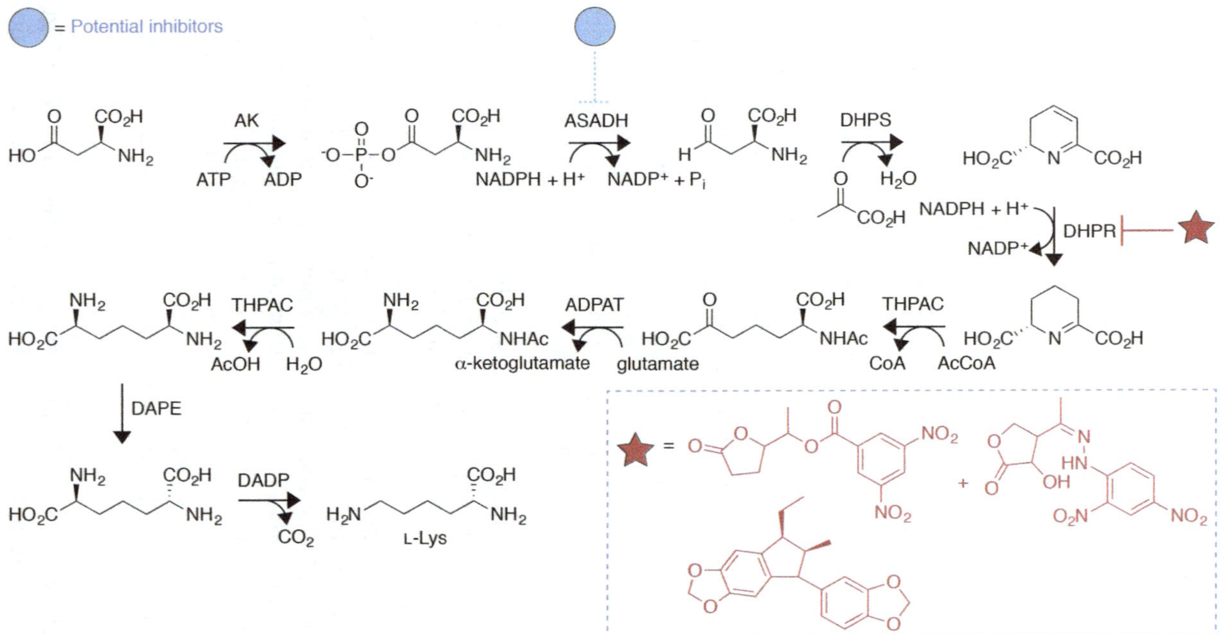

Figure 7.22 The *de novo* synthesis of lysine from L-aspartic acid. The key pathway that has been tested to be "druggable" is indicated with the solid line and the compound or mixture of compounds are noted with the corresponding symbols. Abbreviations: AK = asparto kinase, ASADH = α-aspartate semialdehyde dehydrogenase, DHPS = dihydrodipicolinate synthase, DHPR = dihydrodipicolinate reductase, THPAC = tetrahydrodipicolinate acetyltransferase, ADPAT, acetyldiaminopimelate aminotransferase, ADPA = acetyldiaminopimelate deacetylase, DAPE = diaminopimelate epimerase, DAPD = diaminopimelate decarboxylase.

7.2.2.3(a) CoA Biosynthetic Pathway. CoA is an essential cosubstrate for many biosynthetic pathways, including fatty acid biosynthesis, DNA regulation and transcription, and other useful metabolic pathways. The biosynthesis of CoA starts from aspartate, which is decarboxylated to yield β-alanine, and pantoate (biosynthesized over multiple steps from pyruvic acid). Pantothenate synthetase (PS) catalyzes the first coupling step in the formation of CoA, generating pantothenate from pantoate and β-alanine (Figure 7.23). PS was identified as an antibiotic target of *Mtb* by bioinformatics.[238] This enzyme is an ideal target since it has no homologues in mammalian cells. Studies of *Mtb* lacking the *panC* and *panD* genes were shown to not establish virulence in mice,[241] and this genetic variant is being considered as a possible source of a new vaccine.[242] Actinomycin D was originally identified as a PS inhibitor by HTS.[243] Two unrelated compounds with 10-fold increased PS inhibition were discovered by further *in silico* screening. These molecules were equally active against susceptible and MDR-*Mtb* strains. Analogues of the PS reaction intermediate were shown to bind the enzyme more tightly than the natural substrates.[244]

After the pantothenate is formed, the molecule is phosphorylated by another essential enzyme in the *de novo* generation of CoA, pantothenate kinase (PanK or CoaA), which has also been considered as a target for novel anti-tubercular compounds. Four classes of molecules were found to target CoaA: triazoles, thiazoles, quinoline carboxamides, and biaryl acetic acids.[245] All compounds inhibited CoaA with IC_{50} values ranging from 0.07 to 8.4 μM with various modes of inhibition. These inhibitors were further investigated for optimization with two additional libraries of small molecules, one library of 70 000 compounds focused on triazoles and quinolone scaffolds, known to competitively inhibit the binding of ATP, and the other library of 1 000 000 compounds focused on uncompetitive (thiazoles) and mixed competitive (biaryl acetic acids and quinolone carboxamides) inhibitors.[246] These molecules were even further optimized using SAR studies and co-crystallization with PanK (Figure 7.24A). The crystal structures showed that the triazole and biaryl compounds bind at the site of product formation.[247] While not completely killed, *Mtb* treated with these inhibitors matched the limited growth profile observed for a PanK-knockdown *Mtb* strain. Enzymes acting downstream in the CoA biosynthetic pathway, phosphopantetheine adenylyltransferase (PPAT) and dephospho coenzyme A kinase (DPCK) have also been examined as possible drug targets.[248] Both enzymes were found to be essential for the growth and survival of *Mtb* and no alternative pathways for CoA biogenesis were found.

7.2.2.3(b) Folate Biosynthesis. Folate is biosynthesized from GTP (Figure 7.25), which is converted to dihydroneopterin triphosphate with the removal of formate by GTP cyclohydrolase. The triphosphate is then removed by the action of a phosphatase releasing P_i, inorganic pyrophosphate (PP_i), and dihydroneopterin. This intermediate is degraded to 6-hydroxymethyl-dihydropterin and hydroxyacetaldehyde by dihydroneopterin. The pterin molecule

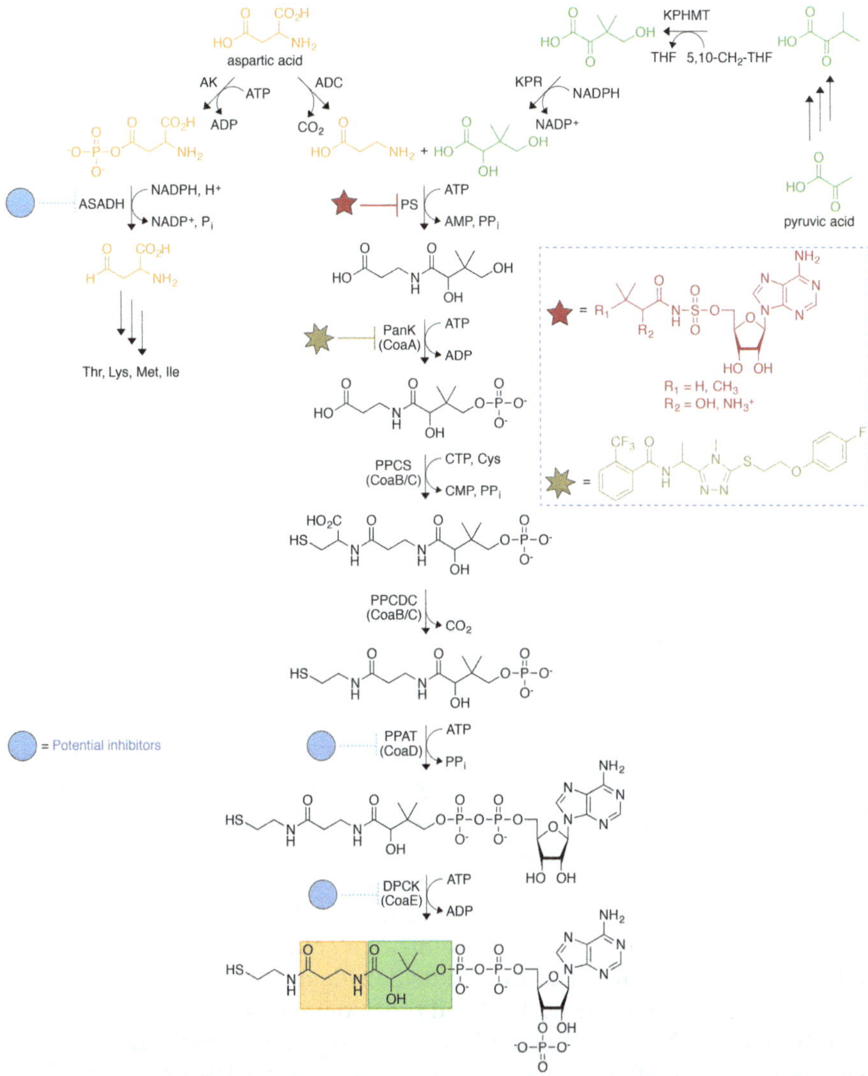

Figure 7.23 The *de novo* synthesis of CoA starting from aspartatic acid and pyruvic acid alongside the *de novo* synthesis of several amino acids originating from aspartic acid. Points in the pathways that have been tested or hypothesized to be "druggable" are indicated, with the solid line indicating compounds have been identified to inhibit the enzyme and dashed lines indicating the necessity of the enzyme and a potential drug target. Abbreviations: AK = asparto kinase, ASADH = α-aspartate semialdehyde dehydrogenase, ADC = aspartyl decarboxylase, KPHMT = ketopantoate hydroxymethyl transferase, KPR = ketopantoate reductase, PS = pantothenate synthase, PanK (CoaA) = pantothenate kinase, PPCS = phoshopantothenoylcysteine synthase, PPCDC = phosphopantothenoylcysteine decarboxylase, PPAT = phosphopantetheine adenylyltransferase, DPCK = dephospho coenzyme A kinase.

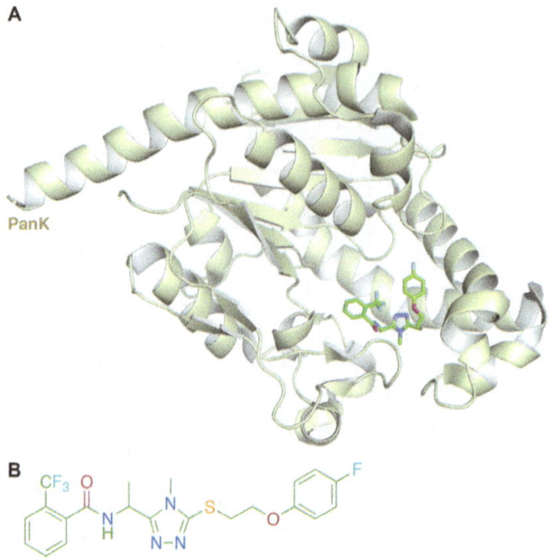

Figure 7.24 (a) The crystal structure of PanK (pale yellow) (PDB ID: 4BFS,[247] at 2.9 Å) and the novel inhibitor shown as sticks and colored by atom. (b) Structure of the novel inhibitor.

is then pyrophosphorylated at the expense of ATP by 6-hydroxymethyl-dihydropterinpyrophosphokinase, activating the alcohol *via* nucleophilic addition. At this point, a product from the shikimate pathway (Section 7.2.2.2(a)), *p*-aminobenzoic acid, is used as the nucleophile in an enzymatic substitution reaction catalyzed by dihydropteroate synthase. Dihydrofolate is then synthesized by dihydrofolate synthase using dihydropteroate, ATP, and glutamate. The production of tetrahydrofolate, a cofactor used in amino acid and nucleic acid metabolism and formaldehyde consumption, is catalyzed by dihydrofolate reductase (DHFR) and the eponymous dihydrofolate. Compounds that have been used as inhibitors for fungal and parasite DHFRs have become the focus of recent studies on mycobacterial DHFRs.[249] Through crystal structure analysis (Figure 7.26), it was found that all inhibitors of DHFR bind in the same pocket and the nitrogenous heterocycles are positioned in the same orientation. A small library of compounds was generated and displayed a wide range of activity against the *Mtb*_DHFR, expressed in *Saccharomyces cerevisiae*.[250] Selected compounds displayed IC_{50} values in a range of 1.6 to 6 μM. The binding sites of human and *Mtb*_DHFR are different enough that a promising and selective inhibitor is a possibility for a novel *Mtb* treatment.

7.2.2.3(c) Biosynthesis of Riboflavin (Vitamin B$_2$). The *de novo* synthesis of flavin requires one molecule of GTP and two molecules of ribulose-5-phosphate, and other cofactors (Figure 7.27). GTP is first converted to

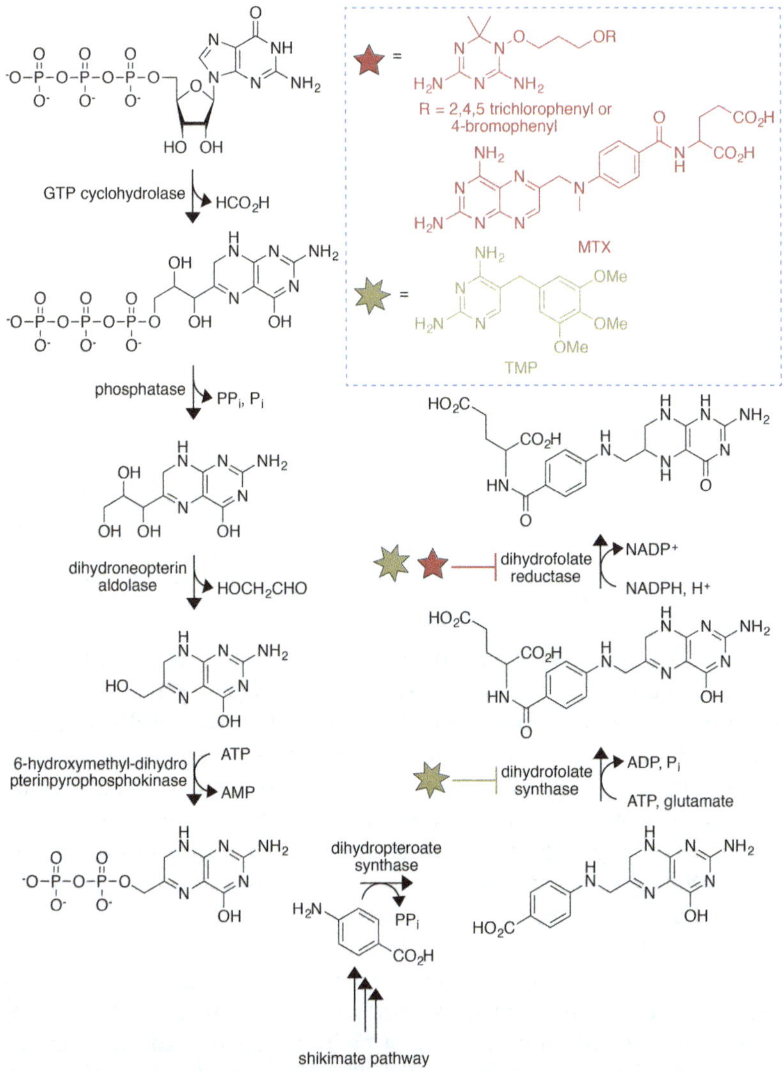

Figure 7.25 The *de novo* synthesis of tetrahydrofolate from GTP. The blocked transition that has been tested to be "druggable" is indicated with the solid line and the compounds are noted with the corresponding symbols.

2,5-diamino-6-ribosylamino-4-(3*H*)-pyrimidone-5′-phosphate by the product of the *ribBA* genes, releasing formate and pyrophosphate. The proteogenic product of *ribDG* then uses water and $NADP^+$ to generate 5-amino-6-ribitylamino-2,3-(1*H*,3*H*)-pyrimidinone-5′-phosphate releasing ammonia and NADPH. Meanwhile, the *ribBA* gene product removes a molecule of formate from ribulose-5-phosphate, creating 3,4-dihydroxy-2-butanone-4-phosphate. This product is then combined with 5-amino-6-ribitylamino-2,3-(1*H*,3*H*)-pyrimidinone-5′-phosphate by 6,7-dimethyl-8-ribityllumazine synthase to

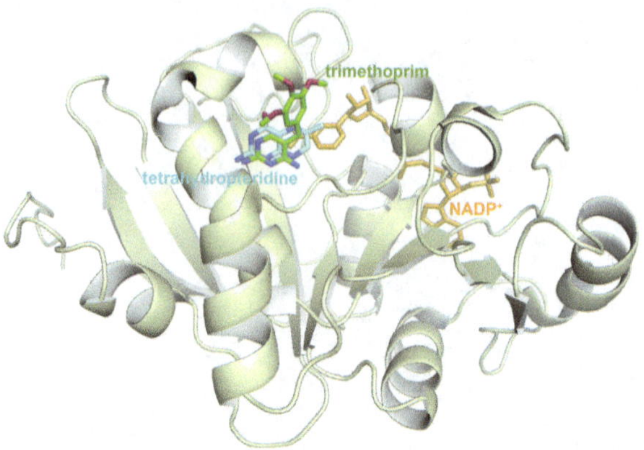

Figure 7.26 An overlay of the crystal structures of dihydrofolate reductase (pale yellow, *note*: only one structure of the enzyme is shown for simplicity) in complex with the novel inhibitor trimethoprim (green sticks with oxygen and nitrogen atoms in red and blue, respectively) (PDB ID: 4XT7,[306] at 2.3 Å), tetrahydropteridine (turquoise sticks with oxygen and nitrogen atoms in red and blue, respectively) (PDB ID: 4XT6,[306] and 1.9 Å), or $NADP^+$ (orange sticks) (PDB ID: 4XT5,[306] at 2.1 Å).

generate 6,7-dimethyl-8-ribityllumazine. Upon the catalytic addition of a second molecule of 3,4-dihydroxy-2-butanone-4-phosphate by riboflavin synthase, the lumazine compound becomes riboflavin and eventually the cofactors FMN and FAD. Inhibition of the lumazine synthase prevents the formation of the vitamin riboflavin and, ergo, the cofactors FMN and FAD. The lack of these molecules can have effects on the electron transport chain and numerous enzymatic transformations. Small molecules have recently been identified to inhibit the *Mtb* lumazine synthase (LS).[251] A new class of LS inhibitors, purinetriones (Figure 7.27 and 7.28B), showed excellent binding to the enzyme (Figure 7.28A). The active site of LS resides at the interface of two subunits of the enzyme pentameric assembly. The inhibitor-LS complex showed a total of twenty hydrogen bonds and two ionic interactions, including seven contacts with one subunit, four with the second subunit, and eleven contacts mediated by water molecules. The binding of the ribityl moiety of the purinetrione is very similar to that of the lumazine molecule, and the phosphate binds in the place of the 3,4-dihydroxy-2-butanone-4-phosphate. The inhibition constants of the inhibitors were 100-fold greater for *Bacillus subtilis* LS than for *Mtb*_LS, showing good selectivity.

7.2.2.3(d) Redox of NADH. The reduction and oxidation of NADH is important for respiration, particularly for *Mtb* in a non-replicating state. The type II NADH-dehydrogenase (NDH-2) enzyme is essential for the survival of *Mtb* and has no mammalian homologues, making it an excellent target for novel drug development.[252] Various studies have identified several inhibitors of NDH-2 (Figure 7.29). Initial studies found that chlorpromazine and

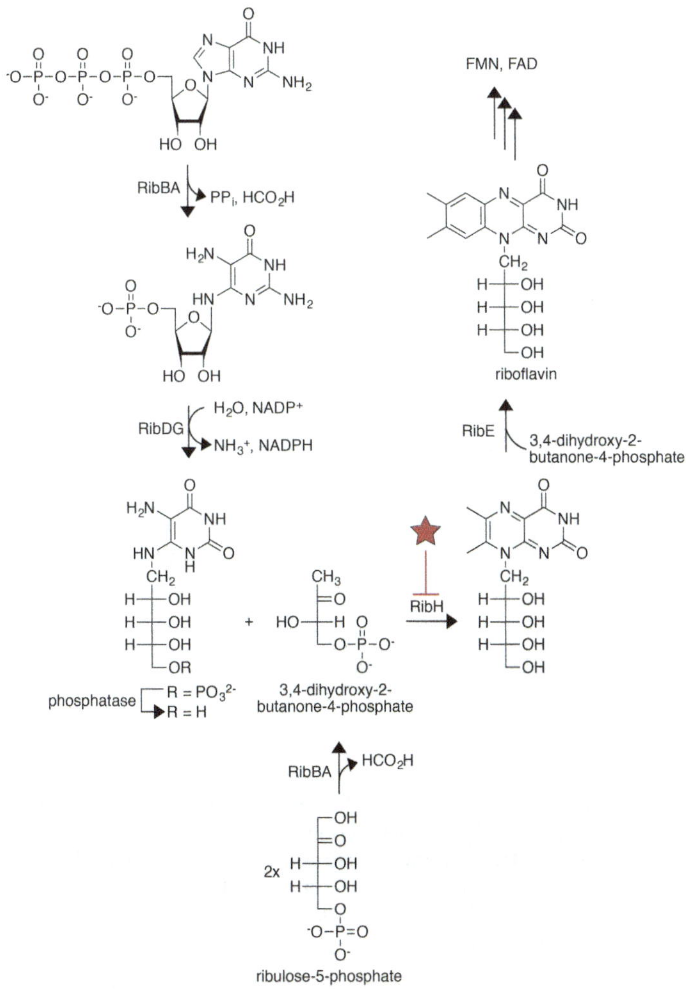

Figure 7.27 The *de novo* synthesis of riboflavin from GTP. The targeted pathway that has been tested to be "druggable" is indicated with the solid line and the compounds are noted with the corresponding symbols shown in Figure 7.28.

trifluoperazine were excellent inhibitors of this enzyme, shutting down electron transport.[253] This scaffold was then subjected to SAR analysis and additional molecules were found to inhibit the enzyme. The model inhibitor, trifluoperazine, even prevented the metabolism of vitamin K_1.[254] For this scaffold, only enzymatic data have been generated and it appears that optimization is needed for development into a therapeutic compound. A second scaffold (without an identifier in Figure 7.29) was also found and subjected to SAR analysis.[252] This scaffold showed activity against *Mtb* cells with best compounds having MIC values in the range of 1–10 μM.

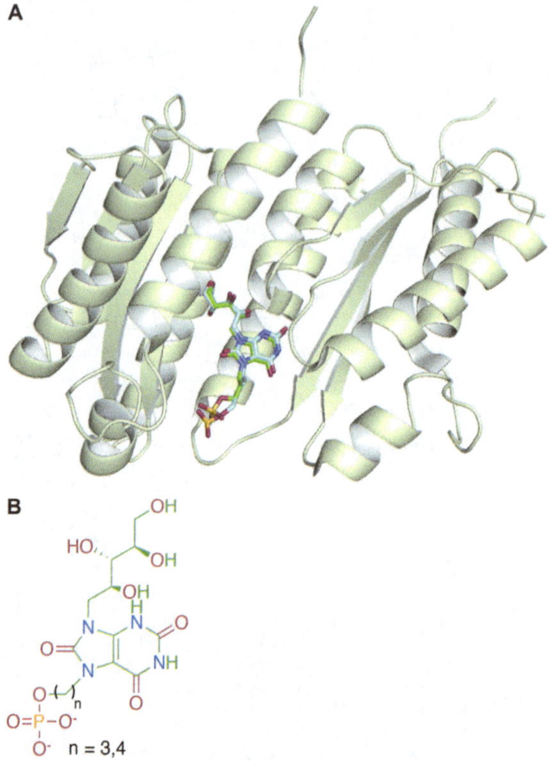

Figure 7.28 (a) The crystal structure of lumazine synthase (pale yellow) in complex with the novel inhibitors shown as green and turquoise sticks (with oxygen, nitrogen, and phosphorous atoms in red, blue, and orange, respectively) (PDB ID: 1W29 and 1W19,[251], at 2.3 Å and 2.0 Å). (b) Chemical structures of the compounds bound to lumazine synthase shown in panel (a).

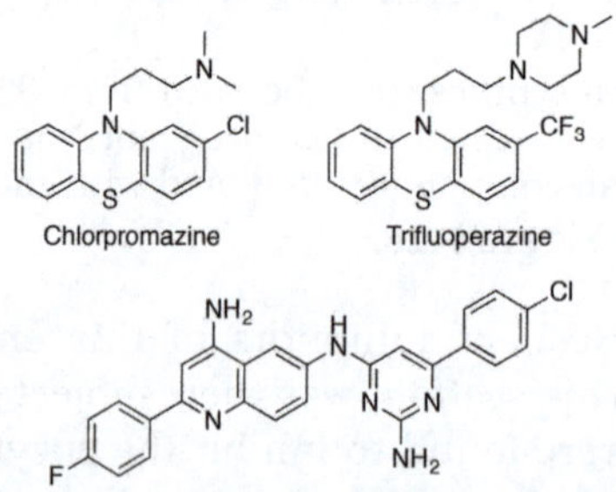

Figure 7.29 Structures of the inhibitors found to be active against NDH-2.

7.2.2.4 DNA Synthesis

So far, we have discussed methods that prevent the synthesis of specific molecules by targeting biosynthetic enzymes in *Mtb*. Precluding these proteins and enzymes from ever being expressed also represent a viable approach for drug development. There are several current anti-tubercular agents that target some point of the DNA-replication, translation, or transcription processes. In this section, a few new strategies that avert the DNA from being maintained, replicated, translated, or transcribed are presented. Recent work has focused on inhibition of (i) deoxythymidine triphosphate (dTTP) generation, (ii) DNA primase (DnaG), and (iii) DNA ligase.

7.2.2.4(a) Deoxythymidine Triphosphate (dTTP) Generation. Thymidylate kinase (TMPK) catalyzes the transformation of deoxythymine monophosphate (dTMP) to the corresponding diphosphate (dTDP) using one molecule of ATP. TMPK is the hub for both the *de novo* synthesis and salvage pathways of deoxythymidine 5′-triphosphate (dTTP) and is the last specific enzyme for both pathways. The *Mtb*_TMPK is particularly interesting as a novel drug target because it has a unique mechanism when compared to other members of the TMPK family, mainly due to its use of a magnesium ion coordinating the dTMP molecule in the active site. Recent studies have identified the phosphorylated form of the anti-HIV drug 3′-azidodeoxythymidine (AZT) (Figure 7.30, top panel) as an inhibitor of *Mtb*_TMPK.[255] This compound showed competitive inhibition with dTMP with a K_i value of ~10 μM. The crystallographic analysis of TMPK in complex with the phosphorylated

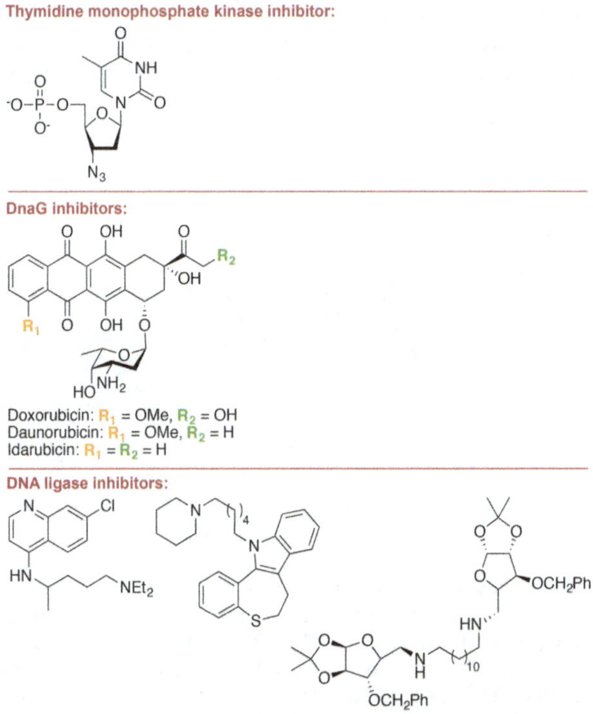

Figure 7.30 The structures of the inhibitors found for thymidylate kinase (top panel), DNA primase (middle panel), and DNA ligase (bottom panel).

AZT demonstrated that it is the azide group that inhibits the phosphorylase (Figure 7.31) by preventing the magnesium ion from binding to the active site of TMPK.

7.2.2.4(b) DNA Primase (DnaG). The bacterial primase DnaG is responsible for synthesizing small RNA oligonucleotides and is essential for chromosomal DNA replication and cell division. Furthermore, the bacterial primase is distinct to that found in humans. Recently, inhibitors of DnaG have been found by using HTS.[256,257] Among the compounds was the anticancer agent doxorubicin (Figure 7.30, middle panel), which showed inhibitory activity against DnaG from *Bacillus anthracis* and *Mtb*. Additional doxorubicin analogues, daunorubicin and idarubicin, were also found to inhibit DnaG.[258] All three compounds showed excellent activity against the purified enzyme and in *M. smegmatis* whole-cell assays. However, idarubicin had a higher MIC value against *Mtb* than doxorubicin or daunorubicin. These studies point to the promise of DnaG and other DNA replication enzymes as novel drug targets.

7.2.2.4(c) DNA Ligase. DNA ligase is a highly modular protein that has distinct domains and architecture, and is either ATP or NAD^+ dependent. Human DNA ligase is ATP dependent, while that of *Mtb* is dependent on NAD^+. A virtual screening library was used to identify inhibitors of the NAD^+-dependent DNA ligase that bind competitively to the cofactor site (Figure 7.30, bottom panel).[259,260] Among these, a series of dimeric furanosyl derivatives was found to show activity against DNA ligase from *Mtb* with IC_{50} values in

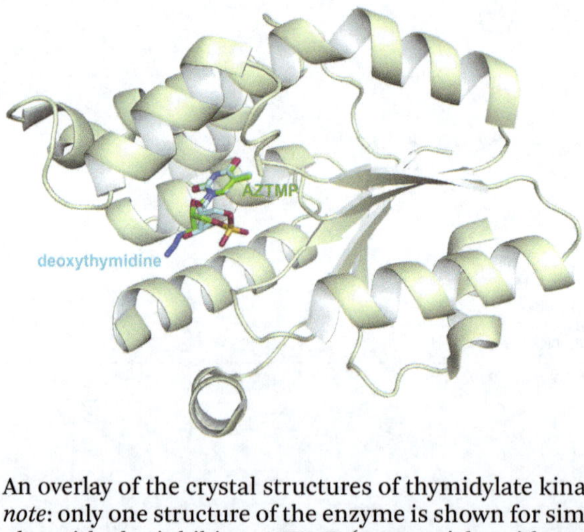

Figure 7.31 An overlay of the crystal structures of thymidylate kinase (pale yellow, *note*: only one structure of the enzyme is shown for simplicity) in complex with the inhibitor AZTMP (green sticks with oxygen, nitrogen, and phosphorous atoms in red, blue, and orange, respectively) (PDB ID: 1W2H,[255] at 2.0 Å) or deoxythymidine (turquoise sticks with oxygen and nitrogen atoms in red and blue, respectively) (PDB ID: 1W2G,[255] at 2.1 Å).

the micromolar range, while displaying no activity against the human ATP-dependent ligase. Some of these compounds also showed good activity in *in vivo* assays.

7.2.2.5 *Other Pathways*

The biosynthesis of molecules essential to the structure and function of *Mtb* is a broad and well-studied area. However, *Mtb* is highly complex for a bacterium and has many processes not directly involved in production of its biomolecules. These pathways have been the topics of recent investigations and are discussed here. These include: (i) cytochrome oxidases, (ii) isocitrate lyase (ICL), (iii) protein tyrosine phosphatase B (PTPB), and (iv) iron acquisition.

7.2.2.5(a) Cytochrome Oxidases. A major drawback of the current anti-tubercular drug therapy is the duration of treatment, which is typically between six months to one year. This long duration of therapy leads to many problems related to patient adherence, which potentially promotes the emergence of drug-resistant strains. Recently, it was observed that *Mtb* cells lacking cytochrome *bd* oxidase died rapidly upon treatment with bedaquiline.[261] This signified that targeting non-essential enzymes such as the cytochrome *bd* oxidase could potentially lead to the development of synergistic combinations with bedaquiline and other anti-tubercular drugs to help shorten duration of therapy. Since bedaquiline inhibits ATP synthase and halts proton influx, increased backpressure is applied to the rest of the proton-pumping machinery of the electron transport chain. It was reported that cytochrome *bd* oxidase was strongly induced in *Mtb* cells treated with bedaquiline to potentially alleviate this backpressure and, thus, allowed the bacteria to survive.[262] Interestingly, lansoprazole, a common gastric proton-pump inhibitor, was recently discovered to possess activity against *Mtb* (Figure 7.32).[263] Mass spectrometry analysis determined that lansoprazole was converted

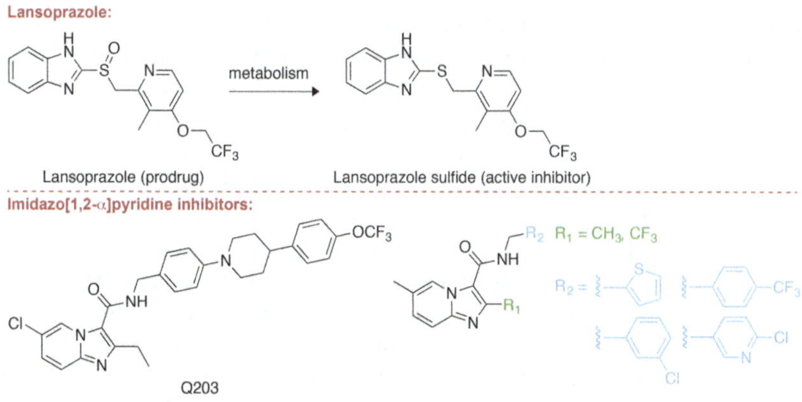

Figure 7.32 Structures of cytochrome oxidase inhibitors.

intracellularly to the active metabolite lansoprazole sulfide. Lansoprazole sulfide was then tested and determined to have MIC_{90} values in the nanomolar range against drug-resistant clinical isolates of *Mtb.* Whole-genome sequencing on lansoprazole-resistant mutants revealed that cytochrome bc_1 was the drug target. It would be interesting to test whether lansoprazole sulfide would also exhibit synergistic effects when used in combination with bedaquiline. Besides lansoprazole sulfide, there are other small-molecule inhibitors of cytochrome bc_1 (*e.g.* the imidazo[1,2-α]pyridine class, which also displayed potent nanomolar MIC values against drug-sensitive and drug-resistant *Mtb* strains (Figure 7.32)).[264,265] Specifically, the imidazo[1,2-α]pyridine Q203 was further validated in a murine model. These findings support cytochrome oxidases as highly valuable drug targets.

7.2.2.5(b) Isocitrate Lyase (ICL). One way to overcome *Mtb* persistence is to target a pathway that is required for a hostile environment. When *Mtb* transitions into its persistent state, its metabolism shifts the carbon source to C_2 substrates generated by β-oxidation of fatty acids. This shift in carbon source causes a decrease in the amount of glycolysis ongoing in *Mtb* and an increase in the glyoxylate shunt, allowing anaplerotic maintenance of the TCA cycle and gluconeogenesis.[266] One such pathway in *Mtb* is the conversion of isocitrate to glyoxylate and succinate by ICL. These compounds are then further processed to generate malate. Elevated levels of ICL are observed when *Mtb* is grown on media containing C_2 carbon sources and shortly after uptake into macrophages.[267,268] Inhibitors of ICL have been identified using a cell-based assay of *M. smegmatis* Δ*Msm*_ICL complemented with the gene expressing *Mtb*_ICL.[269] The inhibitors, 3-bromopyruvate and 3-nitropropionate, mimic the products of the enzyme with the halogenated pyruvate becoming covalently attached to the active site cysteine (Cys191). A large structural change was noted in the crystal structure (Figure 7.33) in two areas that control access to the active site. Since this discovery of ICL inhibitors, many more screens have been performed.[270,271]

7.2.2.5(c) Protein Tyrosine Phosphatase B (PTPB). Many biological processes in mammalian cells are regulated by protein tyrosine phosphorylation, which is controlled by a balance of the actions of protein tyrosine phosphatases (PTPs) and protein tyrosine kinases (PTKs).[272] *Mtb* utilizes a specific PTP called PTPB, which is secreted into the cytoplasm of macrophages during infection, making it an attractive target, as the inhibitors do not need to bypass the complex cell envelope of *Mtb*.[273,274] Deletion of PTPB led to decreased intracellular survival of *Mtb* in macrophages as well as reduction in bacterial load in a guinea pig TB model.[275] Furthermore, PTPB was found to subvert host defense and survival pathways.[276] Targeting virulence factors, such as PTPB, is potentially advantageous due to the lack of strong selective pressure, which is highly valuable in combating *Mtb*. Hence, PTPB has been heavily investigated as a possible drug target and many molecules have been developed as selective PTPB inhibitors. I-A09 is a benzofuran-based salicylic

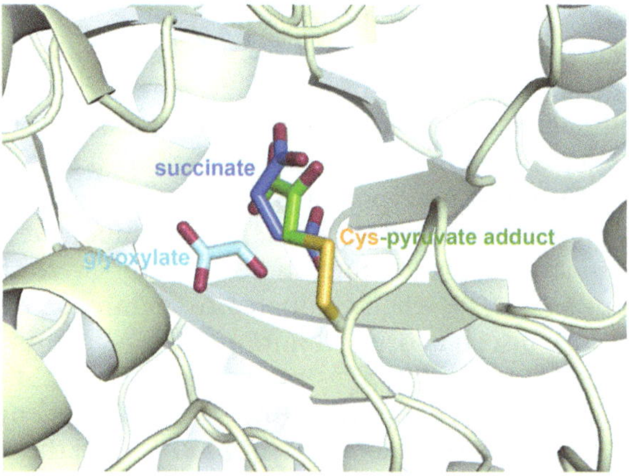

Figure 7.33 A zoomed-in view of the active site of an overlay of the crystal structures of isocitrate lyase (pale yellow, *note*: only one structure of the enzyme is shown for simplicity) in complex with a covalent adduct between Cys191 (orange sticks with the oxygen atom colored red) and 3-bromopyruvate (green sticks with oxygen atoms colored red) (PDB ID: 1F8M,[269] at 1.8 Å) or with glyoxylate (turquoise sticks with oxygen atoms colored red) and succinate (blue sticks with oxygen and nitrogen atoms colored red and blue, respectively) (PDB ID: 1F8I,[269] at 2.3 Å).

acid inhibitor that exhibits non-competitive inhibition of *Mtb*_PTPB at a low micromolar level (Figure 7.34).[276] More importantly, I-A09 is highly cell permeable and displayed similar potency in an infected macrophage model. From a HTS campaign comprised of natural product scaffolds, molecules from the yohimbane family of alkaloids (*e.g.* yohimbine) were identified to be inhibitors of PTPB with moderate IC_{50} values.[277] From the yohimbane scaffold, two indoloquinolizidine analogues were generated from further SAR studies. These compounds displayed IC_{50} values less than 0.43 μM. A major challenge in developing inhibitors against *Mtb*_PTPB is that the active sites of both human PTP and *Mtb*_PTPB are highly conserved.[278] Several potent non-selective human PTP inhibitors were identified and reported *via* NMR-based screening.[279,280] In the literature, a popular strategy to adapt these compounds as TB drug candidates is to generate bidentate inhibitors, where the non-selective PTP inhibitors are used as the "warhead" and coupled with a "secondary-site binder". The secondary site is a unique aromatic binding region adjacent to the active site, which could guide the "warhead" to its intended *Mtb*_PTPB. About 3500 different bidentate inhibitors were generated *via* click chemistry and tested for enzyme inhibition.[278] All the compounds tested were generally cell permeable. The two most potent compounds identified displayed K_i values of 170 and 150 nM.

To further expand the idea of bidentate inhibitors, tridentate inhibitors, which were hypothesized to display additional interactions with the enzyme, were generated (Figure 7.34).[281] The best tridentate inhibitor had a K_i value of 160 nM and a greater than 25-fold selectivity for *Mtb*_PTPB. At the same time, another HTS of 40 000 compounds was used to identify several inhibitors containing the thiazolidiones spiro-fused to indolin-2-ones.[282] A SAR study revealed that the addition of a carboxylic acid group adjacent to the spirothiazolidinone moiety displayed increased solubility while retaining potency. HPLC separation of the enantiomers revealed that the (−)-enantiomers were 7 to 20 times more potent than the corresponding (+)-enantiomers. The inhibitors were found to be reversible competitive inhibitors with respect to the substrate. Furthermore, these inhibitors were also found to be very selective for *Mtb*_PTPB when compared to mammalian PTPs. Recently, cefsulodin, a β-lactam cephalosporin antibiotic, was identified to be a low micromolar *Mtb*_PTPB inhibitor.[283] Interestingly, from the parent cefsulodin, fragment-based screening identified the α-sulfophenylacetic amide (SPAA) moiety to be the most important for binding and the rest of the fragments displayed no enzyme inhibition. SPAA was used to generate multiple analogues with the most potent and selective inhibitor displaying a K_i value of 7.9 nM and more than 10 000-fold selectivity for *Mtb*_PTPB over a panel of 25 other PTPs.

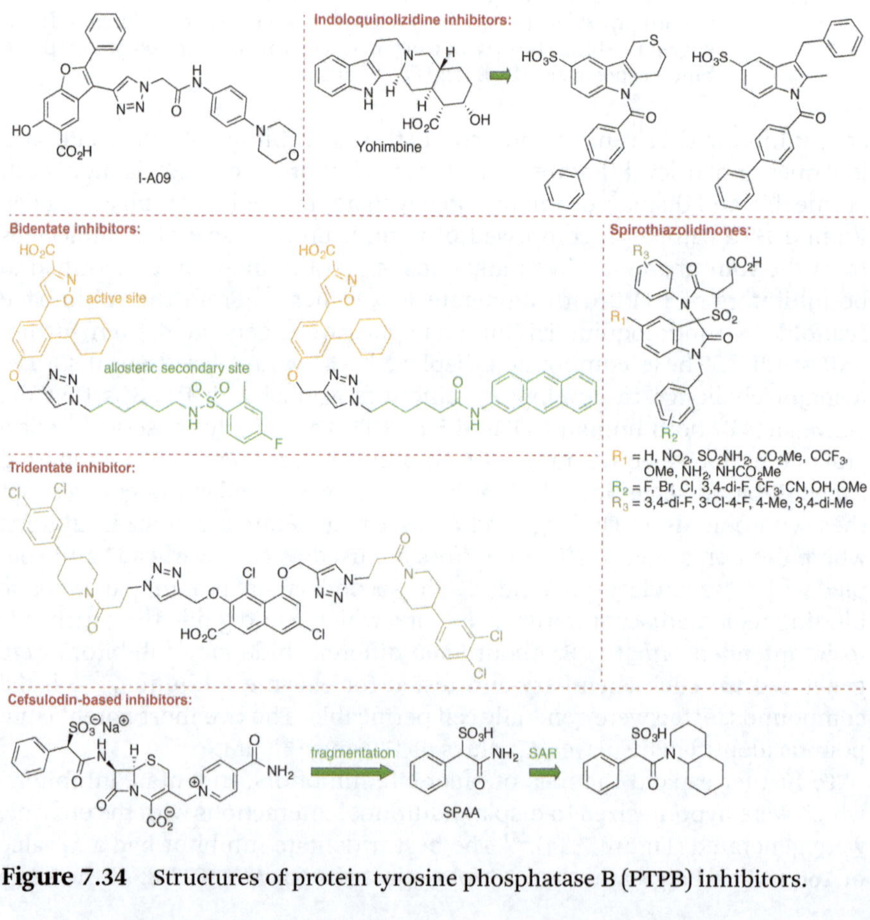

Figure 7.34 Structures of protein tyrosine phosphatase B (PTPB) inhibitors.

7.2.2.5(d) Iron Acquisition Mechanism. Iron is a critical nutrient for most organisms due to its function as a cofactor for many enzymatic processes. *Mtb* is one of these organisms that require iron to survive, especially within the hostile environment of host's macrophages. Ipso facto, targeting iron uptake could be an attractive therapeutic approach.[284] *Mtb* can extract iron from host's macrophages by either siderophore-dependent or siderophore-independent mechanisms.

In the siderophore-mediated mechanism, *Mtb* synthesizes small molecules with a high affinity for iron. These can be divided into two classes called mycobactins and carboxymycobactins (Figure 7.35).[284–286] Both classes contain a 2-hydroxyphenyloxazoline moiety. Mycobactins differ from carboxymycobactins based on the length of their alkyl chain substituents. Alkyl chains of 10–21 carbons are typically found in mycobactins whereas shorter alkyl chains of 2–9 carbons in length are found in carboxymycobactins.[287] The longer alkyl substituents indicate that mycobactins are more lipophilic and, thus, work primarily in the *Mtb* cell envelope.[288,289] Meanwhile, in the cytoplasm of host's macrophages, the less hydrophobic carboxymycobactins actively remove iron from the host iron-binding proteins and then shuttle the iron to mycobactins at the cell envelope. Both siderophores coordinate Fe^{3+}.[290] Both mycobactins and carboxymycobactins have been shown to be crucial for survival of *Mtb* in iron-deprived conditions.[286] The biosynthesis of mycobactins is proposed to result from the action of ten enzymes of polyketide synthase (PKS) and non-ribosomal peptide synthetase (NRPS) origin, designated *mbtA-J*.[291] Inhibition of mycobactins biosynthesis is an attractive strategy and was validated by PAS, which was already known to inhibit mycobactins biosynthesis.[292] In 2005, a mechanism-based inhibitor called 5′-*O*-(*N*-salicylsulfamoyl)adenosine (salicyl-AMS), displaying IC_{50} values in low nanomolar range, was discovered.[293] Salicyl-AMS mimics the intermediate salicyl-adenylate and covalently binds

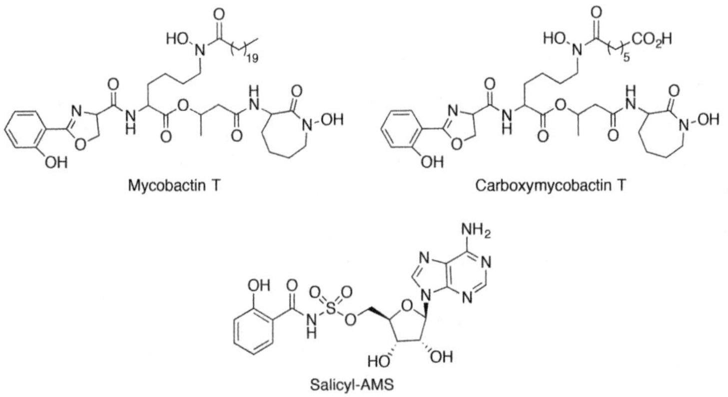

Figure 7.35 Structures of the small-molecule siderophores mycobactin T and carboxymycobactin T used by *Mtb* to extract iron from human macrophages. Salicyl-AMS is the inhibitor of the mycobactin biosynthesis. Salicyl-AMS = 5′-*O*-(*N*-salicylsulfamoyl)adenosine.

to MbtA, preventing adenylation from happening. Salicyl-AMS even displayed potent efficacy against *Mtb* in a bronchial murine model.[294] Other than salicyl-AMS, there have been many ongoing efforts to identify compounds that can interfere with the iron acquisition processes. Other small molecules that inhibit other enzymes during the biosynthesis of mycobactins or synthetic siderophores that compete with the natural siderophores in *Mtb* were summarized in an excellent review recently published.[295] Another interesting idea is to deliver anti-tubercular drugs to *Mtb* using siderophores as drug delivery vehicles.

In the non-siderophore pathway, *Mtb* uptakes heme-iron by three proposed *Mtb* proteins: the heme-binding protein Rv0203, the transmembrane proteins MmpL3 and 11, and the heme-releasing protein MhuD.[135,284] Naturally, MmpL3 inhibitors such as BM212, SQ109, and other various adamantyl urea compounds were hypothesized to be useful in interfering with the non-siderophore pathway. This, however, remains to be evaluated in detail. Inhibition of the non-siderophore pathway has been largely unexplored and would be an interesting area for further research.

7.3 Synergistic Drug Combination Therapy

As proven by current TB drug regimens, the use of a single drug is often not an efficient or wise practice to combat an infectious disease. The current TB treatment, as mentioned above, is usually a combination of up to four compounds with three or four separate biological targets. Combination of drugs lowers treatment costs and duration, and discourages the development of resistance. Combination therapies can consist of (i) unrelated compounds (*e.g.* SQ109 and bedaquiline), (ii) inhibitors of resistance enzymes as adjuvants for therapeutics (*e.g.* AMX and CLV), and (iii) modulators of host immune system (*e.g.* $P2X_7$ receptor agonist) as adjuvants for current drugs.

7.3.1 Unrelated Compounds

In an effort to reduce the emergence of resistance, combinations of small molecules have been explored. Spectinomycin (SPT), an AG structurally similar to STR, was explored as one half of a potential synergistic drug combination. A high-throughput synergy screen was performed with SPT and a library of molecules with known pharmacological properties. Three structural cores were found to work synergistically with SPT and enhance its activity: macrolides, azoles, and butyrophenones.[296] Bromperidol displayed bactericidal activity, was effective against *Mtb* in a macrophage model, and displayed similar enhancement of the activities of RIF, STR, CFZ, and CLA.

7.3.2 Inhibitors of Resistance Enzymes

The use of inhibitors of resistance enzymes along with currently approved anti-tubercular drugs complements the drug combination strategy described above. Perhaps the best-known strategy along this line is the combination of

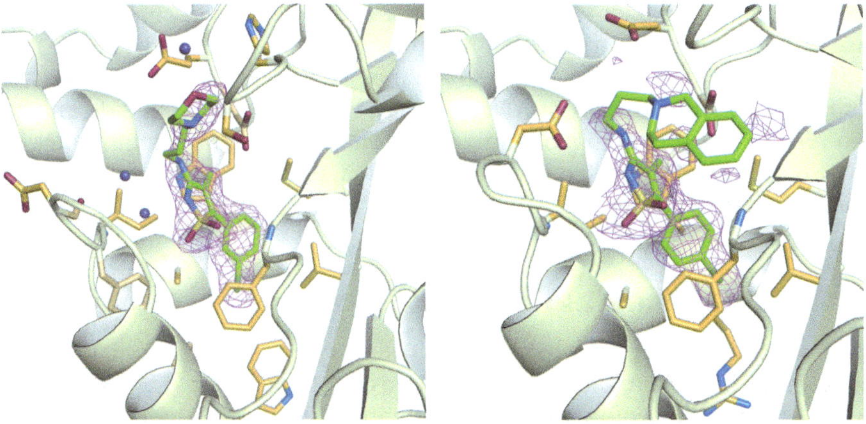

Figure 7.36 Crystal structures of Eis (pale yellow) co-crystallized with inhibitors (PDB ID: 5EBV and 5EC4,[300] both at 2.2 Å) that restore the anti-tubercular activity of KAN.

a β-lactam and a β-lactamase inhibitor. Analogously to the third-line treatments AMX with CLV or IPM with CLN, the combination of CLV and MEM proved to be active against both susceptible and XDR-*Mtb* strains.[140,297] Additionally, a novel β-lactamase inhibitor, NXL104, was recently reported to bind BlaC.[298] Keeping this classical β-lactam/β-lactamase inhibitor combination in mind, inhibitors of the AG-resistance enzyme Eis were discovered.[299] These inhibitors belong to diverse chemical families and display a wide range of *in vitro* activities. Another HTS of the enzyme revealed a series of more than 600 compounds with the same core scaffold that showed a wide range of *in vitro* and *in vivo* activities (Figure 7.36).[300] From the hit scaffold, 18 compounds were found to restore the KAN MIC values to a normal range for a resistant *Mtb* strain. In addition to the HTS, simple metal ions were also explored and found to inhibit the acetyltransferase activity of Eis.[301]

7.3.3 P2X$_7$ Receptor Agonist

Other than utilizing small molecules to target essential pathways in TB, an additional strategy to combat this disease is to modulate the mammalian immune system to eliminate the pathogen. The P2X$_7$ receptors are regulated by ATP, highly expressed in lymphocytes, monocytes, and macrophages, and mediate clearance of *Mtb* during pathophysiological conditions.[302] Different studies have shown that P2X$_7$ receptors play an important role in clearing intracellular *Mtb*. Evidently, macrophages obtained from humans with a 1513A → C polymorphism within the gene sequence of P2X$_7$ receptors failed to induce *Mtb* cell death in the presence of ATP.[303] During TB infection, *Mtb* is able to survive and replicate in macrophages due to inhibition of phagosomal fusion with lysosomes.[304] The treatment of infected macrophages with ATP has been shown to overcome this phagosome-lysosome fusion inhibition and eliminate the intracellular pathogens. Hence, in theory, ATP could be used as a $P2X_7$ receptor agonist to supplement the current TB drugs. Practically, the use of ATP is highly controversial and may not be effective considering that it is a common molecule taking part in many different biological processes and, thus, administration of ATP may lead to many unwanted side effects. However, this leaves an opportunity for the development of a small-molecule $P2X_7$ receptor selective agonist. Potentially, this inhibitor would be a useful compound to act synergistically with the current anti-tubercular drugs.

7.4 Conclusion and Perspectives

We have come a long way from the 19th century when humanity could only cope with TB by "romanticizing" the disease. A large part of this is credited to the contributions of many brilliant scientists who built a solid foundation for today's anti-tubercular drug discovery platform. The recent renaissance in TB research has uncovered many promising drug targets and novel potential anti-tubercular compounds. Evidently, from this TB research renaissance, bedaquiline emerged as a success story. This new compound was recently approved by the FDA for the treatment of MDR-TB. The approval of bedaquiline has certainly been encouraging as far as our battle against drug-resistant TB is concerned. However, besides the ever-growing concerns regarding resistance, there are still other fundamental challenges needing to be addressed, including long duration of treatment, high pill burden, and drug–drug interactions. Other important pitfalls during target/lead validation that we should heed are homologous targets in humans, pan-assay inhibitors, and physicochemical properties of the lead compounds. There is also an important need for the development of improved *in vitro* testing conditions that better simulate the *in vivo* conditions. Accordingly, the war has not been won and there is still much research remaining to be done. Nevertheless, we sincerely hope that some of the targets and compounds covered in this chapter will lead our battle against TB and that one day, the enemy will eventually yield.

Acknowledgements

The work on tuberculosis in our laboratory is supported by NIH AI090048 (to S. G.-T.). S. Y. L. H. is supported by the University of Kentucky Presidential Fellowship Award.

References

1. A. Schatz, E. Bugle and S. A. Waksman, *Exp. Biol. Med.*, 1944, **55**, 66–69.
2. Selman A. Waksman, *Banquet Speech*, 10th December 1952, http://www.nobelprize.org/nobel_prizes/medicine/laureates/1952/waksman-speech.html, accessed April 2016.
3. S. H. Gillespie, *Antimicrob. Agents Chemother.*, 2002, **46**, 267–274.
4. T. M. Daniel, *Respir. Med.*, 2006, **100**, 1862–1870.
5. D. M. Morens, *Emerging Infect. Dis.*, 2002, **8**, 1353–1358.

6. E. Mullins, *How Tuberculosis Shaped Victorian Fashion*, http://www.smithsonianmag.com/science-nature/how-tuberculosis-shaped-victorian-fashion-180959029/.
7. J. Crofton and D. A. Mitchison, *Br. Med. J.*, 1948, **2**, 1009–1015.
8. G. Marshall, J. W. S. Blacklock, C. Cameron, N. B. Capon, R. Cruickshank, J. H. Gaddum, F. R. G. Heaf, A. B. Hill, L. E. Houghton, J. C. Hoyle, H. Raistrick, J. G. Scadding, W. H. Tytler, G. S. Wilson and P. D'Arcy Hart, *Br. Med. J.*, 1948, **2**, 769–782.
9. *World Health Organization Global Tuberculosis Report*, 2015.
10. S. T. Cole, R. Brosch, J. Parkhill, T. Garnier, C. Churcher, D. Harris, S. V. Gordon, K. Eiglmeier, S. Gas, C. E. Barry 3rd, F. Tekaia, K. Badcock, D. Basham, D. Brown, T. Chillingworth, R. Connor, R. Davies, K. Devlin, T. Feltwell, S. Gentles, N. Hamlin, S. Holroyd, T. Hornsby, K. Jagels, A. Krogh, J. McLean, S. Moule, L. Murphy, K. Oliver, J. Osborne, M. A. Quail, M. A. Rajandream, J. Rogers, S. Rutter, K. Seeger, J. Skelton, R. Squares, S. Squares, J. E. Sulston, K. Taylor, S. Whitehead and B. G. Barrell, *Nature*, 1998, **393**, 537–544.
11. G. J. Tortora, B. R. Funke and C. L. Case, *Microbiology: An Introduction*, Peaseson Benjamin Cummings, New York, 8th edn, 2004.
12. K. Todar, *Todar's Online Textbook of Bacteriology*. http://textbookofbacteriology.net/tuberculosis.html.
13. R. Bansal-Mutalik and H. Nikaido, *Proc. Natl. Acad. Sci. U. S. A.*, 2014, **111**, 4958–4963.
14. J. A. Verschoor, M. S. Baird and J. Grooten, *Prog. Lipid Res.*, 2012, **51**, 325–339.
15. S. Garneau-Tsodikova and K. J. Labby, *MedChemComm*, 2016, **7**, 11–27.
16. J. P. Sarathy, V. Dartois and E. J. Lee, *Pharmaceuticals*, 2012, **5**, 1210–1235.
17. M. Niederweis, *Mol. Microbiol.*, 2003, **49**, 1167–1177.
18. C. Mailaender, N. Reiling, H. Engelhardt, S. Bossmann, S. Ehlers and M. Niederweis, *Microbiology*, 2004, **150**, 853–864.
19. P. J. Brennan, *Tuberculosis*, 2003, **83**, 91–97.
20. M. Niederweis, O. Danilchanka, J. Huff, C. Hoffmann and H. Engelhardt, *Trends Microbiol.*, 2010, **18**, 109–116.
21. P. J. Brennan and H. Nikaido, *Annu. Rev. Biochem.*, 1995, **64**, 29–63.
22. D. E. Minnikin, in *The Biology of the Mycobacteria: Physiology, Identification and Classification*, ed. C. Ratledge and J. Stanford, Academic Press, London, 1982, pp. 85–184.
23. I. Yano and K. Saito, *FEBS Lett.*, 1972, **23**, 352–356.
24. I. Yano, K. Kageyama, Y. Ohno, M. Masui, E. Kusunose, M. Kusunose and N. Akimori, *Biomed. Mass Spectrom.*, 1978, **5**, 14–24.
25. L. Favrot, D. H. Lajiness and D. R. Ronning, *J. Biol. Chem.*, 2014, **289**, 25031–25040.
26. H. D. Park, K. M. Guinn, M. I. Harrell, R. Liao, M. I. Voskuil, M. Tompa, G. K. Schoolnik and D. R. Sherman, *Mol. Microbiol.*, 2003, **48**, 833–843.

27. R. L. Leistikow, R. A. Morton, I. L. Bartek, I. Frimpong, K. Wagner and M. I. Voskuil, *J. Bacteriol.*, 2010, **192**, 1662–1670.
28. J. Marrero, C. Trujillo, K. Y. Rhee and S. Ehrt, *PLoS Pathog.*, 2013, **9**, e1003116.
29. D. Schnappinger, S. Ehrt, M. I. Voskuil, Y. Liu, J. A. Mangan, I. M. Monahan, G. Dolganov, B. Efron, P. D. Butcher, C. Nathan and G. K. Schoolnik, *J. Exp. Med.*, 2003, **198**, 693–704.
30. D. G. Muttucumaru, G. Roberts, J. Hinds, R. A. Stabler and T. Parish, *Tuberculosis*, 2004, **84**, 239–246.
31. M. I. Voskuil, K. C. Visconti and G. K. Schoolnik, *Tuberculosis*, 2004, **84**, 218–227.
32. R. L. Gaur, K. Ren, A. Blumenthal, S. Bhamidi, F. D. Gonzalez-Nilo, M. Jackson, R. N. Zare, S. Ehrt, J. D. Ernst and N. Banaei, *PLoS Pathog.*, 2014, **10**, e1004376.
33. A. Venugopal, R. Bryk, S. Shi, K. Rhee, P. Rath, D. Schnappinger, S. Ehrt and C. Nathan, *Cell Host Microbe*, 2011, **9**, 21–31.
34. M. A. Arbex, C. Varella Mde, H. R. Siqueira and F. A. Mello, *J. Bras. Pneumol.*, 2010, **36**, 641–656.
35. G. Di Perri and S. Bonora, *J. Antimicrob. Chemother.*, 2004, **54**, 593–602.
36. L. Davies Forsman, C. G. Giske, J. Bruchfeld, T. Schon, P. Jureen and K. Angeby, *Antimicrob. Agents Chemother.*, 2015, **59**, 3630–3632.
37. A. E. Belanger, G. S. Besra, M. E. Ford, K. Mikusova, J. T. Belisle, P. J. Brennan and J. M. Inamine, *Proc. Natl. Acad. Sci. U. S. A.*, 1996, **93**, 11919–11924.
38. L. Deng, K. Mikusova, K. G. Robuck, M. Scherman, P. J. Brennan and M. R. McNeil, *Antimicrob. Agents Chemother.*, 1995, **39**, 694–701.
39. A. Banerjee, E. Dubnau, A. Quemard, V. Balasubramanian, K. S. Um, T. Wilson, D. Collins, G. de Lisle and W. R. Jacobs Jr, *Science*, 1994, **263**, 227–230.
40. K. Johnsson, D. S. King and P. G. Schultz, *J. Am. Chem. Soc.*, 1995, **117**, 5009–5010.
41. A. R. Baulard, J. C. Betts, J. Engohang-Ndong, S. Quan, R. A. McAdam, P. J. Brennan, C. Locht and G. S. Besra, *J. Biol. Chem.*, 2000, **275**, 28326–28331.
42. A. Alahari, X. Trivelli, Y. Guerardel, L. G. Dover, G. S. Besra, J. C. Sacchettini, R. C. Reynolds, G. D. Coxon and L. Kremer, *PLoS One*, 2007, **2**, e1343.
43. C. Vilcheze, F. Wang, M. Arai, M. H. Hazbon, R. Colangeli, L. Kremer, T. R. Weisbrod, D. Alland, J. C. Sacchettini and W. R. Jacobs Jr, *Nat. Med.*, 2006, **12**, 1027–1029.
44. J. L. Houghton, K. D. Green, W. Chen and S. Garneau-Tsodikova, *ChemBioChem*, 2010, **11**, 880–902.
45. K. J. Labby and S. Garneau-Tsodikova, *Future Med. Chem.*, 2013, **5**, 1285–1309.
46. P. B. Fernandes, D. J. Hardy, D. McDaniel, C. W. Hanson and R. N. Swanson, *Antimicrob. Agents Chemother.*, 1989, **33**, 1531–1534.
47. L. M. Fu and T. M. Shinnick, *J. Infect.*, 2007, **54**, 277–284.

48. L. Vera-Cabrera, E. Gonzalez, A. Rendon, J. Ocampo-Candiani, O. Welsh, V. M. Velazquez-Moreno, S. H. Choi and C. Molina-Torres, *Antimicrob. Agents Chemother.*, 2006, **50**, 3170–3172.
49. S. M. Swaney, H. Aoki, M. C. Ganoza and D. L. Shinabarger, *Antimicrob. Agents Chemother.*, 1998, **42**, 3251–3255.
50. J. J. Champoux, *Annu. Rev. Biochem.*, 2001, **70**, 369–413.
51. J. L. Arbiser and S. L. Moschella, *J. Am. Acad. Dermatol.*, 1995, **32**, 241–247.
52. N. E. Morrison and G. M. Marley, *Int. J. Lepr. Other Mycobact. Dis.*, 1976, **44**, 133–134.
53. V. E. Kagan, *Ann. N. Y. Acad. Sci.*, 1989, **570**, 121–135.
54. H. C. Steel, N. M. Matlola and R. Anderson, *J. Antimicrob. Chemother.*, 1999, **44**, 209–216.
55. V. Nopponpunth, W. Sirawaraporn, P. J. Greene and D. V. Santi, *J. Bacteriol.*, 1999, **181**, 6814–6821.
56. J. Rengarajan, C. M. Sassetti, V. Naroditskaya, A. Sloutsky, B. R. Bloom and E. J. Rubin, *Mol. Microbiol.*, 2004, **53**, 275–282.
57. E. A. Campbell, N. Korzheva, A. Mustaev, K. Murakami, S. Nair, A. Goldfarb and S. A. Darst, *Cell*, 2001, **104**, 901–912.
58. W. Wehrli, F. Knusel, K. Schmid and M. Staehelin, *Proc. Natl. Acad. Sci. U. S. A.*, 1968, **61**, 667–673.
59. H. Engelberg-Kulka, B. Sat, M. Reches, S. Amitai and R. Hazan, *Trends Microbiol.*, 2004, **12**, 66–71.
60. H. I. Boshoff, V. Mizrahi and C. E. Barry 3rd, *J. Bacteriol.*, 2002, **184**, 2167–2172.
61. Y. Zhang, M. M. Wade, A. Scorpio, H. Zhang and Z. Sun, *J. Antimicrob. Chemother.*, 2003, **52**, 790–795.
62. K. D. Green and S. Garneau-Tsodikova, *Front. Microbiol.*, 2013, **4**, 208.
63. P. E. Silva, F. Bigi, M. P. Santangelo, M. I. Romano, C. Martin, A. Cataldi and J. A. Ainsa, *Antimicrob. Agents Chemother.*, 2001, **45**, 800–804.
64. S. K. Banerjee, K. Bhatt, S. Rana, P. Misra and P. K. Chakraborti, *Biochem. Biophys. Res. Commun.*, 1996, **226**, 362–368.
65. M. Singh, G. P. S. Jadaun, R. K. Srivastava, V. Chauhan, R. Mishra, K. Gupta, S. Nair, D. S. Chauhan, V. D. Sharma, K. Venkatesan and V. M. Katoch, *Indian J. Med. Res.*, 2011, **133**, 535–540.
66. A. Telenti, W. J. Philipp, S. Sreevatsan, C. Bernasconi, K. E. Stockbauer, B. Wieles, J. M. Musser and W. R. Jacobs Jr, *Nat. Med.*, 1997, **3**, 567–570.
67. H. E. Takiff, L. Salazar, C. Guerrero, W. Philipp, W. M. Huang, B. Kreiswirth, S. T. Cole, W. R. Jacobs Jr and A. Telenti, *Antimicrob. Agents Chemother.*, 1994, **38**, 773–780.
68. J. Wachino, K. Shibayama, K. Kimura, K. Yamane, S. Suzuki and Y. Arakawa, *FEMS Microbiol. Lett.*, 2010, **311**, 56–60.
69. Y. Suzuki, C. Katsukawa, A. Tamaru, C. Abe, M. Makino, Y. Mizuguchi and H. Taniguchi, *J. Clin. Microbiol.*, 1998, **36**, 1220–1225.
70. S. K. Johansen, C. E. Maus, B. B. Plikaytis and S. Douthwaite, *Mol. Cell*, 2006, **23**, 173–182.

71. F. S. Spies, A. W. Ribeiro, D. F. Ramos, M. O. Ribeiro, A. Martin, J. C. Palomino, M. L. Rossetti, P. E. da Silva and A. Zaha, *J. Clin. Microbiol.*, 2011, **49**, 2625–2630.
72. S. Y. Wong, J. S. Lee, H. K. Kwak, L. E. Via, H. I. Boshoff and C. E. Barry, *Antimicrob. Agents Chemother.*, 2011, **55**, 2515–2522.
73. S. V. Ramaswamy, R. Reich, S. J. Dou, L. Jasperse, X. Pan, A. Wanger, T. Quitugua and E. A. Graviss, *Antimicrob. Agents Chemother.*, 2003, **47**, 1241–1250.
74. M. H. Hazbon, M. Brimacombe, M. Bobadilla del Valle, M. Cavatore, M. I. Guerrero, M. Varma-Basil, H. Billman-Jacobe, C. Lavender, J. Fyfe, L. Garcia-Garcia, C. I. Leon, M. Bose, F. Chaves, M. Murray, K. D. Eisenach, J. Sifuentes-Osornio, M. D. Cave, A. Ponce de Leon and D. Alland, *Antimicrob. Agents Chemother.*, 2006, **50**, 2640–2649.
75. A. E. DeBarber, K. Mdluli, M. Bosman, L. G. Bekker and C. E. Barry 3rd, *Proc. Natl. Acad. Sci. U. S. A.*, 2000, **97**, 9677–9682.
76. G. P. Morlock, B. Metchock, D. Sikes, J. T. Crawford and R. C. Cooksey, *Antimicrob. Agents Chemother.*, 2003, **47**, 3799–3805.
77. K. Stoffels, V. Mathys, M. Fauville-Dufaux, R. Wintjens and P. Bifani, *Antimicrob. Agents Chemother.*, 2012, **56**, 5186–5193.
78. V. Rajendran and R. Sethumadhavan, *J. Biomol. Struct. Dyn.*, 2014, **32**, 209–221.
79. M. H. Larsen, C. Vilcheze, L. Kremer, G. S. Besra, L. Parsons, M. Salfinger, L. Heifets, M. H. Hazbon, D. Alland, J. C. Sacchettini and W. R. Jacobs Jr, *Mol. Microbiol.*, 2002, **46**, 453–466.
80. C. Vilcheze, A. D. Baughn, J. Tufariello, L. W. Leung, M. Kuo, C. F. Basler, D. Alland, J. C. Sacchettini, J. S. Freundlich and W. R. Jacobs Jr, *Antimicrob. Agents Chemother.*, 2011, **55**, 3889–3898.
81. J. E. Hugonnet and J. S. Blanchard, *Biochemistry*, 2007, **46**, 11998–12004.
82. M. S. Ramirez and M. E. Tolmasky, *Drug Resist. Updates*, 2010, **13**, 151–171.
83. S. S. Hegde, F. Javid-Majd and J. S. Blanchard, *J. Biol. Chem.*, 2001, **276**, 45876–45881.
84. M. A. Zaunbrecher, R. D. Sikes Jr, B. Metchock, T. M. Shinnick and J. E. Posey, *Proc. Natl. Acad. Sci. U. S. A.*, 2009, **106**, 20004–20009.
85. W. Chen, T. Biswas, V. R. Porter, O. V. Tsodikov and S. Garneau-Tsodikova, *Proc. Natl. Acad. Sci. U. S. A.*, 2011, **108**, 9804–9808.
86. J. L. Houghton, T. Biswas, W. Chen, O. V. Tsodikov and S. Garneau-Tsodikova, *ChemBioChem*, 2013, **14**, 2127–2135.
87. J. L. Houghton, K. D. Green, R. E. Pricer, A. S. Mayhoub and S. Garneau-Tsodikova, *J. Antimicrob. Chemother.*, 2013, **68**, 800–805.
88. O. V. Tsodikov, K. D. Green and S. Garneau-Tsodikova, *PLoS One*, 2014, **9**, e92370.
89. B. C. Jennings, K. J. Labby, K. D. Green and S. Garneau-Tsodikova, *Biochemistry*, 2013, **52**, 5125–5132.
90. W. Chen, K. D. Green and S. Garneau-Tsodikova, *Antimicrob. Agents Chemother.*, 2012, **56**, 5831–5838.

91. L. Duan, M. Yi, J. Chen, S. Li and W. Chen, *Biochem. Biophys. Res. Commun.*, 2016, **473**, 1229–1234.
92. D. M. Shin, B. Y. Jeon, H. M. Lee, H. S. Jin, J. M. Yuk, C. H. Song, S. H. Lee, Z. W. Lee, S. N. Cho, J. M. Kim, R. L. Friedman and E. K. Jo, *PLoS Pathog.*, 2010, **6**, e1001230.
93. K. H. Kim, D. R. An, J. Song, J. Y. Yoon, H. S. Kim, H. J. Yoon, H. N. Im, J. Kim, J. Kim do, S. J. Lee, K. H. Kim, H. M. Lee, H. J. Kim, E. K. Jo, J. Y. Lee and S. W. Suh, *Proc. Natl. Acad. Sci. U. S. A.*, 2012, **109**, 7729–7734.
94. K. D. Green, T. Biswas, C. Chang, R. Wu, W. Chen, B. K. Janes, D. Chalupska, P. Gornicki, P. C. Hanna, O. V. Tsodikov, A. Joachimiak and S. Garneau-Tsodikova, *Biochemistry*, 2015, **54**, 3197–3206.
95. R. E. Pricer, J. L. Houghton, K. D. Green, A. S. Mayhoub and S. Garneau-Tsodikova, *Mol. BioSyst.*, 2012, **8**, 3305–3313.
96. W. Chen, K. D. Green, O. V. Tsodikov and S. Garneau-Tsodikova, *Biochemistry*, 2012, **51**, 4959–4967.
97. K. D. Green, R. E. Pricer, M. N. Stewart and S. Garneau-Tsodikova, *ACS Infect. Dis.*, 2015, **1**, 272–283.
98. K. M. George, Y. Yuan, D. R. Sherman and C. E. Barry 3rd, *J. Biol. Chem.*, 1995, **270**, 27292–27298.
99. H. Marrakchi, M. A. Laneelle and M. Daffe, *Chem. Biol.*, 2014, **21**, 67–85.
100. K. Takayama, C. Wang and G. S. Besra, *Clin. Microbiol. Rev.*, 2005, **18**, 81–101.
101. D. Portevin, C. De Sousa-D'Auria, C. Houssin, C. Grimaldi, M. Chami, M. Daffe and C. Guilhot, *Proc. Natl. Acad. Sci. U. S. A.*, 2004, **101**, 314–319.
102. Y. Yuan, R. E. Lee, G. S. Besra, J. T. Belisle and C. E. Barry 3rd, *Proc. Natl. Acad. Sci. U. S. A.*, 1995, **92**, 6630–6634.
103. R. A. Slayden, R. E. Lee and C. E. Barry 3rd, *Mol. Microbiol.*, 2000, **38**, 514–525.
104. M. R. Kuo, H. R. Morbidoni, D. Alland, S. F. Sneddon, B. B. Gourlie, M. M. Staveski, M. Leonard, J. S. Gregory, A. D. Janjigian, C. Yee, J. M. Musser, B. Kreiswirth, H. Iwamoto, R. Perozzo, W. R. Jacobs Jr, J. C. Sacchettini and D. A. Fidock, *J. Biol. Chem.*, 2003, **278**, 20851–20859.
105. T. Matviiuk, F. Rodriguez, N. Saffon, S. Mallet-Ladeira, M. Gorichko, A. L. de Jesus Lopes Ribeiro, M. R. Pasca, C. Lherbet, Z. Voitenko and M. Baltas, *Eur. J. Med. Chem.*, 2013, **70**, 37–48.
106. X. He, A. Alian and P. R. Ortiz de Montellano, *Bioorg. Med. Chem.*, 2007, **15**, 6649–6658.
107. M. Isaka, N. Rugseree, P. Maithip, P. Kongsaeree, S. Prabpai and Y. Thebtaranonth, *Tetrahedron*, 2005, **61**, 5577–5583.
108. C. Freiberg, N. A. Brunner, G. Schiffer, T. Lampe, J. Pohlmann, M. Brands, M. Raabe, D. Habich and K. Ziegelbauer, *J. Biol. Chem.*, 2004, **279**, 26066–26073.
109. L. Slepikas, G. Chiriano, R. Perozzo, S. Tardy, A. Kranjc-Pietrucci, O. Patthey-Vuadens, H. Ouertatani-Sakouhi, S. Kicka, C. F. Harrison, T. Scrignari, K. Perron, H. Hilbi, T. Soldati, P. Cosson, E. Tarasevicius and L. Scapozza, *J. Med. Chem.*, 2016, **59**, 10917–10928.

110. C. Menendez, S. Gau, C. Lherbet, F. Rodriguez, C. Inard, M. R. Pasca and M. Baltas, *Eur. J. Med. Chem.*, 2011, **46**, 5524–5531.
111. C. Menendez, A. Chollet, F. Rodriguez, C. Inard, M. R. Pasca, C. Lherbet and M. Baltas, *Eur. J. Med. Chem.*, 2012, **52**, 275–283.
112. S. Sachdeva, F. N. Musayev, M. M. Alhamadsheh, J. N. Scarsdale, H. T. Wright and K. A. Reynolds, *Chem. Biol.*, 2008, **15**, 402–412.
113. F. Musayev, S. Sachdeva, J. N. Scarsdale, K. A. Reynolds and H. T. Wright, *J. Mol. Biol.*, 2005, **346**, 1313–1321.
114. H. Oishi, T. Noto, H. Sasaki, K. Suzuki, T. Hayashi, H. Okazaki, K. Ando and M. Sawada, *J. Antibiot.*, 1982, **35**, 391–395.
115. H. Sasaki, H. Oishi, T. Hayashi, I. Matsuura, K. Ando and M. Sawada, *J. Antibiot.*, 1982, **35**, 396–400.
116. T. Noto, S. Miyakawa, H. Oishi, H. Endo and H. Okazaki, *J. Antibiot.*, 1982, **35**, 401–410.
117. S. Miyakawa, K. Suzuki, T. Noto, Y. Harada and H. Okazaki, *J. Antibiot.*, 1982, **35**, 411–419.
118. C. A. Machutta, G. R. Bommineni, S. R. Luckner, K. Kapilashrami, B. Ruzsicska, C. Simmerling, C. Kisker and P. J. Tonge, *J. Biol. Chem.*, 2010, **285**, 6161–6169.
119. Q. Al-Balas, N. G. Anthony, B. Al-Jaidi, A. Alnimr, G. Abbott, A. K. Brown, R. C. Taylor, G. S. Besra, T. D. McHugh, S. H. Gillespie, B. F. Johnston, S. P. Mackay and G. D. Coxon, *PLoS One*, 2009, **4**, e5617.
120. X. He, A. M. Reeve, U. R. Desai, G. E. Kellogg and K. A. Reynolds, *Antimicrob. Agents Chemother.*, 2004, **48**, 3093–3302.
121. M. M. Alhamadsheh, N. C. Waters, S. Sachdeva, P. Lee and K. A. Reynolds, *Bioorg. Med. Chem. Lett.*, 2008, **18**, 6402–6405.
122. M. M. Alhamadsheh, N. C. Waters, D. P. Huddler, M. Kreishman-Deitrick, G. Florova and K. A. Reynolds, *Bioorg. Med. Chem. Lett.*, 2007, **17**, 879–883.
123. Y. Liu, W. Zhong, R. J. Li and S. Li, *Molecules*, 2012, **17**, 4770–4781.
124. C. Vilcheze, V. Molle, S. Carrere-Kremer, J. Leiba, L. Mourey, S. Shenai, G. Baronian, J. Tufariello, T. Hartman, R. Veyron-Churlet, X. Trivelli, S. Tiwari, B. Weinrick, D. Alland, Y. Guerardel, W. R. Jacobs Jr and L. Kremer, *PLoS Pathog.*, 2014, **10**, e1004115.
125. L. Kremer, L. G. Dover, S. Carrere, K. M. Nampoothiri, S. Lesjean, A. K. Brown, P. J. Brennan, D. E. Minnikin, C. Locht and G. S. Besra, *Biochem. J.*, 2002, **364**, 423–430.
126. A. K. Brown, S. Sridharan, L. Kremer, S. Lindenberg, L. G. Dover, J. C. Sacchettini and G. S. Besra, *J. Biol. Chem.*, 2005, **280**, 32539–32547.
127. K. Kapilashrami, G. R. Bommineni, C. A. Machutta, P. Kim, C. T. Lai, C. Simmerling, F. Picart and P. J. Tonge, *J. Biol. Chem.*, 2013, **288**, 6045–6052.
128. L. Kremer, J. D. Douglas, A. R. Baulard, C. Morehouse, M. R. Guy, D. Alland, L. G. Dover, J. H. Lakey, W. R. Jacobs Jr, P. J. Brennan, D. E. Minnikin and G. S. Besra, *J. Biol. Chem.*, 2000, **275**, 16857–16864.
129. A. K. Brown, R. C. Taylor, A. Bhatt, K. Futterer and G. S. Besra, *PLoS One*, 2009, **4**, e6306.

130. E. Martens and A. L. Demain, *J. Antibiot.*, 2011, **64**, 705–710.
131. M. Daffe and P. Draper, *Adv. Microb. Physiol.*, 1998, **39**, 131–203.
132. P. Draper, *Front. Biosci.*, 1998, **3**, D1253–D1261.
133. D. Kaur, M. E. Guerin, H. Skovierova, P. J. Brennan and M. Jackson, *Adv. Appl. Microbiol.*, 2009, **69**, 23–78.
134. W. Li, A. Upadhyay, F. L. Fontes, E. J. North, Y. Wang, D. C. Crans, A. E. Grzegorzewicz, V. Jones, S. G. Franzblau, R. E. Lee, D. C. Crick and M. Jackson, *Antimicrob. Agents Chemother.*, 2014, **58**, 6413–6423.
135. C. Varela, D. Rittmann, A. Singh, K. Krumbach, K. Bhatt, L. Eggeling, G. S. Besra and A. Bhatt, *Chem. Biol.*, 2012, **19**, 498–506.
136. C. P. Owens, N. Chim, A. B. Graves, C. A. Harmston, A. Iniguez, H. Contreras, M. D. Liptak and C. W. Goulding, *J. Biol. Chem.*, 2013, **288**, 21714–21728.
137. K. Tahlan, R. Wilson, D. B. Kastrinsky, K. Arora, V. Nair, E. Fischer, S. W. Barnes, J. R. Walker, D. Alland, C. E. Barry 3rd and H. I. Boshoff, *Antimicrob. Agents Chemother.*, 2012, **56**, 1797–1809.
138. K. Li, L. A. Schurig-Briccio, X. Feng, A. Upadhyay, V. Pujari, B. Lechartier, F. L. Fontes, H. Yang, G. Rao, W. Zhu, A. Gulati, J. H. No, G. Cintra, S. Bogue, Y. L. Liu, K. Molohon, P. Orlean, D. A. Mitchell, L. Freitas-Junior, F. Ren, H. Sun, T. Jiang, Y. Li, R. T. Guo, S. T. Cole, R. B. Gennis, D. C. Crick and E. Oldfield, *J. Med. Chem.*, 2014, **57**, 3126–3139.
139. A. E. Grzegorzewicz, H. Pham, V. A. Gundi, M. S. Scherman, E. J. North, T. Hess, V. Jones, V. Gruppo, S. E. Born, J. Kordulakova, S. S. Chavadi, C. Morisseau, A. J. Lenaerts, R. E. Lee, M. R. McNeil and M. Jackson, *Nat. Chem. Biol.*, 2012, **8**, 334–341.
140. A. Zumla, P. Nahid and S. T. Cole, *Nat. Rev. Drug Discovery*, 2013, **12**, 388–404.
141. S. Lun, H. Guo, O. K. Onajole, M. Pieroni, H. Gunosewoyo, G. Chen, S. K. Tipparaju, N. C. Ammerman, A. P. Kozikowski and W. R. Bishai, *Nat. Commun.*, 2013, **4**, 2907.
142. G. Poce, R. H. Bates, S. Alfonso, M. Cocozza, G. C. Porretta, L. Ballell, J. Rullas, F. Ortega, A. De Logu, E. Agus, V. La Rosa, M. R. Pasca, E. De Rossi, B. Wae, S. G. Franzblau, F. Manetti, M. Botta and M. Biava, *PLoS One*, 2013, **8**, e56980.
143. V. La Rosa, G. Poce, J. O. Canseco, S. Buroni, M. R. Pasca, M. Biava, R. M. Raju, G. C. Porretta, S. Alfonso, C. Battilocchio, B. Javid, F. Sorrentino, T. R. Ioerger, J. C. Sacchettini, F. Manetti, M. Botta, A. De Logu, E. J. Rubin and E. De Rossi, *Antimicrob. Agents Chemother.*, 2012, **56**, 324–331.
144. S. A. Stanley, S. S. Grant, T. Kawate, N. Iwase, M. Shimizu, C. Wivagg, M. Silvis, E. Kazyanskaya, J. Aquadro, A. Golas, M. Fitzgerald, H. Dai, L. Zhang and D. T. Hung, *ACS Chem. Biol.*, 2012, **7**, 1377–1384.
145. J. T. Belisle, V. D. Vissa, T. Sievert, K. Takayama, P. J. Brennan and G. S. Besra, *Science*, 1997, **276**, 1420–1422.
146. A. K. Sanki, J. Boucau, D. R. Ronning and S. J. Sucheck, *Glycoconjugate J.*, 2009, **26**, 589–596.
147. D. R. Ronning, T. Klabunde, G. S. Besra, V. D. Vissa, J. T. Belisle and J. C. Sacchettini, *Nat. Struct. Biol.*, 2000, **7**, 141–146.

148. D. H. Anderson, G. Harth, M. A. Horwitz and D. Eisenberg, *J. Mol. Biol.*, 2001, **307**, 671–681.
149. D. R. Ronning, V. Vissa, G. S. Besra, J. T. Belisle and J. C. Sacchettini, *J. Biol. Chem.*, 2004, **279**, 36771–36777.
150. L. Y. Armitige, C. Jagannath, A. R. Wanger and S. J. Norris, *Infect. Immun.*, 2000, **68**, 767–778.
151. C. Abou-Zeid, T. L. Ratliff, H. G. Wiker, M. Harboe, J. Bennedsen and G. A. Rook, *Infect. Immun.*, 1988, **56**, 3046–3051.
152. L. S. Schlesinger and M. A. Horwitz, *J. Immunol.*, 1991, **147**, 1983–1994.
153. J. S. Schorey, M. C. Carroll and E. J. Brown, *Science*, 1997, **277**, 1091–1093.
154. G. Harth, B. Y. Lee, J. Wang, D. L. Clemens and M. A. Horwitz, *Infect. Immun.*, 1996, **64**, 3038–3047.
155. J. Kraut, *Annu. Rev. Biochem.*, 1977, **46**, 331–358.
156. D. A. Ibrahim, J. Boucau, D. H. Lajiness, S. K. Veleti, K. R. Trabbic, S. S. Adams, D. R. Ronning and S. J. Sucheck, *Bioconjugate Chem.*, 2012, **23**, 2403–2416.
157. C. Gould and P. I. Folb, *Nonproliferation Rev.*, 2000, **7**, 10–23.
158. S. Gobec, I. Plantan, J. Mravljak, U. Svajger, R. A. Wilson, G. S. Besra, S. L. Soares, R. Appelberg and D. Kikelj, *Eur. J. Med. Chem.*, 2007, **42**, 54–63.
159. J. Wang, B. Elchert, Y. Hui, J. Y. Takemoto, M. Bensaci, J. Wennergren, H. Chang, R. Rai and C. W. Chang, *Bioorg. Med. Chem.*, 2004, **12**, 6397–6413.
160. J. D. Rose, J. A. Maddry, R. N. Comber, W. J. Suling, L. N. Wilson and R. C. Reynolds, *Carbohydr. Res.*, 2002, **337**, 105–120.
161. C. S. Barry, K. M. Backus, C. E. Barry 3rd and B. G. Davis, *J. Am. Chem. Soc.*, 2011, **133**, 13232–13235.
162. V. Puech, N. Bayan, K. Salim, G. Leblon and M. Daffe, *Mol. Microbiol.*, 2000, **35**, 1026–1041.
163. A. K. Sanki, J. Boucau, P. Srivastava, S. S. Adams, D. R. Ronning and S. J. Sucheck, *Bioorg. Med. Chem.*, 2008, **16**, 5672–5682.
164. A. K. Sanki, J. Boucau, F. E. Umesiri, D. R. Ronning and S. J. Sucheck, *Mol. BioSyst.*, 2009, **5**, 945–956.
165. C. Scheich, V. Puetter and M. Schade, *J. Med. Chem.*, 2010, **53**, 8362–8367.
166. T. Warrier, M. Tropis, J. Werngren, A. Diehl, M. Gengenbacher, B. Schlegel, M. Schade, H. Oschkinat, M. Daffe, S. Hoffner, A. N. Eddine and S. H. Kaufmann, *Antimicrob. Agents Chemother.*, 2012, **56**, 1735–1743.
167. L. Favrot, A. E. Grzegorzewicz, D. H. Lajiness, R. K. Marvin, J. Boucau, D. Isailovic, M. Jackson and D. R. Ronning, *Nat. Commun.*, 2013, **4**, 2748.
168. H. X. Ngo, S. K. Shrestha, K. D. Green and S. Garneau-Tsodikova, *Bioorg. Med. Chem.*, 2016, **24**, 6298–6306.
169. M. J. Parnham and H. Sies, *Biochem. Pharmacol.*, 2013, **86**, 1248–1253.
170. T. Yamaguchi, K. Sano, K. Takakura, I. Saito, Y. Shinohara, T. Asano and H. Yasuhara, *Stroke*, 1998, **29**, 12–17.
171. E. Lynch and J. Kil, *Semin. Hear.*, 1009, **30**, 47–55.
172. G. K. Azad and R. S. Tomar, *Mol. Biol. Rep.*, 2014, **41**, 4865–4879.

173. T. R. Ioerger, T. O'Malley, R. Liao, K. M. Guinn, M. J. Hickey, N. Mohaideen, K. C. Murphy, H. I. Boshoff, V. Mizrahi, E. J. Rubin, C. M. Sassetti, C. E. Barry 3rd, D. R. Sherman, T. Parish and J. C. Sacchettini, *PLoS One*, 2013, **8**, e75245.
174. S. A. Stanley, T. Kawate, N. Iwase, M. Shimizu, A. E. Clatworthy, E. Kazyanskaya, J. C. Sacchettini, T. R. Ioerger, N. A. Siddiqi, S. Minami, J. A. Aquadro, S. S. Grant, E. J. Rubin and D. T. Hung, *Proc. Natl. Acad. Sci. U. S. A.*, 2013, **110**, 11565–11570.
175. W. Li, S. Gu, J. Fleming and L. Bi, *Sci. Rep.*, 2015, **5**, 15493.
176. S. Galandrin, V. Guillet, R. S. Rane, M. Leger, N. Radha, N. Eynard, K. Das, T. S. Balganesh, L. Mourey, M. Daffe and H. Marrakchi, *J. Biomol. Screen.*, 2013, **18**, 576–587.
177. S. Thanna and S. J. Sucheck, *MedChemComm*, 2016, **7**, 69–85.
178. K. A. De Smet, A. Weston, I. N. Brown, D. B. Young and B. D. Robertson, *Microbiology*, 2000, **146**(pt 1), 199–208.
179. L. Shi, H. Zhang, Y. Qiu, Q. Wang, X. Wu, H. Wang, X. Zhang and D. Lin, *Acta Biochim. Biophys. Sin.*, 2013, **45**, 837–844.
180. H. N. Murphy, G. R. Stewart, V. V. Mischenko, A. S. Apt, R. Harris, M. S. McAlister, P. C. Driscoll, D. B. Young and B. D. Robertson, *J. Biol. Chem.*, 2005, **280**, 14524–14529.
181. H. N. Murphy, G. R. Stewart, V. V. Mischenko, A. S. Apt, R. Harris, M. S. McAlister, P. C. Driscoll, D. B. Young and B. D. Robertson, *J. Biol. Chem.*, 2005, **280**, 14524–14529.
182. J. Neres, F. Pojer, E. Molteni, L. R. Chiarelli, N. Dhar, S. Boy-Rottger, S. Buroni, E. Fullam, G. Degiacomi, A. P. Lucarelli, R. J. Read, G. Zanoni, D. E. Edmondson, E. De Rossi, M. R. Pasca, J. D. McKinney, P. J. Dyson, G. Riccardi, A. Mattevi, S. T. Cole and C. Binda, *Sci. Transl. Med.*, 2012, **4**, 150ra121.
183. I. Bhutani, S. Loharch, P. Gupta, R. Madathil and R. Parkesh, *PLoS One*, 2015, **10**, e0119771.
184. C. M. Sassetti, D. H. Boyd and E. J. Rubin, *Proc. Natl. Acad. Sci. U. S. A.*, 2001, **98**, 12712–12717.
185. C. M. Sassetti, D. H. Boyd and E. J. Rubin, *Mol. Microbiol.*, 2003, **48**, 77–84.
186. S. M. Batt, T. Jabeen, V. Bhowruth, L. Quill, P. A. Lund, L. Eggeling, L. J. Alderwick, K. Futterer and G. S. Besra, *Proc. Natl. Acad. Sci. U. S. A.*, 2012, **109**, 11354–11359.
187. T. Christophe, M. Jackson, H. K. Jeon, D. Fenistein, M. Contreras-Dominguez, J. Kim, A. Genovesio, J. P. Carralot, F. Ewann, E. H. Kim, S. Y. Lee, S. Kang, M. J. Seo, E. J. Park, H. Skovierova, H. Pham, G. Riccardi, J. Y. Nam, L. Marsollier, M. Kempf, M. L. Joly-Guillou, T. Oh, W. K. Shin, Z. No, U. Nehrbass, R. Brosch, S. T. Cole and P. Brodin, *PLoS Pathog.*, 2009, **5**, e1000645.
188. S. Magnet, R. C. Hartkoorn, R. Szekely, J. Pato, J. A. Triccas, P. Schneider, C. Szantai-Kis, L. Orfi, M. Chambon, D. Banfi, M. Bueno, G. Turcatti, G. Keri and S. T. Cole, *Tuberculosis*, 2010, **90**, 354–360.

189. L. J. Alderwick, M. Seidel, H. Sahm, G. S. Besra and L. Eggeling, *J. Biol. Chem.*, 2006, **281**, 15653–15661.
190. H. Huang, M. S. Scherman, W. D'Haeze, D. Vereecke, M. Holsters, D. C. Crick and M. R. McNeil, *J. Biol. Chem.*, 2005, **280**, 24539–24543.
191. L. Kremer, L. G. Dover, C. Morehouse, P. Hitchin, M. Everett, H. R. Morris, A. Dell, P. J. Brennan, M. R. McNeil, C. Flaherty, K. Duncan and G. S. Besra, *J. Biol. Chem.*, 2001, **276**, 26430–26440.
192. F. Pan, M. Jackson, Y. Ma and M. McNeil, *J. Bacteriol.*, 2001, **183**, 3991–3998.
193. W. Li, Y. Xin, M. R. McNeil and Y. Ma, *Biochem. Biophys. Res. Commun.*, 2006, **342**, 170–178.
194. M. Lavollay, M. Arthur, M. Fourgeaud, L. Dubost, A. Marie, N. Veziris, D. Blanot, L. Gutmann and J. L. Mainardi, *J. Bacteriol.*, 2008, **190**, 4360–4366.
195. R. Gupta, M. Lavollay, J. L. Mainardi, M. Arthur, W. R. Bishai and G. Lamichhane, *Nat. Med.*, 2010, **16**, 466–469.
196. J. L. Mainardi, M. Fourgeaud, J. E. Hugonnet, L. Dubost, J. P. Brouard, J. Ouazzani, L. B. Rice, L. Gutmann and M. Arthur, *J. Biol. Chem.*, 2005, **280**, 38146–38152.
197. M. Lavollay, M. Fourgeaud, J. L. Herrmann, L. Dubost, A. Marie, L. Gutmann, M. Arthur and J. L. Mainardi, *J. Bacteriol.*, 2011, **193**, 778–782.
198. S. B. Erdemli, R. Gupta, W. R. Bishai, G. Lamichhane, L. M. Amzel and M. A. Bianchet, *Structure*, 2012, **20**, 2103–2115.
199. H. S. Kim, J. Kim, H. N. Im, J. Y. Yoon, D. R. An, H. J. Yoon, J. Y. Kim, H. K. Min, S. J. Kim, J. Y. Lee, B. W. Han and S. W. Suh, *Acta Crystallogr., Sect. D: Biol. Crystallogr.*, 2013, **69**, 420–431.
200. J. R. Silva, T. Govender, G. E. Maguire, H. G. Kruger, J. Lameira, A. E. Roitberg and C. N. Alves, *Chem. Commun.*, 2015, **51**, 12560–12562.
201. J. B. Billones, M. C. Carrillo, V. G. Organo, S. J. Macalino, J. B. Sy, I. A. Emnacen, N. A. Clavio and G. P. Concepcion, *Drug Des., Dev. Ther.*, 2016, **10**, 1147–1157.
202. J. Coyette and G. D. Shockman, *J. Bacteriol.*, 1973, **114**, 34–41.
203. J. Allignet, S. Aubert, K. G. Dyke and N. El Solh, *Infect. Immun.*, 2001, **69**, 712–718.
204. L. L. Deng, D. E. Humphries, R. D. Arbeit, L. E. Carlton, S. C. Smole and J. D. Carroll, *Biochim. Biophys. Acta*, 2005, **1747**, 57–66.
205. D. M. Prigozhin, D. Mavrici, J. P. Huizar, H. J. Vansell and T. Alber, *J. Biol. Chem.*, 2013, **288**, 31549–31555.
206. A. Padhi, S. K. Naik, S. Sengupta, G. Ganguli and A. Sonawane, *Microbes Infect.*, 2016, **18**, 224–236.
207. L. G. Wayne and L. G. Hayes, *Infect. Immun.*, 1996, **64**, 2062–2069.
208. J. C. Betts, P. T. Lukey, L. C. Robb, R. A. McAdam and K. Duncan, *Mol. Microbiol.*, 2002, **43**, 717–731.
209. C. D. Sohaskey and M. I. Voskuil, *Methods Mol. Biol.*, 2015, **1285**, 201–213.
210. S. P. Rao, S. Alonso, L. Rand, T. Dick and K. Pethe, *Proc. Natl. Acad. Sci. U. S. A.*, 2008, **105**, 11945–11950.

211. D. R. Sherman, M. Voskuil, D. Schnappinger, R. Liao, M. I. Harrell and G. K. Schoolnik, *Proc. Natl. Acad. Sci. U. S. A.*, 2001, **98**, 7534–7539.
212. K. Andries, P. Verhasselt, J. Guillemont, H. W. Gohlmann, J. M. Neefs, H. Winkler, J. Van Gestel, P. Timmerman, M. Zhu, E. Lee, P. Williams, D. de Chaffoy, E. Huitric, S. Hoffner, E. Cambau, C. Truffot-Pernot, N. Lounis and V. Jarlier, *Science*, 2005, **307**, 223–227.
213. A. Koul, L. Vranckx, N. Dendouga, W. Balemans, I. Van den Wyngaert, K. Vergauwen, H. W. Gohlmann, R. Willebrords, A. Poncelet, J. Guillemont, D. Bald and K. Andries, *J. Biol. Chem.*, 2008, **283**, 25273–25280.
214. M. Gengenbacher, S. P. Rao, K. Pethe and T. Dick, *Microbiology*, 2010, **156**, 81–87.
215. A. H. Diacon, A. Pym, M. Grobusch, R. Patientia, R. Rustomjee, L. Page-Shipp, C. Pistorius, R. Krause, M. Bogoshi, G. Churchyard, A. Venter, J. Allen, J. C. Palomino, T. De Marez, R. P. van Heeswijk, N. Lounis, P. Meyvisch, J. Verbeeck, W. Parys, K. de Beule, K. Andries and D. F. Mc Neeley, *N. Engl. J. Med.*, 2009, **360**, 2397–2405.
216. V. M. Reddy, L. Einck, K. Andries and C. A. Nacy, *Antimicrob. Agents Chemother.*, 2010, **54**, 2840–2846.
217. P. Santos, A. Gordillo, L. Osses, L. M. Salazar and C. Y. Soto, *Peptides*, 2012, **36**, 121–128.
218. P. A. Mak, S. P. Rao, M. Ping Tan, X. Lin, J. Chyba, J. Tay, S. H. Ng, B. H. Tan, J. Cherian, J. Duraiswamy, P. Bifani, V. Lim, B. H. Lee, N. Ling Ma, D. Beer, P. Thayalan, K. Kuhen, A. Chatterjee, F. Supek, R. Glynne, J. Zheng, H. I. Boshoff, C. E. Barry 3rd, T. Dick, K. Pethe and L. R. Camacho, *ACS Chem. Biol.*, 2012, **7**, 1190–1197.
219. J. R. Coggins, C. Abell, L. B. Evans, M. Frederickson, D. A. Robinson, A. W. Roszak and A. P. Lapthorn, *Biochem. Soc. Trans.*, 2003, **31**, 548–552.
220. K. M. Herrmann and L. M. Weaver, *Annu. Rev. Plant Physiol. Plant Mol. Biol.*, 1999, **50**, 473–503.
221. B. D. Davis, *Adv. Enzymol. Relat. Subj. Biochem.*, 1955, **16**, 247–312.
222. S. Reichau, W. Jiao, S. R. Walker, R. D. Hutton, E. N. Baker and E. J. Parker, *J. Biol. Chem.*, 2011, **286**, 16197–16207.
223. P. H. Ray, *J. Bacteriol.*, 1980, **141**, 635–644.
224. F. Kona, X. Xu, P. Martin, P. Kuzmic and D. L. Gatti, *Biochemistry*, 2007, **46**, 4532–4544.
225. T. Garbe, S. Servos, A. Hawkins, G. Dimitriadis, D. Young, G. Dougan and I. Charles, *Mol. Gen. Genet.*, 1991, **228**, 385–392.
226. A. T. Tran, N. P. West, W. J. Britton and R. J. Payne, *ChemMedChem*, 2012, **7**, 1031–1043.
227. R. J. Payne, F. Peyrot, O. Kerbarh, A. D. Abell and C. Abell, *ChemMedChem*, 2007, **2**, 1015–1029.
228. J. H. Pereira, J. S. de Oliveira, F. Canduri, M. V. Dias, M. S. Palma, L. A. Basso, D. S. Santos and W. F. de Azevedo Jr, *Acta Crystallogr., Sect. D: Biol. Crystallogr.*, 2004, **60**, 2310–2319.
229. T. Parish and N. G. Stoker, *Microbiology*, 2002, **148**, 3069–3077.

230. V. S. Rajput, R. Mehra, S. Kumar, A. Nargotra, P. P. Singh and I. A. Khan, *Appl. Microbiol. Biotechnol.*, 2016, **100**, 5415–5426.
231. J. E. Griffin, J. D. Gawronski, M. A. Dejesus, T. R. Ioerger, B. J. Akerley and C. M. Sassetti, *PLoS Pathog.*, 2011, **7**, e1002251.
232. T. Lonhienne, A. Nouwens, C. M. Williams, J. A. Fraser, Y. T. Lee, N. P. West and L. W. Guddat, *Angew. Chem.*, 2016, **55**, 4247–4251.
233. J. A. Grandoni, P. T. Marta and J. V. Schloss, *J. Antimicrob. Chemother.*, 1998, **42**, 475–482.
234. Y. Lv, A. Kandale, S. J. Wun, R. P. McGeary, S. J. Williams, B. Kobe, V. Sieber, M. A. Schembri, G. Schenk and L. W. Guddat, *FEBS J.*, 2016, **283**, 1184–1196.
235. R. J. Cox, A. Sutherland and J. C. Vederas, *Bioorg. Med. Chem.*, 2000, **8**, 843–871.
236. M. S. Pavelka Jr and W. R. Jacobs Jr, *J. Bacteriol.*, 1996, **178**, 6496–6507.
237. A. M. Paiva, D. E. Vanderwall, J. S. Blanchard, J. W. Kozarich, J. M. Williamson and T. M. Kelly, *Biochim. Biophys. Acta*, 2001, **1545**, 67–77.
238. K. Mdluli and M. Spigelman, *Curr. Opin. Pharmacol.*, 2006, **6**, 459–467.
239. S. Pedreno, J. P. Pisco, G. Larrouy-Maumus, G. Kelly and L. P. de Carvalho, *Biochemistry*, 2012, **51**, 8027–8038.
240. J. Meng, Y. Yang, C. Xiao, Y. Guan, X. Hao, Q. Deng and Z. Lu, *Int. J. Mol. Sci.*, 2015, **16**, 23572–23586.
241. V. K. Sambandamurthy, X. Wang, B. Chen, R. G. Russell, S. Derrick, F. M. Collins, S. L. Morris and W. R. Jacobs Jr, *Nat. Med.*, 2002, **8**, 1171–1174.
242. V. K. Sambandamurthy, S. C. Derrick, T. Hsu, B. Chen, M. H. Larsen, K. V. Jalapathy, M. Chen, J. Kim, S. A. Porcelli, J. Chan, S. L. Morris and W. R. Jacobs Jr, *Vaccine*, 2006, **24**, 6309–6320.
243. Y. Yang, P. Gao, Y. Liu, X. Ji, M. Gan, Y. Guan, X. Hao, Z. Li and C. Xiao, *Bioorg. Med. Chem. Lett.*, 2011, **21**, 3943–3946.
244. A. Ciulli, D. E. Scott, M. Ando, F. Reyes, S. A. Saldanha, K. L. Tuck, D. Y. Chirgadze, T. L. Blundell and C. Abell, *ChemBioChem*, 2008, **9**, 2606–2611.
245. J. Venkatraman, J. Bhat, S. M. Solapure, J. Sandesh, D. Sarkar, S. Aishwarya, K. Mukherjee, S. Datta, K. Malolanarasimhan, B. Bandodkar and K. S. Das, *J. Biomol. Screening*, 2012, **17**, 293–302.
246. B. K. Reddy, S. Landge, S. Ravishankar, V. Patil, V. Shinde, S. Tantry, M. Kale, A. Raichurkar, S. Menasinakai, N. V. Mudugal, A. Ambady, A. Ghosh, R. Tunduguru, P. Kaur, R. Singh, N. Kumar, S. Bharath, A. Sundaram, J. Bhat, V. K. Sambandamurthy, C. Bjorkelid, T. A. Jones, K. Das, B. Bandodkar, K. Malolanarasimhan, K. Mukherjee and V. Ramachandran, *Antimicrob. Agents Chemother.*, 2014, **58**, 3312–3326.
247. C. Bjorkelid, T. Bergfors, A. K. Raichurkar, K. Mukherjee, K. Malolanarasimhan, B. Bandodkar and T. A. Jones, *J. Biol. Chem.*, 2013, **288**, 18260–18270.
248. A. Ambady, D. Awasthy, R. Yadav, S. Basuthkar, K. Seshadri and U. Sharma, *Tuberculosis*, 2012, **92**, 521–528.

249. R. Li, R. Sirawaraporn, P. Chitnumsub, W. Sirawaraporn, J. Wooden, F. Athappilly, S. Turley and W. G. Hol, *J. Mol. Biol.*, 2000, **295**, 307–323.
250. A. B. Gerum, J. E. Ulmer, D. P. Jacobus, N. P. Jensen, D. R. Sherman and C. H. Sibley, *Antimicrob. Agents Chemother.*, 2002, **46**, 3362–3369.
251. E. Morgunova, W. Meining, B. Illarionov, I. Haase, G. Jin, A. Bacher, M. Cushman, M. Fischer and R. Ladenstein, *Biochemistry*, 2005, **44**, 2746–2758.
252. P. S. Shirude, B. Paul, N. Roy Choudhury, C. Kedari, B. Bandodkar and B. G. Ugarkar, *ACS Med. Chem. Lett.*, 2012, **3**, 736–740.
253. E. A. Weinstein, T. Yano, L. S. Li, D. Avarbock, A. Avarbock, D. Helm, A. A. McColm, K. Duncan, J. T. Lonsdale and H. Rubin, *Proc. Natl. Acad. Sci. U. S. A.*, 2005, **102**, 4548–4553.
254. T. Yano, L. S. Li, E. Weinstein, J. S. Teh and H. Rubin, *J. Biol. Chem.*, 2006, **281**, 11456–11463.
255. E. Fioravanti, V. Adam, H. Munier-Lehmann and D. Bourgeois, *Biochemistry*, 2005, **44**, 130–137.
256. T. Biswas, K. D. Green, S. Garneau-Tsodikova and O. V. Tsodikov, *Biochemistry*, 2013, **52**, 6905–6910.
257. A. H. Pang, S. Garneau-Tsodikova and O. V. Tsodikov, *In vitro* assays to identify antibiotics targeting DNA metabolism, in *Antibiotics: Methods and Protocols*, ed. Peter Sass, Springer, 2017, pp. 175–200.
258. C. Gajadeera, M. J. Willby, K. D. Green, P. Shaul, M. Fridman, S. Garneau-Tsodikova, J. E. Posey and O. V. Tsodikov, *J. Antibiot.*, 2015, **68**, 153–157.
259. S. K. Srivastava, D. Dube, N. Tewari, N. Dwivedi, R. P. Tripathi and R. Ramachandran, *Nucleic Acids Res.*, 2005, **33**, 7090–7101.
260. S. K. Srivastava, D. Dube, V. Kukshal, A. K. Jha, K. Hajela and R. Ramachandran, *Proteins*, 2007, **69**, 97–111.
261. M. Berney, T. E. Hartman and W. R. Jacobs Jr, *mBio*, 2014, **5**, e01275–14.
262. A. Koul, L. Vranckx, N. Dhar, H. W. Gohlmann, E. Ozdemir, J. M. Neefs, M. Schulz, P. Lu, E. Mortz, J. D. McKinney, K. Andries and D. Bald, *Nat. Commun.*, 2014, **5**, 3369.
263. J. Rybniker, A. Vocat, C. Sala, P. Busso, F. Pojer, A. Benjak and S. T. Cole, *Nat. Commun.*, 2015, **6**, 7659.
264. K. A. Abrahams, J. A. Cox, V. L. Spivey, N. J. Loman, M. J. Pallen, C. Constantinidou, R. Fernandez, C. Alemparte, M. J. Remuinan, D. Barros, L. Ballell and G. S. Besra, *PLoS One*, 2012, **7**, e52951.
265. K. Pethe, P. Bifani, J. Jang, S. Kang, S. Park, S. Ahn, J. Jiricek, J. Jung, H. K. Jeon, J. Cechetto, T. Christophe, H. Lee, M. Kempf, M. Jackson, A. J. Lenaerts, H. Pham, V. Jones, M. J. Seo, Y. M. Kim, M. Seo, J. J. Seo, D. Park, Y. Ko, I. Choi, R. Kim, S. Y. Kim, S. Lim, S. A. Yim, J. Nam, H. Kang, H. Kwon, C. T. Oh, Y. Cho, Y. Jang, J. Kim, A. Chua, B. H. Tan, M. B. Nanjundappa, S. P. Rao, W. S. Barnes, R. Wintjens, J. R. Walker, S. Alonso, S. Lee, J. Kim, S. Oh, T. Oh, U. Nehrbass, S. J. Han, Z. No, J. Lee, P. Brodin, S. N. Cho, K. Nam and J. Kim, *Nat. Med.*, 2013, **19**, 1157–1160.

266. P. R. Wheeler and C. Ratledge, *J. Gen. Microbiol.*, 1988, **134**, 2111–2121.
267. K. Honer Zu Bentrup, A. Miczak, D. L. Swenson and D. G. Russell, *J. Bacteriol.*, 1999, **181**, 7161–7167.
268. J. E. Graham and J. E. Clark-Curtiss, *Proc. Natl. Acad. Sci. U. S. A.*, 1999, **96**, 11554–11559.
269. V. Sharma, S. Sharma, K. Hoener zu Bentrup, J. D. McKinney, D. G. Russell, W. R. Jacobs Jr and J. C. Sacchettini, *Nat. Struct. Biol.*, 2000, **7**, 663–668.
270. B. Bai, J.-P. Xie, J.-F. Yan, H.-H. Wang and C.-H. Hu, *Drug Dev. Res.*, 2006, **67**, 818–823.
271. L. Ji, Q. Long, D. Yang and J. Xie, *Int. J. Biol. Sci.*, 2011, **7**, 376–382.
272. N. K. Tonks and B. G. Neel, *Curr. Opin. Cell Biol.*, 2001, **13**, 182–195.
273. A. Koul, A. Choidas, M. Treder, A. K. Tyagi, K. Drlica, Y. Singh and A. Ullrich, *J. Bacteriol.*, 2000, **182**, 5425–5432.
274. A. Koul, T. Herget, B. Klebl and A. Ullrich, *Nat. Rev. Microbiol.*, 2004, **2**, 189–202.
275. R. Singh, V. Rao, H. Shakila, R. Gupta, A. Khera, N. Dhar, A. Singh, A. Koul, Y. Singh, M. Naseema, P. R. Narayanan, C. N. Paramasivan, V. D. Ramanathan and A. K. Tyagi, *Mol. Microbiol.*, 2003, **50**, 751–762.
276. B. Zhou, Y. He, X. Zhang, J. Xu, Y. Luo, Y. Wang, S. G. Franzblau, Z. Yang, R. J. Chan, Y. Liu, J. Zheng and Z. Y. Zhang, *Proc. Natl. Acad. Sci. U. S. A.*, 2010, **107**, 4573–4578.
277. A. Noren-Muller, I. Reis-Correa Jr, H. Prinz, C. Rosenbaum, K. Saxena, H. J. Schwalbe, D. Vestweber, G. Cagna, S. Schunk, O. Schwarz, H. Schiewe and H. Waldmann, *Proc. Natl. Acad. Sci. U. S. A.*, 2006, **103**, 10606–10611.
278. L. P. Tan, H. Wu, P. Y. Yang, K. A. Kalesh, X. Zhang, M. Hu, R. Srinivasan and S. Q. Yao, *Org. Lett.*, 2009, **11**, 5102–5105.
279. G. Liu, Z. Xin, H. Liang, C. Abad-Zapatero, P. J. Hajduk, D. A. Janowick, B. G. Szczepankiewicz, Z. Pei, C. W. Hutchins, S. J. Ballaron, M. A. Stashko, T. H. Lubben, C. E. Berg, C. M. Rondinone, J. M. Trevillyan and M. R. Jirousek, *J. Med. Chem.*, 2003, **46**, 3437–3440.
280. G. Liu, Z. Xin, Z. Pei, P. J. Hajduk, C. Abad-Zapatero, C. W. Hutchins, H. Zhao, T. H. Lubben, S. J. Ballaron, D. L. Haasch, W. Kaszubska, C. M. Rondinone, J. M. Trevillyan and M. R. Jirousek, *J. Med. Chem.*, 2003, **46**, 4232–4235.
281. R. He, Z. Yu, Y. He, L. F. Zeng, J. Xu, L. Wu, A. M. Gunawan, L. Wang, Z. X. Jiang and Z. Y. Zhang, *ChemMedChem*, 2010, **5**, 2051–2056.
282. V. V. Vintonyak, K. Warburg, H. Kruse, S. Grimme, K. Hubel, D. Rauh and H. Waldmann, *Angew. Chem.*, 2010, **49**, 5902–5905.
283. R. He, Z. H. Yu, R. Y. Zhang, L. Wu, A. M. Gunawan and Z. Y. Zhang, *ACS Med. Chem. Lett.*, 2015, **6**, 1231–1235.
284. S. Hameed, R. Pal and Z. Fatima, *Open Microbiol. J.*, 2015, **9**, 91–97.
285. M. Luo, E. A. Fadeev and J. T. Groves, *Nat. Chem. Biol.*, 2005, **1**, 149–153.
286. J. J. De Voss, K. Rutter, B. G. Schroeder, H. Su, Y. Zhu and C. E. Barry 3rd, *Proc. Natl. Acad. Sci. U. S. A.*, 2000, **97**, 1252–1257.
287. C. Ratledge and L. G. Dover, *Annu. Rev. Microbiol.*, 2000, **54**, 881–941.

288. J. Gobin and M. A. Horwitz, *J. Exp. Med.*, 1996, **183**, 1527–1532.
289. C. A. Golden, I. Kochan and D. R. Spriggs, *Infect. Immun.*, 1974, **9**, 34–40.
290. J. Gobin, C. H. Moore, J. R. Reeve Jr, D. K. Wong, B. W. Gibson and M. A. Horwitz, *Proc. Natl. Acad. Sci. U. S. A.*, 1995, **92**, 5189–5193.
291. L. E. Quadri, J. Sello, T. A. Keating, P. H. Weinreb and C. T. Walsh, *Chem. Biol.*, 1998, **5**, 631–645.
292. K. A. Brown and C. Ratledge, *Biochim. Biophys. Acta*, 1975, **385**, 207–220.
293. J. A. Ferreras, J. S. Ryu, F. Di Lello, D. S. Tan and L. E. Quadri, *Nat. Chem. Biol.*, 2005, **1**, 29–32.
294. S. Lun, H. Guo, J. Adamson, J. S. Cisar, T. D. Davis, S. S. Chavadi, J. D. Warren, L. E. Quadri, D. S. Tan and W. R. Bishai, *Antimicrob. Agents Chemother.*, 2013, **57**, 5138–5140.
295. J.-L. He and J.-P. Xie, *Acta Pharmacol. Sin.*, 2011, **1**, 8–13.
296. S. Ramon-Garcia, C. Ng, H. Anderson, J. D. Chao, X. Zheng, T. Pfeifer, Y. Av-Gay, M. Roberge and C. J. Thompson, *Antimicrob. Agents Chemother.*, 2011, **55**, 3861–3869.
297. J. E. Hugonnet, L. W. Tremblay, H. I. Boshoff, C. E. Barry 3rd and J. S. Blanchard, *Science*, 2009, **323**, 1215–1218.
298. H. Xu, S. Hazra and J. S. Blanchard, *Biochemistry*, 2012, **51**, 4551–4557.
299. K. D. Green, W. Chen and S. Garneau-Tsodikova, *ChemMedChem*, 2012, **7**, 73–77.
300. M. J. Willby, K. D. Green, C. S. Gajadeera, C. Hou, O. V. Tsodikov, J. E. Posey and S. Garneau-Tsodikova, *ACS Chem. Biol.*, 2016, **11**, 1639–1646.
301. Y. Li, K. D. Green, B. R. Johnson and S. Garneau-Tsodikova, *Antimicrob. Agents Chemother.*, 2015, **59**, 4148–4156.
302. C. M. Miller, N. R. Boulter, S. J. Fuller, A. M. Zakrzewski, M. P. Lees, B. M. Saunders, J. S. Wiley and N. C. Smith, *PLoS Pathog.*, 2011, **7**, e1002212.
303. S. L. Fernando, B. M. Saunders, R. Sluyter, K. K. Skarratt, H. Goldberg, G. B. Marks, J. S. Wiley and W. J. Britton, *Am. J. Respir. Crit. Care Med.*, 2007, **175**, 360–366.
304. I. Vergne, J. Chua, H. H. Lee, M. Lucas, J. Belisle and V. Deretic, *Proc. Natl. Acad. Sci. U. S. A.*, 2005, **102**, 4033–4038.
305. W. J. Li, D. F. Li, Y. L. Hu, X. E. Zhang, L. J. Bi and D. C. Wang, *Cell Res.*, 2013, **23**, 728–731.
306. Y. S. Cheng and J. C. Sacchettini, *Biochemistry*, 2016, **55**, 1107–1119.

CHAPTER 8

Antibacterial Leads Targeting Isoprenoid Biosynthesis

CARL J. BALIBAR[a]

[a]Merck Research Laboratories, 2015 Galloping Hill Road, K15-D409-12, Kenilworth, NJ 07033, USA
*E-mail: carl.balibar@merck.com

8.1 Introduction

Found in all living organisms, isoprenoids represent a ubiquitous class of compounds serving a myriad of vital biological functions. The diverse array of chemical structures classified as isoprenoids all derive from the common five-carbon precursor isopentenyl diphosphate (IPP) and its isomer dimethylallyl diphosphate (DMAPP). The regulation of IPP biosynthesis has been most extensively studied in eukaryotes, where it was first discovered in the 1960s,[1,2] and as a result, serves as an important therapeutic target in several human diseases. Targeting the rate-limiting step in IPP biosynthesis, HMG-CoA reductase, the statin class of molecules was first discovered in 1976[3] and by 1990 three drugs were on the market to treat hypercholesterolemia.[4] The bisphosphonate class of drugs used to treat diseases of excessive bone resorption, including osteoporosis and Paget's disease, targets farnesyl pyrophosphate synthase, an enzyme that catalyzes two consecutive condensations between IPP and DMAPP.[5] Finally, several inhibitors of farnesyl transferase, an enzyme that post-translationally appends the isoprenoid farnesyl

Drug Discovery Series No. 58
Antibiotic Drug Discovery: New Targets and Molecular Entities
Edited by Steven M. Firestine and Troy Lister

Published by the Royal Society of Chemistry, www.rsc.org

to proteins to facilitate proper membrane localization, are in clinical development to treat various cancers and Hutchinson-Gilford Progeria syndrome.[6]

Although focusing on eukaryotes has had an undeniable impact, targeting isoprenoid biosynthesis in pathogenic bacteria represents a significant opportunity to further advance human health by treating infectious diseases. In bacteria isoprenoids are essential for viability and virulence, as they are utilized for cell wall biogenesis, electron transport, protein synthesis, maintaining membrane architecture, and biosynthesis of secondary metabolites. Undecaprenyl phosphate, otherwise known as bactoprenol, is a C_{55} lipid carrier essential for synthesis and transport of GlcNAc-MurNAc-peptide monomers that ultimately comprise the peptidoglycan, a structure essential for maintaining cell shape and resisting high osmotic pressure. Bactoprenol also serves as a carrier for the synthesis of additional cell surface associated polysaccharides including teichoic acids, lipopolysaccharide, capsule polysaccharides, and common antigens.[7] The bacterial respiratory electron-transport chain involves the compounds ubiquinone, menaquinone, and heme A/O, which all contain redox active head groups linked to isoprenoid sidechains. Whereas ubiquinone and menaquinone are lipid-soluble quinones capable of interacting with membrane-bound protein complexes,[8] heme A/O is a prosthetic group found in certain membrane-bound proteins termed cytochromes.[9] Transfer RNAs (tRNAs) that recognize codons beginning with uridine are isopentylated at adenosine 37 adjacent to the anticodon. This modification improves the efficiency of translation, affects translational fidelity, controls expression of stress-responsive sigma factors, and its absence increases the rate of spontaneous mutations.[10,11] Hopanoids are pentacyclic isoprenoids derived from squalene that serve as functional homologs of eukaryotic sterols to modulate membrane fluidity. These molecules order lipids, provide the basis for membrane lateral segregation, and promote formation of fluid, biochemically active, yet mechanically durable membranes.[12] Many bacteria also utilize isoprenoid precursors to produce a variety of secondary metabolites, including carotenoids, which contribute to virulence of certain pathogenic bacteria.[13] The array of diverse and critical biological processes reliant upon isoprenoids illustrates the potential of targeting involved biosynthetic pathways in treating bacterial infections.

There are two biosynthetic pathways leading to formation of the common five-carbon isoprenoid precursors IPP and DMAPP (Scheme 8.1). The first is the classical mevalonate (MVA) pathway first discovered in eukaryotes and at one point thought to be universal in all organisms.[14] In the MVA pathway, acetyl-CoA and acetoacetyl-CoA, which itself is formed from two molecules of acetyl-CoA, undergo condensation to form 3-hydroxy-3-methylglutaryl (HMG)-CoA through action of HMG-CoA synthase (HMGS). HMG-CoA is then irreversibly reduced utilizing two equivalents of NADPH by HMG-CoA reductase (HMGR) to form the intermediate mevalonate for which the pathway is named. Subsequently, mevalonate is sequentially phosphorylated by mevalonate kinase (MK) and phosphomevalonate kinase (PMK) and then decarboxylated, with concomitant dehydration, by diphosphomevalonate

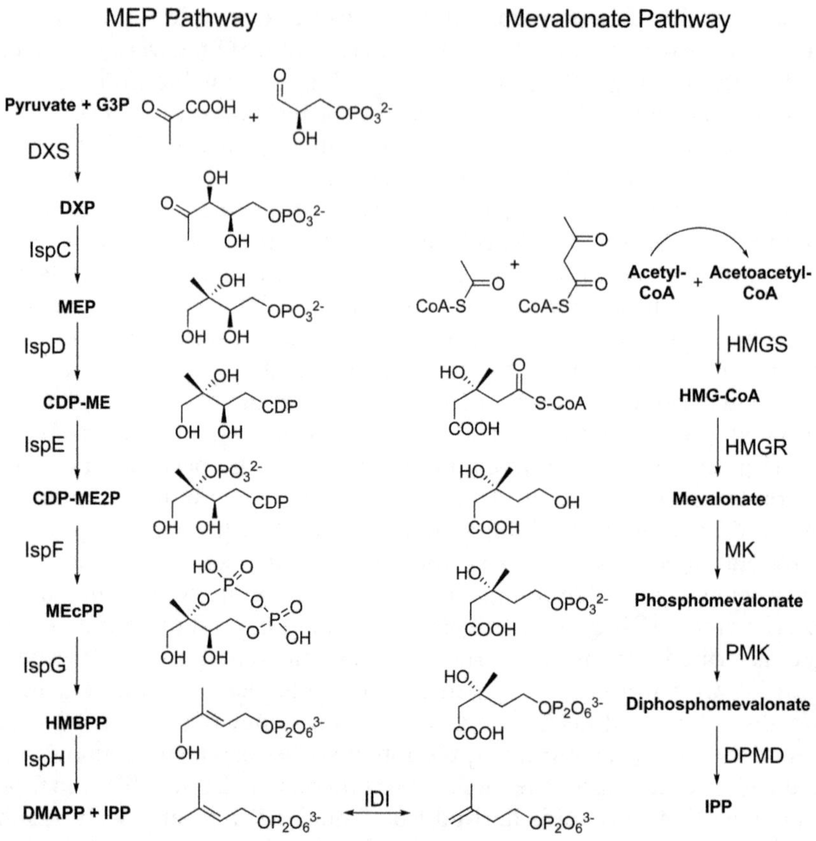

Scheme 8.1 The MEP and mevalonate pathways.

decarboxylase (DPMD) to yield IPP. Conversion of IPP to DMAPP is carried out by either a type I or type II IPP isomerase (IDI)[15,16] to generate the other essential isomer. In the late 1990s a second route, known as the methylerythritol 4-phosphate (MEP) pathway, was discovered to be responsible for IPP and DMAPP biosynthesis in various organisms.[17] In the MEP pathway, D-glyceraldehyde 3-phosphate (G3P) and pyruvate are condensed to form 1-deoxy-D-xylulose 5-phosphate (DXP), *via* the enzyme DXP synthase (DXS). DXP undergoes intramolecular rearrangement and reduction by DXP reductoisomerase (IspC) to form 2-C-methyl-D-erythritol 4-phosphate (MEP) for which the pathway is named. MEP is subsequently phosphocytidylylated, phosphorylated, and cyclized by the sequential actions of MEP cytidylyltransferase (IspD), 4-(cytidine 5′-diphospho)-3-C-methyl-D-erythritol kinase (IspE), and 2-C-methyl-D-erythritol 2,4-cyclodiphosphate synthase (IspF) to form 2-C-methyl-D-erythritol 2,4-cyclodiphosphate (MEcPP). Reductive ring opening of MEcPP by 4-hydroxy-3-methylbut-2-enyl diphosphate (HMBPP) synthase (IspG) followed by reductive dehydroxyaltion by HMBPP reductase (IspH) ultimately yields both IPP and DMAPP.

With the global threat of multidrug resistance ever growing, there is a dire need to develop novel agents against bacterial pathogens. Bacteria causing the most frequent and difficult to treat US hospital infections have been termed the ESKAPE pathogens and include the Gram-positives *Enterococcus faecium* and *Staphylococcus aureus*, and Gram-negatives *Klebsiella pneumoniae* (along with other *Enterobacteriacae* like *Escherichia coli*), *Acinetobacter baumannii*, *Pseudomonas aeruginosa*, and *Enterobacter* spp.[18,19] Isoprenoid biosynthesis represents a large and underexplored target space for discovery and development of antibacterials for these organisms. Of the ESKAPE pathogens, the Gram-positives synthesize isoprenoids *via* the MVA pathway and the Gram-negatives produce isoprenoids *via* the MEP pathway.[20–22] The success in targeting the human MVA pathway augurs well for the ability to intervene in the homologous pathway in Gram-positives, while the unique MEP pathway constituting enzymes with little homology to mammalian proteins offers the possibility of creating antibiotics with safer toxicity profiles and narrower spectrum that could spare some of the beneficial bacteria that reside in the human microbiota. This chapter will discuss recent advances in discovery and development of inhibitors targeting isoprenoid pathways from pathogenic bacteria.

8.2 Targeting the MVA Pathway

8.2.1 Historic Compounds Inhibiting the MVA Pathway

The first forays into exploring the potential for developing antibacterial compounds targeting the MVA pathway stemmed from the work done on the homologous pathway in humans. Observational studies that found patients on statin treatment had better outcomes of severe bacterial infections[23,24] led to proposals that statins may have a direct antibacterial effect on pathogens such as *S. aureus* and *Streptococcus pneumoniae* through inhibition of bacterial HMGR. Ultimately, it was found that only certain statins, such as simvastatin but not fluvastatin or prevastatin, had measurable direct antibacterial activity, though at concentrations far exceeding peak plasma concentrations attained in human serum after dosing.[25,26] The direct antibacterial activity of simvastatin is not a consequence of inhibiting HMGR because growth cannot be rescued by supplementing with exogenous mevalonate, a downstream pathway intermediate, but mostly likely is a result of disruption of bacterial membranes due to its hydrophobicity.[26] As a class, the beneficial effects of statins during infection are probably due to anti-inflammatory properties targeting the host.[27] Notwithstanding a lack of activity for the statin class of molecules on prokaryotic HMGR, there is a broad-spectrum natural product which does inhibit the MVA pathway. Hymeglusin **1** (Figure 8.1) was first discovered in 1971[28] and demonstrated to decrease cholesterogenesis in cell-free rat liver enzyme system by inhibiting HMGS.[29] Subsequent studies revealed that **1** had antimicrobial properties against several fungi and bacteria,[30] but unlike in the case of the statins, the mechanism of action remained inhibition of HMGS.[31] Detailed biochemical and structural studies showed

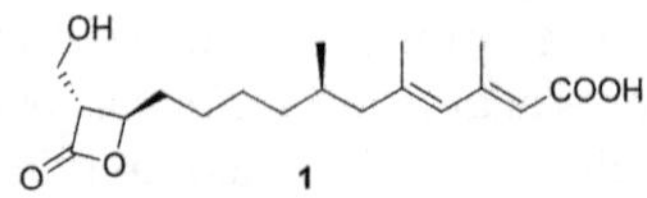

1

Figure 8.1 Structure of hymeglusin (Compound **1**).

that **1** inhibits *Enterococcus faecalis* HMGS through covalent inactivation of the active site cysteine in a similar fashion to HMGS of plant and animal origin, however, with different affinities and kinetics. Although *E. faecalis* HMGS was less sensitive to inhibition by **1** than human HMGS, with a 10-fold higher K_i, inactivation of the bacterial HMGS occurred faster and was less transient. Differences in the binding kinetics and affinities may be explained by a more restricted and less solvent-accessible binding pocket providing access to the catalytic cysteine in the prokaryotic enzyme,[31] which may provide sufficient differentiation to design inhibitors selective for bacterial HMGS.

8.2.2 Screening for Inhibitors of the MVA Pathway in Bacteria

Findings that all five genes of the MVA pathway are essential for growth of *S. pneumoniae*[22] and the discovery of natural products inhibiting bacterial isoprenoid biosynthetic enzymes have elicited interest in discovering and developing antibacterial compounds specifically targeting the prokaryotic pathway. In order to validate the MVA pathway as a target in the common human pathogen *S. aureus*, detailed experiments were carried out in which HMGS, HMGR, MK, DMK, and DPMD—encoded by the genes *mvaS*, *mvaA*, *mvaK1*, *mvaK2*, and *mvaD*, respectively—were placed under control of a regulatable Pspac promoter.[32] Growth, transcriptional, and metabolic profiles were generated for these strains under induced and uninduced conditions in order to understand the cellular responses to perturbations in the pathway. As expected, all strains required induction in order to grow, further validating the essentiality of all steps in MVA pathway. Transcriptional profiles derived from inhibiting any step of the pathway yielded similar results: genes involved in non-essential isoprenoid biosynthesis, such as the carotenoid staphyloxanthin, were two-fold downregulated, whereas genes involved in synthesis and recycling of the essential lipid carrier bactoprenol, were two-fold upregulated; the cell wall stress stimulon, a set of genes that responds to inhibition of peptidoglycan biosynthesis, was dramatically upregulated between 3–500 fold; and primary metabolism was severely attenuated with genes required for nucleotide and amino acid biosynthesis, respiration, ribosome assembly, and β-oxidation downregulated by as much as 20-fold. Bolstered by these dramatic findings that illustrated the central role the MVA pathway plays in *S. aureus* physiology, a high-throughput screen was performed on the Pspac *mvaS* and *mvaK1* regulated strains.[33] Decreased expression of the target of interest should increase sensitivity to specific inhibitors, allowing for identification of hits biased towards a desired mechanism of action. From a screen of 1.3 million compounds, 1201 confirmed hits were identified with an IC_{50} shift of at least five-fold between high and low expression conditions of *mvaS* and 277 confirmed hits were identified with an IC_{50} shift of at least 2.5-fold between high and low expression conditions of *mvaK1*. Further characterization of hits in downstream assays ultimately resulted in identification of two compounds that selectively showed increased activity when *mvaS* was downregulated, had bioactivity that could be suppressed with exogenous mevalonate supplementation, and were demonstrated to inhibit HMGS in an *in vitro* biochemical assay with IC_{50}s of 0.7 μM and 0.4 μM. Although the structures of these two HMGS inhibitors were not disclosed, they were reported as having progressed to an active hit-to-lead program.[33]

8.2.3 Targeting the GHMP Kinase Family Members of the MVA Pathway

The final three enzymes in the MVA pathway all belong to the GHMP (galactokinase, homoserine kinase, mevalonate kinase, phosphomevalonate kinase) kinase family.[34] Interestingly, in addition to their structural homology and substrate similarity, it was shown that *S. pneumoniae* MK is subject to feedback inhibition by the product of PMK, diphosphomevalonate.[35] Diphosphomevalonate was shown to bind non-competitively with the MK substrates ATP and mevalonate, with a K_i of 0.5 μM, indicating there may be a separate allosteric binding site at the interface of the MK dimer. The lack of inhibition of human MK by diphosphomevalonate spurred the exploration of diphosphomevalonate analogues that could simultaneously act as substrate inhibitors of the downstream DPMD as well as allosteric inhibitors of the upstream MK.[36,37] Critical to the utility of such a strategy is the ability of the inhibitor to enter the cell, but the high net negative charge on diphosphomevalonate would preclude passage through the cell membrane. As a result, an antimetabolite strategy was used in which non-phosphorylated mevalonate analogues were created in the hopes that they would enter cells, act as substrates of MK and DMK, and generate the desired inhibitor *in situ*. Such an approach was previously shown to work for 6-fluoromevalonate, which can be converted to 6-fluoromevalonate 5-pyrophosphate,[38] which in turn acts an inhibitor of eukaryotic DPMD[39] and downstream isoprenoid biosynthesis in cells.[40,41] Ten mevalonate analogues (**2–11**) were synthesized in which the C3-methyl position was elaborated with hydrocarbons of variable length (Figure 8.2). Studies on mammalian DPMD have shown that the C3-methyl is important for inductively stabilizing the carbocation that forms prior to

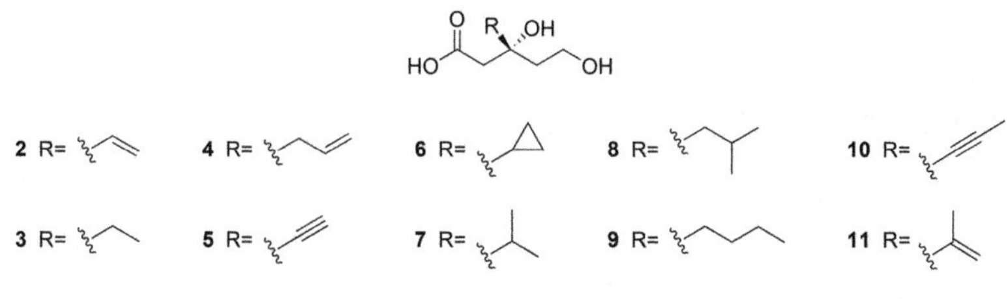

Figure 8.2 MK substrate analogues that could potentially inhibit MVA pathway enzymes (Compounds **2–11**).

decarboxylation,[42] and so it was hypothesized that modifying this substituent could delocalize the positive charge through resonance to transiently form an electrophile that could be captured by an adjacent nucleophile in the enzyme active site. Of the ten analogues tested, only **2–6** were acceptable substrates for *S. pneumoniae* MK, though the kinetic efficiency of the best analogue, **2**, was decreased 67-fold compared to mevalonate. Conversely, the diphosphorylated versions of **2–11**, were all acceptable substrates of *S. pneumoniae* DPMD, with **2** demonstrating only a two-fold decrease in catalytic efficiency. Furthermore, assessment of allosteric inhibition of MK by diphosphorylated versions of the mevalonate analogues revealed that all were substantially less active than diphosphomevalonate, with IC_{50}s that were 50-fold higher for analogues **2** and **6**, 250-fold higher for analogues **3**, **5**, and **10**, 900-fold higher for analogue **4**, or had no detectable inhibition for analogues **7**, **8**, **9** and **11**. In all, mevalonate analogues substituted at the C3-methyl position were poor substrates of MK, which has a binding pocket that appears limited to accepting substrates with no more than a three-carbon substituent, were all accepted as substrates for decarboxylation by DPMD rather than inactivating the enzyme, and exhibit poor non-competitive inhibition of MK in *S. pneumoniae*.[36,37]

As alluded to previously, eukaryotic DPMD can be inhibited by substrate analogues 6-fluoromevalonate diphosphate **12** and 6,6,6-trifluoromevalonate diphosphate **13**,[39] as well as the transition state mimic diphosphoglycolylproline **14** (Figure 8.3).[43] Biochemical, structural biology and molecular docking experiments[44–46] have demonstrated that these inhibitors are capable of binding and inhibiting prokaryotic DPMD with K_i values for **12** and **14** of 49 nM and 4 µM for *Staphylococcus epidermidis* enzyme and 230 nM and 34 µM for *S. aureus* enzyme, respectively. Co-crystal structures of *S. epidermidis* DMPD bound to **12** and **14** revealed a key difference between eukaryotic and prokaryotic enzymes; Arg-193, which interacts with the β-phosphate moiety

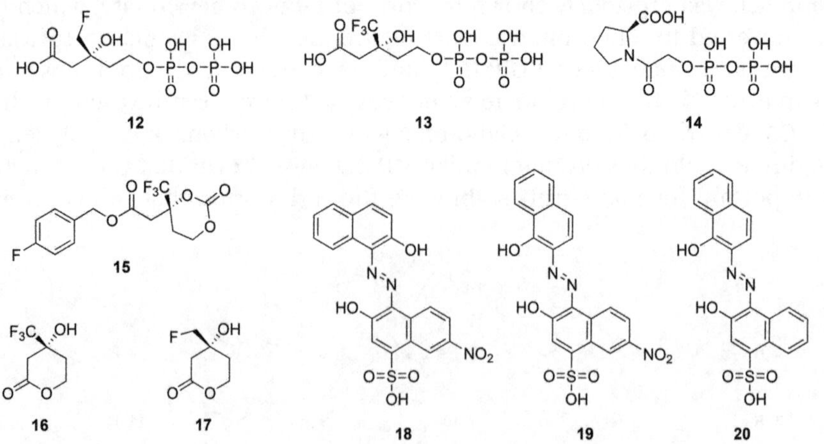

Figure 8.3 Inhibitors of DPMD (Compounds **12–20**).

of the inhibitors, is substituted by a Thr in the corresponding position of the human enzyme.[46] Perhaps encouraged by the low K_i against prokaryotic DPMD and the potential for taking advantage of structural differences in the active site to make selective inhibitors, a series of fluorinate mevalonate antimetabolite prodrugs were synthesized in order to increase cellular permeability and bioavailability.[47] Over 40 cyclic carbonate, acetal, ketal, amide, ester, and lactone prodrugs were synthesized and were shown to possess a wide range of stabilities in human plasma. Whereas cyclic carbonates decomposed with half-lives on the order of minutes, cyclic acetals and ketals were stable for over 48 h. Despite demonstrating that such prodrugs had tunable stabilities, only three compounds, **15**, **16**, and **17**, displayed any, albeit weak, cellular antibacterial activity. When tested against *S. pneumoniae*, *Streptococcus pyogenes*, *Streptococcus agalactiae*, *S. aureus*, *Enterococcus*, and *E. coli*, **15** and **16** only had activity against *S. pneumoniae* with minimal inhibitory concentrations (MICs) of 1.6 mM and 0.8 mM, respectively, and **17** only had activity against *S. pneumoniae*, *S. pyogenes*, and *S. agalactiae* with MICs of 0.2 mM, 0.2 mM, and 0.8 mM, respectively. Given the poor membrane permeation of phosphorylated inhibitors, the narrow substrate parameters required for turnover of mevalonate analogues by MK and DMK, and the poor antibacterial activity of antimetabolites, a biochemical screen was initiated to identify small, more drug-like molecules that could inhibit DPMD. The mechanistic diversity set from the National Cancer Institute was screened against *S. epidermidis* DPMD, and one compound, Eriochrome Black A **18**, was identified as a hit with an IC_{50} < 5 μM.[48] Virtual screening also identified **18** as a top hit, and biochemical experiments ultimately demonstrated that **18**, along with the structurally related Eriochrome Black T **19** and Eriochrome Black B **20**, were competitive inhibitors with respect to diphosomevalonate with K_i values of 0.58 μM, 1.2 μM, and 2.7 μM, respectively, and non-competitive inhibitors with respect to ATP with K_i values of 1.8 μM, 0.59 μM, and 6.8 μM, respectively. Despite that fact that the Eriochrome Black compounds are substantially more hydrophobic and lack the high negative charge of diphosphomevalonate analogues, these compounds still had poor cellular activity, with **18**, **19**, and **20** only inhibiting growth of *S. epidermidis* by 50% at concentrations of 250 μM, 31 μM, and 63 μM, respectively.

8.3 Targeting the MEP Pathway

8.3.1 Historic Compounds Inhibiting the MEP Pathway

Even before the discovery of the alternate MEP pathway for isoprenoid biosynthesis, a natural product targeting IspC was isolated from *Streptomyces lavendulae* in the late 1970s.[49,50] Studies on the mechanism of action of fosmidomycin **21** (Figure 8.4) demonstrated that treatment of *E. coli* cells led to formation of spheroplasts or swollen cells without inhibiting cell wall formation or impacting the amount of envelope lipoprotein, phospholipid, or DNA produced.[51] Instead, **21** reduced the amount of menaquinones and ubiquinones in growing *E. coli* as well as reduced carotenoids and menaquinones in *Micrococcus luteus*, an effect attributed to inhibition of isoprenoid biosynthesis. When the MEP pathway was ultimately delineated, a search for antibiotics active against the MEP pathway-containing bacteria *E. coli* and *Bacillus subtilis* but inactive against the MVA pathway-containing bacteria *S. aureus* identified **21** as a potential selective inhibitor.[52] Indeed, **21** was determined to be a mixed-type inhibitor of *E. coli* IspC in an *in vitro* enzyme assay with an IC_{50} of 8.2 nM and a K_i of 38 nM. Furthermore, cellular activity of **21** could be suppressed by supplementation with 0.025% 2-C-methylerythritol, the free alcohol of a downstream MEP pathway intermediate, and **21** had no effect on *E. coli ispC* knockout mutants maintained on 2-C-methylerythritol, supporting that **21** was an IspC inhibitor. Complementary studies in various plants reached the same conclusion.[53] Whereas **21** was first identified as an antibacterial and later shown to possess herbicidal activity, clomazone **22** was first discovered as an herbicide for use in growing soybeans and later demonstrated to possess antibacterial activity. Like **21**, **22** was discovered prior to elucidation of the MEP pathway, but it was known to affect isoprenoid biosynthesis in plants because it inhibited production of carotenoids and chlorophylls, leading to leaf bleaching.[54] After recognition of the existence of the MEP pathway, various studies indicated that perhaps not clomazone, but rather its metabolite ketoclomazone **23**, was responsible for the herbicidal activity, and that the mechanism of action of **23** was inhibition of DXS. Whereas **23** inhibited DXS from *Chlamydomonas*[55] and *Catharanthus roseus*[56] with IC_{50}s of 100 μM and 20 μM, respectively, **22** had no measurable inhibitory activity. When tested against *E. coli* and *Haemophilus influenzae*, **23** had MICs of 800 and 12.5 μM, respectively, and this inhibitory activity could be suppressed by supplementation with 1-deoxyxylulose, the free alcohol form of the product of DXS.[57] *In vitro* enzyme assays directly demonstrated that **23** was an uncompetitive inhibitor with respect to pyruvate with K_is of 28 μM and 75 μM and a mixed-type inhibitor with respect to G3P with K_is of 23 μM and 460 μM for *H. influenzae* and *E. coli* DXS, respectively. Clearly chemical biology approaches were validating the effectiveness of targeting the MEP pathway decades before its discovery, and the distinct divergence

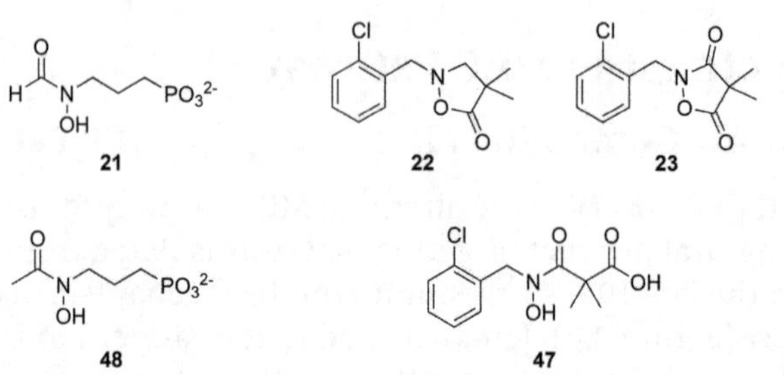

Figure 8.4 Historic MEP pathway inhibitors identified in phenotypic screens (Compounds **21–23**, **47** and **48**).

from the human MVA pathway only serves to strengthen interest in identifying inhibitors that would be nontoxic to humans but efficacious in treating infections.

8.3.2 Inhibiting DXS

DXS is a thiamin diphosphate and Mg^{2+}-dependent enzyme that catalyzes the decarboxylative condensation of pyruvate and G3P in the first and rate-limiting step of the MEP pathway. Because the product DXP is also a precursor of vitamins B_1 [58] and B_6,[59] this enzyme has been considered a particularly attractive target for antibacterial drug discovery. Despite high homology in the thiamin diphosphate binding sites between DXS and human enzymes such as transketolase and pyruvate dehydrogenase E1 subunit,[60] there have been several attempts made to create thiamin diphosphate competitive inhibitors. A known transketolase inhibitor **24** (Figure 8.5) was demonstrated to have poor activity against *Mycobacterium tuberculosis* DXS, with an IC_{50} of 114.1 μM.[61] This lack of activity was thought to be the result of structural differences in the domain arrangements of the active site between DXS and the transketolase superfamily of enzymes, leading to speculation that inhibitory specificity could be achieved. After creating approximately 30 analogues, the two most potent compounds **25** and **26**, were found to have IC_{50}s of 10.9 μM and 10.6 μM, respectively, against *M. tuberculosis* DXS. Although **25** and **26** did show anti-tuberculosis activity with MICs of 7.6 μM and 7.7 μM, respectively, they also were cytotoxic against Vero cells, with IC_{50}s of 16.0 μM and 35.6 μM, respectively, demonstrating that although activity could be improved,

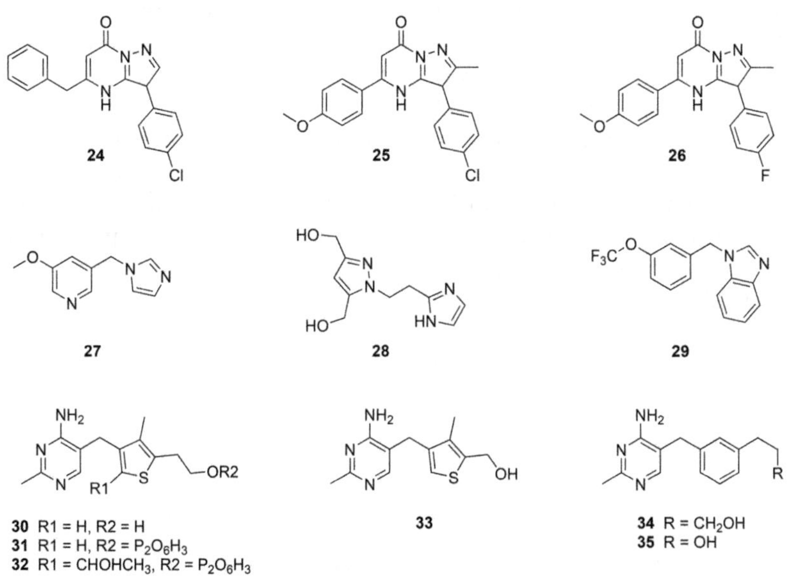

Figure 8.5 Thiamin-based DXS inhibitors (Compounds **24–35**).

selectively remained a challenge. Rather than starting with a known inhibitor of thiamin diphosphate-dependent enzymes, another approach was taken to utilize the crystal structure of the *Deinococcus radiodurans* DXS enzyme to design fragment-based inhibitors *de novo*. Focusing on the thiamin aminopyrimidine and thiazolium pockets because the hydrophilic diphosphate binding pocket was deemed less druggable and the substrate binding pocket was poorly defined, two fragments, **27** and **28**, were designed that when tested biochemically yielded IC_{50}s of 1810 μM and 762 μM and K_is of 448 μM and 183 μM, respectively. A combination of three NMR methods were used to confirm the binding mode of **27** and the fragment was further pursued for growth and optimization, ultimately leading to synthesis of **29**, which had an IC_{50} of 595 μM and a K_i of 151 μM. This fragment had approximately equivalent activity to the known thiamin mimic deazathiamine[62] **30**, though clearly additional moieties need to be added in order to take advantage of the diphosphate binding site because the diphosphate derivative of deazathiamine, **31**, is far more potent with an IC_{50} of 34 nM and a K_i of 7 nM. Activity of **30** and **31**, as well as the additional analogues **32–35** were also tested against *M. tuberculosis* DXS, and although **32–35** had similar activity as **30** against the *D. radiodurans* enzyme, only compounds **31** and **32** retained activity against the *M. tuberculosis* enzyme, with IC_{50}s of 0.74 and 1.4 μM, respectively.[63] These results reveal the difficulty in targeting a cofactor binding site where the natural ligand is tightly bound and there are only subtle differences in a pocket highly conserved across protein families that can be exploited to build in selectivity.

Rather than targeting the thiamin diphosphate cofactor binding site, an alternate approach has been taken to create substrate analogues that can inhibit DXS. DXS is unique among thiamin diphosphate utilizing enzymes in that it forms a ternary complex during catalysis[64] instead of following typical ping-pong kinetics,[65] and substrates bind independently and reversibly in a random sequential mechanism.[66] Furthermore, flexibility in the *E. coli* DXS active site allows for promiscuity where more non-polar acceptor substrates can substitute for G3P.[67] On this basis, several acetylphosphonates were synthesized which were either modified at the phosphonate (compounds **36–38**) to mimic a non-polar acceptor substrate, for which promiscuity had previously been observed, or modified at the acyl position (compounds **39–41**) to mimic alternative donor substrates, derivatives of which the enzyme was previously shown to poorly tolerate (Figure 8.6).[68] As was hypothesized based on previous mechanistic work, acetylphosphonates **36–38** that exploited the

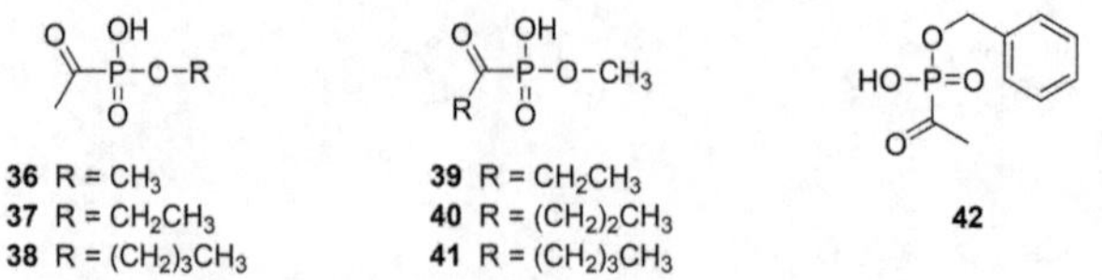

Figure 8.6 Acetylphosphonate DXS inhibitors (Compounds **36–42**).

promiscuity of the acceptor substrate binding site were demonstrated to be potent inhibitors of *E. coli* DXS, with K_is between 1–7 μM, whereas acetylphosphonates **39–41** that poorly fit the pyruvate donor site were mostly inactive, with only **39** having a measurable K_i of 258 μM.[69] Importantly, unlike thiamin diphosphate DXS inhibitors, **36–38** demonstrated selective inhibition over other thiamin diphosphate utilizing enzymes. They were completely inactive against *Saccharomyces cerevisiae* transketolase, and they were less active against the pyruvate-utilizing porcine pyruvate dehydrogenase E1 subunit, with K_is from seven-fold to 60-fold higher than against *E. coli* DXS. Additional studies on the substrate promiscuity of *E. coli* DXS identified a series of aromatic nitroso analogues that could act as donor substrates, yielding hydroxamate products through C–N bond formation with pyruvate.[70] The reasonably low K_ms for several of these aromatic nitroso compounds motivated testing of these alternate substrates for inhibitory activity, and intriguingly, all were found to be weak inhibitors with the best compound having an IC_{50} of 208 μM. Mechanistic studies and evaluation of activity with mutant protein variants determined that the nitrosonaphthol analogues were competitive inhibitors with respect to G3P, although they adopt a binding mode distinct from the natural substrate during turnover. These studies proving that the DXS acceptor substrate binding site can accommodate larger aromatic scaffolds that could also act as inhibitors at higher concentrations led to the synthesis of the hybrid compound benzylacetylphosphonate **42**, which combines the bisubstrate scaffold of the aforementioned inhibitory acetylphosphonates with a benzyl group discovered in the aromatic nitroso analogues. **42** exhibited competitive inhibition with respect to pyruvate with a K_i of 10.4 μM and noncompetitive inhibition with respect to G3P with a K_i of 70 μM, further validating that the aromatic scaffold adopts a different binding mode than the natural substrate. As was the case with **36–38**, **42** was selective for *E. coli* DXS, with an 85-fold higher K_i measured for porcine pyruvate dehydrogenase E1 subunit. The acetylphosphonates display broad spectrum activity because **38** can inhibit DXS from multiple prokaryotes, with single digit μM K_is measured for enzymes isolated from *Yersinia pestis*, *M. tuberculosis*, and *Salmonella enterica*.[71] Testing for antimicrobial activity revealed that **38** could inhibit growth of several pathogens, including *E. coli*, *S. enterica*, *Micrococcus* sp., *Bacilus anthracis*, and *P. aeruginosa* with MICs ranging from 1 to 4 mg mL^{-1}. Supplementing the growth media with 1-deoxy-D-xylulose, the free alcohol of the downstream MEP-metabolite DXP, or thiamin, an essential cofactor that also uses DXP as a precursor for biosynthesis, was able to rescue growth of *E. coli* treated with **38**. Growth inhibition could also be suppressed by overexpression of wild-type DXS but not a catalytically dead mutant. These results support inhibition of DXS as the mechanism of action for the antibacterial effect of **38**.

Attempting to identify unique chemotypes that could act as bisubstrate inhibitors of DXS, a series of compounds tethering the donor mimic glyoxylate to an aryl aldehyde library through an oxime-based linker was synthesized.[72] Initial testing of the library identified compounds derived from

2,4,5-trihydroxybenaldehye that displayed potent inhibition of *E. coli* DXS, with **43** and **44** showing noncompetitive or uncompetitive inhibition *versus* pyruvate and competitive inhibition *versus* G3P, with K_is of 18.4 μM and 1.0 μM, respectively (Figure 8.7). The more potent inhibition by symmetrical **44** and the uncompetitive mode of inhibition with respect to pyruvate, indicated that the glyoxylate moiety was not acting as a donor substrate mimic and that these compounds had a different mechanism of inhibition from the acetylphosphonates. Ultimately it was determined that neither the linker nor the donor mimic were required for inhibition; a single oximine, either substituted or unsubstituted at the oxygen, linked to the trihydroxybenzoyl moiety yielded inhibition, with the most potent compounds **45** and **46** having K_is of 2.6 μM and 3.9 μM, respectively. Inhibition by **46** was reversible and occurred even under reducing conditions, so although the multiple hydroxyl substituents were found to enhance inhibitory activity, electrophilic quinone formation is not required.

Despite efforts to create novel compounds inhibiting DXS, none have demonstrated superior activity to ketoclomazone **23**, a compound found through phenotypic screening before the MEP pathway was ever elucidated. Even the hydrolyzed derivative **47** is capable of inhibiting *H. influenzae* DXS with an IC_{50} of 1.0 μM and displays antibacterial activity with an MIC of 32 μg mL^{-1} against *H. influenzae*, which is within four-fold of the activity observed for **23**.[73] Supplementation of the growth media with 1-deoxy-D-xylulose, which is presumably converted to the downstream MEP intermediate DXP by D-xylulokinase,[74] suppresses the growth inhibitory effect of **23** and **47**, supporting the notion that antibacterial activity can be attributed to inhibition of DXS. Although the modest antibacterial activity of **23** may inspire continued pursuit of novel DXS inhibitors in the hopes of one day creating compounds with improved cellular activity, a serious question arises as to the validity of this target for developing new antibiotics. Although DXS plays a critical role in primary metabolism because it is responsible for synthesizing G3P, an essential precursor of isoprenoids and the cofactors thiamin diphosphate (vitamin B1) and pyridoxyl phosphate (vitamin B6), studies have shown that in *E. coli* suppressor mutations can arise that bypass the essentiality of this first step in the MEP pathway.[75,76] By expressing a synthetic MVA pathway operon containing the *mvaK1*, *mvaK2*, and *mvaD* genes, *E. coli* are able to synthesize IPP and DMAPP from exogenously provided mevalonate,

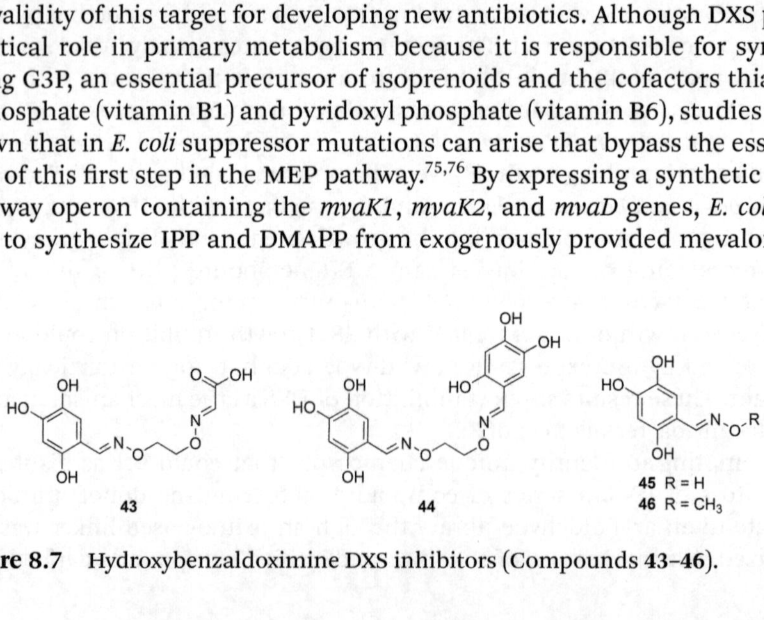

Figure 8.7 Hydroxybenzaldoximine DXS inhibitors (Compounds **43–46**).

completely bypassing the MEP pathway.[77,78] Utilizing such a strain, knockouts in the individual MEP pathway genes were made and assessed for conditional essentiality. For most of the MEP pathway knockouts, no suppressors of the mevalonate auxotrophy could be isolated, but for DXS, suppressors were isolated at a frequency of 6.4×10^{-9}.[76] Subsequent genetic mapping and sequencing determined that suppressors harbored one of four missense mutations in *aceE*, which encodes pruvate dehydrogenase E1 subunit, or one of two missense mutations in *ribB*, which encodes 3,4-dihydroxy-2-butanone 4-phosphate synthase.[75] All of these mutations appeared to be gain-of-function and allowed for synthesis of DXP *in vivo* in the absence of DXS. Given how immensely difficult it is to discover and develop new antibiotics, it is perhaps imprudent to focus on a target for which bypass mutations are known to exist. Such mutations could give rise to resistance in the clinic, shortening the timeframe of utility for a new antibiotic.

8.3.3 Inhibiting IspC

IspC catalyzes the first committed step in the MEP pathway, utilizing an NADPH cofactor in the presence of a divalent metal cation to isomerize and reduce DXP to MEP. Despite over 30 structures of IspC having been reported,[79] dramatic conformation changes in IspC upon ligand binding, which involve the repositioning of a flexible loop to shield a reactive intermediate in the active site from bulk solvent,[80] have made it challenging to capitalize on crystallographic information to design rational inhibitors. Instead, most inhibitors to date are based on the natural product fosmidomycin **21**[50] and its acetyl analogue FR-900098 **48** isolated from *Streptomyces rubellomurinus*.[81] These compounds have garnered broad interest because of their antibacterial activity against a variety of pathogens as well as their low toxicity, with LD_{50}s greater than 5 g kg^{-1} when administered intravenously to mice.[50] Optimization of IspC inhibition and antimicrobial activity has been most extensively explored in the pathogens *E. coli*, *M. tuberculosis*, and *Plasmodium falciparum* for which there is vast biochemical and structural information reported. Activity of **21** and **48** against *P. falciparum* is impressive, with an IC_{50} against IspC of 10 nM, the ability to inhibit parasite growth in cell culture at submicromolar concentrations, and sustained cure in a lethal *Plasmodium vinckei* mouse model of infection when treated with 30 mg kg^{-1} for eight days.[82] In fact, **21** has been under clinical investigation for treatment of malaria in combination with clindamycin,[83–87] artemisinin,[88,89] and piperaquine.[90] However, this section will focus primarily on efforts directed towards targeting bacteria, such as *E. coli* and *M. tuberculosis*. As with the polar antibiotic fosfomycin, uptake of **21** and **48** into *E. coli* is facilitated by the glucose-3-phosphate transporter, GlpT, mutation of which confers resistance to these compounds.[91–93] Notably, *M. tuberculosis* is resistant to **21** and **48** despite demonstrated activity against IspC from this organism,[94,95] presumably because *M. tuberculosis* lacks GlpT and these compounds cannot cross the mycobacterial cell envelope.[96] Therefore, optimization of **21** and **48** is needed to increase cellular penetration because in *M. tuberculosis* permeability limits antibacterial activity and in *E. coli* specific uptake by the non-essential transporter GlpT provides a facile high-frequency mechanism of resistance.

In order to directly assess whether increased lipophilicity could promote cellular penetration of **21** and **48,** a series of lipophilic phosphonate ester prodrugs was synthesized.[97] When tested against *E. coli* and *M. tuberculosis*, antibacterial activity was found to increase with the size of the ester, and primary esters were found to outperform their secondary counterparts. Although all of the analogues demonstrated some activity against *M. tuberculosis*, with MICs ranging from 25 μg mL^{-1} for larger esters like **49** to 400 μg mL^{-1} for smaller esters like **50**, most of the analogues were inactive against *E. coli* (Figure 8.8). The best activity was observed for the largest of the prodrugs **49** and **51**, with an MIC of 50 μg mL^{-1}, eight-fold higher than the parent **48**. Intriguingly, **51** was demonstrated to have approximately equivalent antibacterial activity as **48** against *Francisella novicida* when tested directly, during intracellular infection of eukaryotic cells, and in a *Galleria mellonella* caterpillar infection model.[98] Furthermore, the activity of **51**, but not **48,** was shown to be GlpT-independent in *F. novicida*, validating the strategy of modulating lipophilicity to increase membrane penetration and circumvent transporter-specific entry. Although these prodrugs had antibacterial activity, the mechanism of action cannot be definitively attributed to inhibition of IspC because conversion of the prodrug to the active compound in cells was not demonstrated and some of the lipophilic esters had *S. aureus* activity, which utilizes the MVA pathway for isoprenoid biosynthesis and does not encode *ispC*. Rather than modulate lipophilicity with a prodrug, the co-crystal structure of *E. coli* IspC bound to **21** was used to design inhibitors which either replaced the phosphonate with a sulfone or sulfonamide moiety to maintain the hydrogen bonding network or added additional lipophilic substituents tethered to the phosphonate to potentially take advantage of a small hydrophobic binding pocket in the vicinity.[99] Although none of the sulfone or sulfonamide analogues demonstrated inhibition of *E. coli* IspC, a sulfonic acid derivative and a series of extended phosphonate monoesters did inhibit the

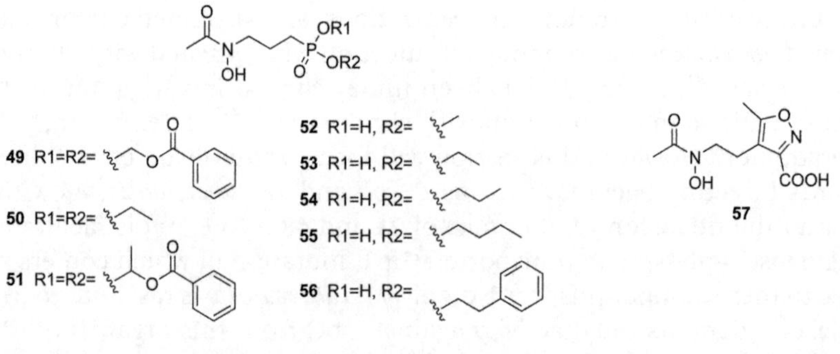

Figure 8.8 Phosphonate-modified analogues of fosmidomycin as inhibitors of IspC (Compounds **49–57**).

enzyme. Activity of the phosphonate monoesters increased with the length of the alkyl chain, with IC_{50}s of 50 μM, 23 μm, 16 μM, and 3.9 μM for methyl **52**, ethyl **53**, propyl **54**, and butyl **55** derivatives, respectively. The best compound **56** had an IC_{50} of 0.49 μM, which was only fourteen-fold higher than that for **21** and **48**, and it was reported to possess some weak cellular activity against *M. tuberculosis*, although subsequent studies only measured 37% inhibition of *M. tuberculosis* IspC at 100 μM.[100] Addition studies exploring replacement of the phosphonate group demonstrated a lack of activity for the bioisosters carboxylic acid, tetrazole, and imidazolidinedione, and a weak IC_{50} of 151 μM for isoxazole carboxylic acid **57** against *M. tuberculosis* IspC.[100]

Other attempts to modulate the lipophilicity of **21** and **48** have focused on altering the hydroxamate moiety. Although hydroxamates are very strong inhibitors of metalloenzymes due to their ability to chelate active site metals as bidentate ligands, they have significant liabilities with respect to bioavailability, metabolism, pharmacokinetics and toxicity,[101] properties which may be preventing development of other hydroxamate-containing antibiotics targeting LpxC, the essential UDP-(3-*O*-(*R*-3-hydroxymyristoyl))-*N*-acetylglucosamine deacetylase involved in lipopolysaccharide biosynthesis,[102] and peptide deformylase, an enzyme involved in processing nascent peptides during protein synthesis.[103] Generally, most modifications to the hydroxamate have led to significant loss of enzymatic and/or cellular potency. A series of heteroarylcarboxamides, synthesized based on *in silico* analysis of the *E. coli* IspC binding site and computer simulated docking studies suggesting that derivatives of **21** could potentially bind in the reverse orientation with the phosphonate chelating the active site divalent cation, were demonstrated to be poor enzyme inhibitors, with the best compound **58** having an IC_{50} of 1 mM (Figure 8.9).[104] Subsequent studies interrogating additional *N*-aryl and *N*-heteroarylcarboxamides with varied spacer lengths between the phosphonate and carboxamide yielded similar results, with the best compounds **59** and **60** only showing 26.8% and 49.2% inhibition of *E. coli* IspC at 250 μM, respectively. Since the ability to chelate metal is critical to the activity of **21**, as evidenced by an 87 000-fold increase in K_i for the methylated analogue **61** measured on the *Synechocystis* sp.

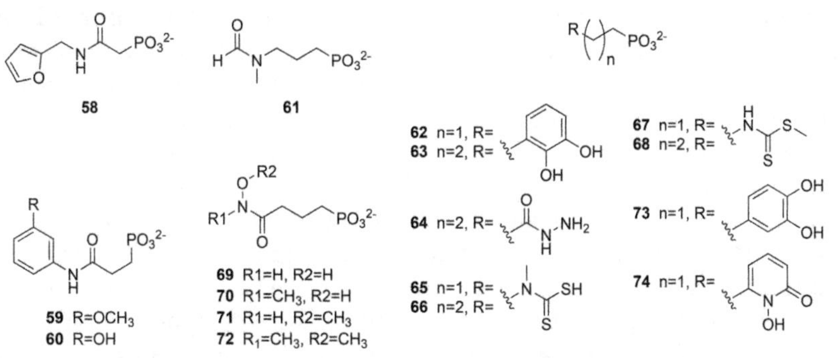

Figure 8.9 Hydroxamate-replaced analogues of fosmidomycin as inhibitors of IspC (Compound **58–74**).

PCC6803 IspC,[105] a series of amidopropylphosphonates were synthesized in an effort to identify an alternative bidentate ligand. None of the fourteen analogs with various chelating moieties were found to inhibit IspC from either *E. coli* or *M. tuberculosis*, though this may be attributed to the use of a suboptimal three-carbon linker between the phosphonate and amide.[106] Other bidentate chelators have been incorporated into analogues of **21**, yielding catechols **62** and **63**, hydrazide **64**, *N*-methyldithiocarbamates **65** and **66**, *S*-methyldithiocarbamates **67** and **68**, reverse hydroxamates **69** and **70**, and *O*-methyl reverse hydroxamates **71** and **72**. No detectable inhibition of *E. coli* IspC was observed for compounds **63**, **65**, **66**, and **68**, and only weak activity was seen with compounds **62**, **64**, **67**, **71**, and **72**, with IC_{50}s in the range of 1–4 mM, well above the 32 nM IC_{50} measured for **21**. A 3,4-substituted catechol **73**, as well as a related hydroxypyridinone **74** were demonstrated to have activity against *M. tuberculosis* IspC, with IC_{50}s of 41 µM and 53 µM, respectively, corroborating the inhibition observed with catechol **62** in *E. coli*.[100] Intriguingly, **69** and **70** were very active against *E. coli* IspC and were shown to follow a slow-binding inhibition mechanism with K_i values of 169 nM and 54 nM and K_i* values of 68 nM and 3 nM, respectively, activity that is comparable to that of **21**, which has a K_i and K_i* of 40 nM and 10 nM, respectively.[107] Furthermore, **70**, had *E. coli* antibacterial activity similar to that of **21** and **48**, and retained some activity against a fosmidomycin resistant strain.

Although replacement of the hydroxamate moiety proved rather unsuccessful, with the only substitution retaining similar activity to **21** or **48** being the highly related reverse hydroxamates **69** and **70**, an alternative strategy to increase lipophilicity focused on elongating **21** beyond the formyl group of the hydroxamate. Initially these efforts were exploratory and designed to aid in building a ligand-based QSAR model. Using a glycine spacer to facilitate synthesis and provide a biologically relevant moiety that could potentially form hydrogen bond interactions, a series of compounds were synthesized containing large aryl or alkyl substituents.[108] Two additional derivatives lacking the glycine spacer with simple phenyl substituents were also made. These molecules were found to be active on *E. coli* IspC with IC_{50}s between 12–20 µM for compounds **75–80**, between 1–7 µM for compounds **81–83**, and 0.13 µM for compound **84**, which is only 2.5-fold greater than that observed for **21** and **48** (Figure 8.10). Additional acyl analogues made without the glycine spacer also demonstrated activity towards *E. coli* IspC, with IC_{50}s between 10–20 µM for compounds **85–87** and 1–9 µM for compounds **88–92**.[109] Compounds with extremely long extensions or charged hydrophilic groups did not show activity. Docking of these molecules into the co-crystal structure of *E. coli* IspC bound to **21** revealed that these acyl groups could extend into the adjacent NADPH binding site, providing an explanation for how such large inhibitors can be accommodated by the enzyme. Using the crystal structure of *M. tuberculosis* IspC, bisubstrate inhibitors were designed to take advantage of the finding that the nicotinamide ring of NADPH lies within 3.5 Å from the formyl carbon atom of **21**.[110] In addition to creating amide linked derivatives as had been done in previous studies, *O*-linked modifications

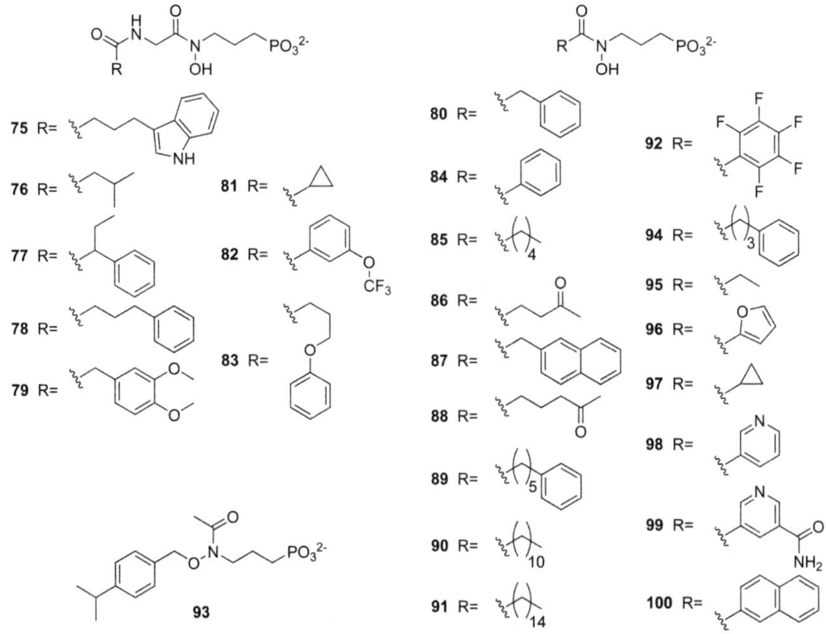

Figure 8.10 Hydroxamate-derivatized analogues of fosmidomycin as inhibitors of IspC (Compounds **75–100**).

were also investigated. Most of the *O*-linked derivatives were inactive when tested at 100 μM, though **93** did have an IC_{50} of 48.4 μM. Amide-linked derivatives were more active and the best compound **94** had an IC_{50} of 17.8 μM. Mechanistic studies demonstrated that **94** was competitive with respect to DXP, but not competitive with respect to NADPH, indicating that despite being made to be a bisubstrate inhibitor, the phenylpropyl substituent of **94** was likely binding in an alternate orientation rather than competing with the nicotinamide ring of NADPH. Furthermore, the diethyl phosphonate ester prodrug of **94** displayed antitubercular activity with an MIC of 200 μg mL^{-1}, which is actually better than the MIC of 250 μg mL^{-1} measured for the equivalent prodrug of **21**. Additional acyl derivatives tested against *M. tuberculosis* IspC identified similarly potent compounds **80** and **95–97** with IC_{50}s between 49 μM and 156 μM, as well as significantly more potent analogues **84** and **98–100** with IC_{50}s between 1.6 μM and 3.6 μM, although none possessed antitubercular activity.[111] It appears that altering the lipophilicity of the hydroxamate of **21** or **48** is insufficient to balance the polarity of the phosphonate moiety and confer cellular activity on *M. tuberculosis*.

The length of the carbon linker between the phosphonate and hydroxamate substituents of **21** and **48** is critical. From initial work done after isolation of these natural products, it was noted that shortening the propyl linker to an ethyl led to complete loss of antibacterial activity.[112] More recent studies have confirmed that the propyl linker is optimal, as increasing the chain length to

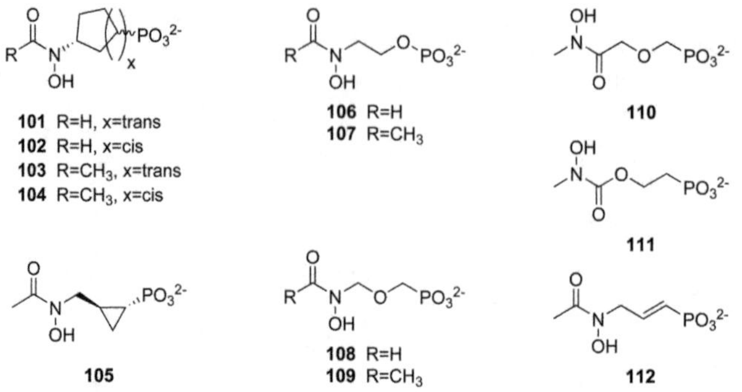

Figure 8.11 Linker modified analogues of fosmidomycin as inhibitors of IspC (Compounds **101–112**).

four or five carbons also results in dramatic loss of enzymatic and cellular activity.[113] Restraining the rotational freedom of the propyl linker is tolerated as conformationally constrained analogues retain fairly potent enzymatic activity. For cyclopentyl derivatives **101–104** the *trans*-isomers **101** and **103** had a measured IC_{50}s on *E. coli* IspC of 0.2 μM and 2.3 μM, respectively, and were an order of magnitude more potent than the *cis*-isomers **102** and **104** (Figure 8.11).[114] This is in agreement with findings for cyclopropyl derivative **105**, which with an IC_{50} of 0.05 μM on *E. coli* IspC, was found to be three-fold more potent than a racemic mixture.[115] The propyl linker is also amenable to incorporating heteroatoms. Studies in which oxaisosteres of **21**, **48**, and **70** were created by replacement of the carbon α (**106**, **107**), β (**108**, **109**, **110**), or γ (**111**) to the phosphonate revealed that proximity of an electron rich atom to the phosphonate increases activity of these compounds. Compounds **106** and **107** were an order of magnitude more potent than **21** and **48** at inhibiting Synechocystis sp. PCC6803 IspC with K_is of 19 nM and 2 nM, respectively,[105] **108–110** were essentially equipotent to their parental compounds with IC_{50}s of 1.1 μM, 0.09 μM, and 0.07 μM on *E. coli* IspC, respectively, and **111** lost measurable enzymatic activity.[116] Despite the dramatic enzymatic potency observed for **106** and **107** and studies demonstrating antibacterial activity,[117] these compounds are unlikely to progress for development as novel antibiotics because the phosphate, which is critical for high affinity binding to the target,[99] is highly labile to hydrolysis. Given that both α and β oxa derivatives of 48 are active, it is perhaps unsurprising that the unsaturated analogue **112** demonstrates activity on *M. tuberculosis* IspC, with an IC_{50} of 1 μM.[113]

Beyond simply modifying the propyl linker, many compounds have been synthesized incorporating an aryl moiety at the α-position. Interest in this substitution first came when a small library investigating the effects of increased lipophilicity near the phosphonate revealed that several phenyl derivatives maintained enzymatic activity on *E. coli* IspC within a log of **21** and **48**, but actually had vastly improved antimalarial activity.[118] The best compounds from this series, **113** and **114** (Figure 8.12), had IC_{50}s

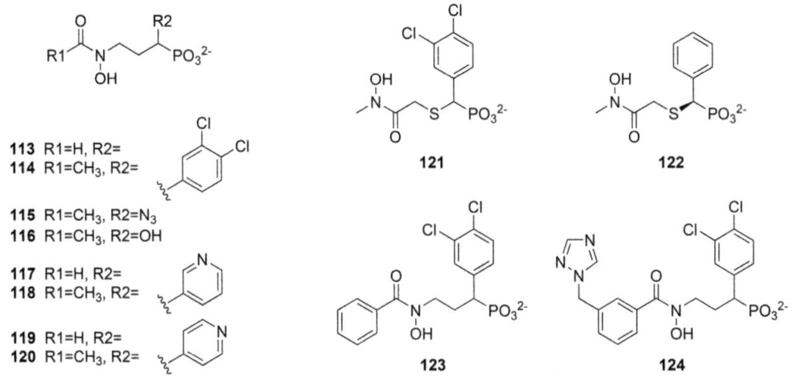

Figure 8.12 α-Substituted analogues of fosmidomycin as inhibitors of IspC (Compounds **113–124**).

of 0.06 μM and 0.12 μM against *E. coli* IspC, within two-fold of compounds **21** and **48**, respectively. Additional studies demonstrated that **113** and **114** also had activity on *M. tuberculosis* IspC, with IC_{50}s of 0.15 μM and 0.7 μM, respectively. Co-crystal structures revealed that the phosphonate and hydroxamate moieties adopt similar conformations when **21, 48, 113**, and **114** are bound, but the propyl linker for **21** and **48** lies deeper in the active site than for **113** and **114**. Furthermore, the α-phenyl of the derivatives displaces a tryptophan side chain as it reaches out towards the solvent, causing the active site to become disordered. Since this α-substituent makes limited protein contacts, this position appears to tolerate many substitutions. Generally, biaryl and fused ring derivatives lead to a ten-fold reduction in enzyme inhibition,[119] *ortho* substitutions on the phenyl ring are poorly tolerated,[94] and phenyl rings substituted with electron withdrawing groups were more active than those substituted with electron donating groups.[120] Small substituents such as azido **115** and hydroxy **116** yield potent inhibitors, with IC_{50}s against *E. coli* IspC of 75 nM and 52 nM, respectively, though α-triazoles are inactive.[121] Pyridinyl α-analogues are exquisitely potent as well, and compounds **117–120** are approximately 20-fold more potent than their phenyl analogues, with K_is on *E. coli* IspC between 35–87 nM, which is within two-fold of **21.**[122] Utilizing structure activity relationships (SAR) from multiple studies, hybrid molecules were constructed using the best modifications at various positions. Utilizing the reverse hydroxamate identified in compound **70**, β-heteroatom containing linkers like those found in compound **110**, and various α-aryl substituents like that in compound **114**, a small library of oxa, thia, and carba isosteres was synthesized and evaluated for inhibition of both *E. coli* and *M. tuberculosis* IspC.[123] *N*-methyl hydroxamates were more active than their non-methylated counterparts. Of the five α-substituents tested, the best activity was found for analogues containing 3,4-dichlorophenyl followed by 3,4-difluorophenyl, phenyl, 4-methylphenyl, and naphthalene-1-yl. Comparing substitution at the β-position, thia isosteres were more active than carba isosteres, which in turn were more active than oxa isoteres. Taken together the most potent compound synthesized was **121**, which had an IC_{50}s of 5.9 nM and 9.2 nM against *E. coli* and *M. tuberculosis* IspC, which was over 30-fold more potent that **21** on both enzymes. Moreover, the co-crystal structure and chiral separation of one derivative demonstrated that the *S*-(+)-enantiomer **122** was the active compound, with 800-fold more potency than the other enantiomer. Another hybrid molecule looked at combining the α-substituent of compound **114** with various acyl hydroxamates like those in compound **84**.[111] The lead compound **123** had an IC_{50} of 0.32 μM when tested against *M. tuberculosis* IspC, Substitutions on the phenyl moiety of the hydroxamate generally led to reduced activity, with *ortho* substitution decreasing activity between twenty- to seventy-fold and a single *para* substituted derivative losing three-fold activity. However, *meta* substituted derivative **124** was more inhibitory with an IC_{50} of 0.14 μM, which is comparable to **48**. By integrating SAR determined from systematic exploration of the fosmidomycin **21** scaffold, it is clearly possible to create improved inhibitors. Although enzymatic activity has been improved significantly, the hurdle of improving cellular antibacterial activity remains.

Rather than modifying the natural product inhibitors **21** and **48**, an alternative approach to targeting IspC utilized modification of the substrate DXP. During the course of investigating the reaction mechanism of IspC, two substrate mimics lacking either the C4 (**125**) or C3 (**126**) hydroxyl were synthesized to determine whether α-ketol rearrangement was occurring (Figure 8.13).[124] Neither compound was accepted by *E. coli* IspC as a substrate, although they did reversibly inhibit the enzyme with K_is of 120 μM and 800 μM, respectively. Complementary results were obtained investigating the enzyme mechanism using fluorinated analogues of DXP, wherein C3 (**127**) and C4 (**128**) fluoro derivatives of DXP were not found to be substrates for catalysis by *E. coli* IspC but rather noncompetitive inhibitors with K_is of 444 μM and 733 μM, respectively.[125] When tested against *Synechocystis* PCC6803 IspC, **125** and **126**, as well as additional substrate analogues, were found to be competitive inhibitors, with **125** having the lowest measured K_i of 30 μM.[126] Four additional substrate analogues **129–132** with modification to the terminal acetyl of DXP but retaining both C3 and C4 hydroxyls have been demonstrated to inhibit *E. coli* IspC with IC_{50} values between 0.25–1.0 mM.[127] Although they retain their ability to coordinate with the active site divalent cation, these substrate analogues were orders of magnitude less potent than **48**.

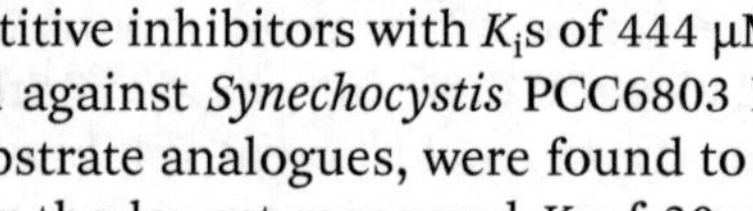

125 R1=OH, R2=H
126 R1=H, R2=OH
127 R1=F, R2=OH
128 R1=OH, R2=F

129 R=$CONH_2$
130 R=COOH
131 R=$COOCH_3$
132 R=CH_2OH

Figure 8.13 DXP analogues as substrate-based inhibitors of IspC (Compounds **125–132**).

A few compounds unrelated to the scaffolds of DXP and **21** have been found that inhibit IspC. Testing a series of bisphosphonates against *E. coli* IspC identified compounds **133** and **134** with IC_{50}s of 4 μM and 7 μM, respectively (Figure 8.14).[128] Co-crystal structures revealed that the bisphosphonates adopt a unique binding mode in which the phosphonates coordinate the active site metal and the aromatic side chain binds in a hydrophobic cleft enclosed by Trp-211. Intriguingly, neither **133** nor **134** overlap the phosphonate binding site of **21**, which in the crystal structure is instead occupied by a sulfate ion. Using a coordination chemistry approach, other lipophilic chelators were designed and evaluated for their ability to inhibit *E. coli* IspC.[129] Of seventeen compounds tested, six had IC_{50}s below 100 μM. Whereas compounds **135–138** had IC_{50}s between 25–75 μM, compounds **139** and **140** were ten-fold more potent with IC_{50}s of 1.4 μM and 4.5 μM. Furthermore, **139** demonstrated antibacterial activity similar to **21** against *E. coli*, *P. aeruginosa*, *B. anthracis*, and *M. luteus* with MICs between 3.7–19 μg mL^{-1}. Derivatives of **140** incorporating a more electron deficient pyridinyl substituent yielded compounds **141–143** with IC_{50}s between 5–10 μM, and an additional 5-phenyl substituent yielded the most potent analogue **144** with an IC_{50} of 0.8 μM.[130] Synthesis of an additional 41 analogues demonstrated that a one-carbon linker between the pyridinyl and phosphonate was optimal, the pyridin-2-yl was more potent than the pyridine-3-yl or pyrindin-4-yl, and replacement of the phosphonate with a carboxylate, hydroxamate, or sulfate resulted in a complete loss of activity.[131] Although more potent analogues were not identified, **143** and **144** were found to inhibit *M. tuberculosis* IspC with K_is of 1.6 μM and 3.2 μM, which is similar to the K_is of 2.3 μM and 4.2 μM determined for *E. coli* IspC, respectively. Co-crystal structures of *E. coli* IspC in complex with **143** and **144** revealed that these compounds adopt a binding mode more similar to **21** than that of the bisphosphonates **133** and **134**.[130] Rather than chelate the active site divalent cation, the phosphonate of **143** and **144** overlaps with the phosphonate binding site of **21**. Furthermore, the lipophilic side chain binds in a pocket exploited by α-substituted fosmidomycin derivatives like **113**. Interestingly, Trp-112 which forms a hydrophobic pocket and participates in π–π stacking interactions with

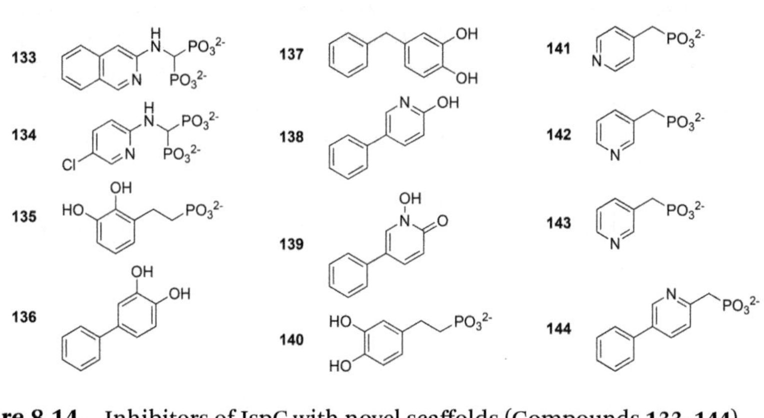

Figure 8.14 Inhibitors of IspC with novel scaffolds (Compounds **133–144**).

the side chains of **133** and **134**, rotates almost 180° and now forms an adjacent hydrophobic pocket where it participates in π–π stacking interactions with the side chains of **143** and **144**. Identification of these novel chemotypes and elucidation of their binding modes may provide a path forward for discovering non-substrate based inhibitors IspC with cellular activity.

8.3.4 Inhibiting IspD

In the third step of the MEP pathway, IspD catalyzes the transfer of phosphocytidyl derived from CTP onto the phosphate of MEP yielding 4-diphosphocytidyl-2-C-methyl-D-erythritol (CDP-ME). The reaction requires the presence of an active site Mg^{2+} and proceeds in an ordered sequential mechanism wherein CTP binds first followed by MEP.[132] Despite several crystal structures having been reported from various bacteria, few inhibitors have been discovered. Screening campaigns have identified several inhibitors of IspD from *Arabidopsis thaliana*[133–135] and *P. falciparum*,[133,136,137] which may prove useful in developing herbacides and antimalarials, but none of these have been demonstrated to have activity on bacterial IspD. Significant differences exist between prokaryotic and eukaryotic IspD homologues at both a sequence[138] and structural[139] level, perhaps explaining the lack of spectrum for identified inhibitors. Rational design of inhibitors is difficult because bacterial IspD has a relatively polar active site[79] and the CTP binding pocket is solvent-exposed. Structure-based drug design targeting the active site of *M. tuberculosis* IspD was performed to generate a series of virtual hits that were scored by *in silico* docking experiments, however none of these compounds were actually synthesized or tested for biological activity.[140] The only substrate-competitive inhibitor reported for *E. coli* IspD is **145** (Figure 8.15), which has a very weak IC_{50} of 1.36 mM.[141] High throughput screening of *M. tuberculosis* IspD against a small 3200 compound library at 10 μg mL^{-1} identified a single hit, domiphen bromide **146**, with reproducible biochemical inhibition and a measured IC_{50} of 33 μg mL^{-1}.[142] Overexpression of IspD in *Mycobacterium smegmatis* led to a five-fold decrease in the MIC of **146**, knockdown of IspD by antisense resulted in growth kinetics similar to treatment with 0.5 μg mL^{-1} **146**, and the combination of antisense knockdown and **146** treatment resulted in a complete loss of growth. Additionally, *M. smegmatis* grown in the presence of **146** had significantly decreased levels of cell wall-bound mycolic acids, a cellular component that requires isoprenoids for biosynthesis. Although these data would suggest that **146** is targeting IspD, MICs against drug resistant strains of *M. tuberculosis* were approximately 8 μg mL^{-1}, which is four-fold lower than the IC_{50} measured for the enzyme. This coupled with the detergent-like structure of **146** might

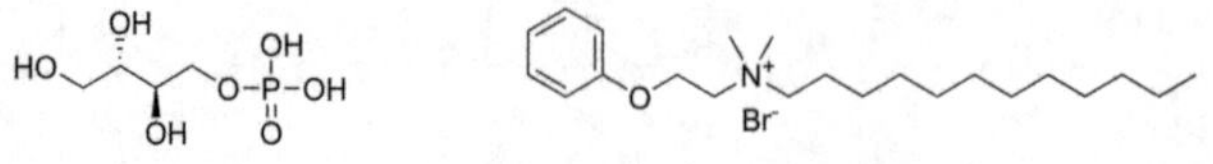

145 **146**

Figure 8.15 Inhibitors of IspD (Compounds **145–146**).

suggest that the antitubercular mechanism of action is not elicited through specific inhibition of IspD.

8.3.5 Inhibiting IspE

IspE is a Mg^{2+}- and ATP-dependent kinase that phosphorylates the 2-hydroyl group of CDP-ME to yield 4-diphosphocytidyl-2-C-methyl-D-erythritol 2-phosphate (CDP-ME2P). Although IspE belongs to the GHMP kinase superfamily that possess a high degree of sequence and structural similarity, there are differences in the active site between IspE and other kinases which can potentially be exploited to confer selectivity. Since IspE binds two nucleotide substrates, there are a variety of possibilities for targeting this enzyme. One approach focused on the substrate binding pocket and sought to make derivatives that could take advantage of a small adjacent hydrophobic subpocket that appeared well conserved across species based on sequence alignment.[143] Utilizing cytosine as a core, various small heteroalicyclic and aromatic rings were appended to approximate the ribose moiety and a propargylicsulfonamide linker was used to access the unique subpocket for placement of various small hydrophobic substituents. Compounds **147–157** were synthesized using structure based drug design and all were found to be viable inhibitors of *E. coli* IspE, most of which followed a competitive mechanism (Figure 8.16). Compounds **147** and **148** had IC_{50}s of 400 and 100 μM,

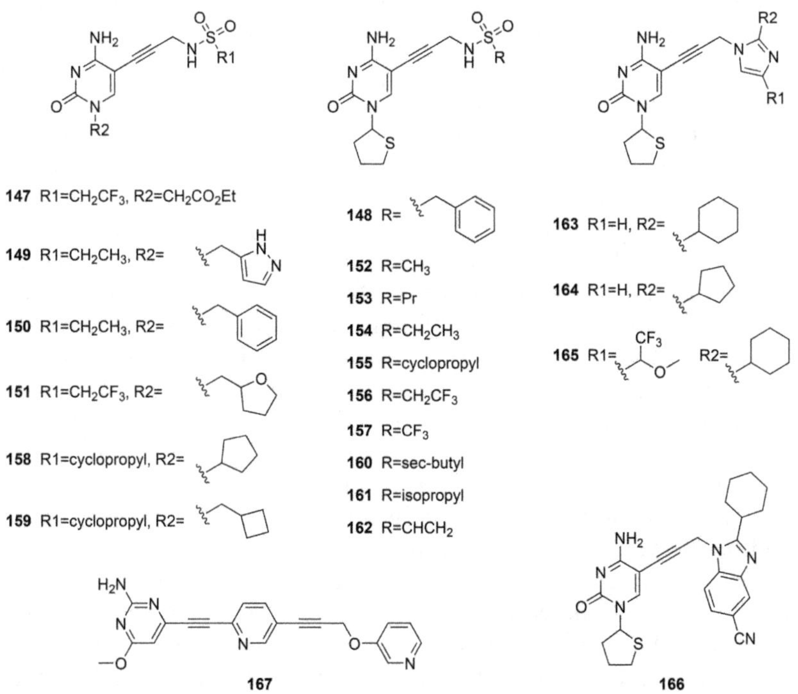

Figure 8.16 Cytosine-based inhibitors of IspE (Compounds **147–167**).

respectively, compounds **149–153** had IC_{50}s in the range of 19–79 μM, and compounds **154–157** had IC_{50}s of 6 μM or 8 μM. Tetrahydrothiophenyl substitution off of N1 of the cytosine was found to be optimal as additional studies demonstrated that although cyclohexyl **158** and cyclobutyl **159** were tolerated and had IC_{50} values of approximately 40 μM, any larger, more polar, or aryl substitution yielded poorly active compounds with IC_{50}s at least an order of magnitude higher.[144,145] Replacement of the sulfonamide linker with 4-morpholinyl, 1-piperidinyl, or 1-pyrrolidinyl groups led to a drastic reduction in activity with IC_{50}s in the range of 0.2–1.3 mM, and small moieties no larger than four carbons appended to the sulfonamide linker afforded the most potent compounds. No analogues more potent than **154–157** were identified, although *sec*-butyl and isopropyl derivatives **160** and **161** were of equal potency with IC_{50}s of 8 μM.[144] A co-crystal structure of *Aquifex aeolicus* IspE bound to a water soluble derivative of **155** confirmed the binding mode of this class of inhibitors was as had been proposed during structure-based design efforts. The inhibitor cytosine overlays with that of the natural substrate, the N1 substituent occupies the ribose binding site, and the propargylicsulfonamide linker extends towards the MEP binding pocket, placing the cyclopropyl substituent in the newly defined hydrophobic subpocket. Interestingly, in *A. aeolicus*, as well as in *M. tuberculosis* and *P. falciparum*, a key phenylalanine sidechain that lines the subpocket is replaced by tyrosine, leading to an unfavorable interaction between the tyrosine hydroxyl and the inhibitor cyclopropyl. A series of analogues were synthesized replacing the cyclopropyl with more polar vinyl, methyl ether, oxetane, and furan substituents in order to optimize contacts to this tyrosine in *A. aeolicus* IspE, as well as larger ribosyl and mannosyl moieties that could extend into the adjacent MEP binding pocket and displace a water cluster that was observed in the crystal structure.[146] Surprisingly, none of these new analogues inhibited *A. aeolicus* IspE, however they did possess some activity towards *E. coli* IspE. As had been previously observed, the larger sugar-substituted ligands showed weak if any inhibitory activity, but compounds with smaller substituents were active, displaying IC_{50}s of 30 μM for the methyl ether and oxetane analogues and a more potent IC_{50} of 3 μM for vinylsulfonamide **162**, which presumably is still able to interact with the phenylalanine side chain of *E. coli* IspE much in the same way as the original cyclopropyl moiety. Additional structure based drug design was employed to expand the interaction of these novel inhibitors with the expansive *E. coli* IspE substrate binding site.[147] Replacing the sulfonamide of **155** with an imidazole to facilitate growth of the molecule, branching at the C2 position with propyl, butyl, phenyl, cyclohexyl and cyclopentyl groups to probe the MEP binding pocket identified compounds **163** and **164** with IC_{50}s of 12 μM and 15 μM. Further derivatization at the C4 position to access the glycine-rich phosphate-binding loop of the enzyme yielded **165** with an IC_{50} of 9.9 μM. Finally, fused benzimidazoles were also shown to be tolerated with the best compound **166** displaying an IC_{50} of 13 μM. An additional bisubstrate inhibitor that occupies both the 4-diphosphocytidyl-2-C-methyl-D-erythritol and ATP-binding pockets has also been described.[148] Although **167** is extremely rigid and highly insoluble, use of a peptide carrier allowed measurement of an IC_{50} of 8.7 μM. Despite impressive work in structure-based *de novo* design of *E. coli* IspE inhibitors that grew out of the cytosine binding-pocket to interact with the ribosyl binding site, a hydrophobic subpocket, the MEP binding site, the phosphate binding loop and ultimately the ATP-binding site, the most active inhibitors were some of the earlier compounds which only occupied the cytidine binding pocket and hydrophobic subpocket. This may underscore the difficulty in effectively drugging polar binding sites with effective ligands and may suggest why previous inhibitors of the MEP pathway hitting IspC required polar hydroxamate and phosphonate moieties to achieve antibacterial activity. These cytosine derivatives have yet to be tested for cellular activity.

In an effort to find inhibitors distinct from those hitting the CDP-ME binding site, the structural similarity of IspE with other kinases in the superfamily led to testing existing inhibitors of GHMP kinases. From a library of 150 compounds found to inhibit galactokinase,[149] 24 were tested on *E. coli* IspE, leading to identification of seven compounds demonstrating cross-inhibition, of which three compounds had increased activity.[150] Although there is sequence and structural similarity amongst the GHMP family, the fact that 70% of the hits tested on IspE were inactive at concentrations 10-fold higher than the IC_{50} measured on galactokinase, and the identification of three hits that were more potent on IspE than on galactokinase, demonstrates that it is possible design inhibitors with selectivity among related kinases. From these hits, two compounds, **168** and **169** were selected for additional follow-up (Figure 8.17). Not only did **168** and **169** inhibit *E. coli* IspE with IC_{50}s of 18 μM and 5.5 μM, they also inhibited IspE from *Yersinia pestis* with IC_{50}s of 9 μM and 6 μM, and **168** was able to slow bacterial growth in culture. Intriguingly, although these compounds were selected due to their inhibition of a family of kinases, *in silico* docking predicted that both would occupy the substrate cytosine binding pocket as opposed to the ATP binding pocket. Follow-up SAR studies focused on altering the substituents bound to the core thiazinecarbonitrile

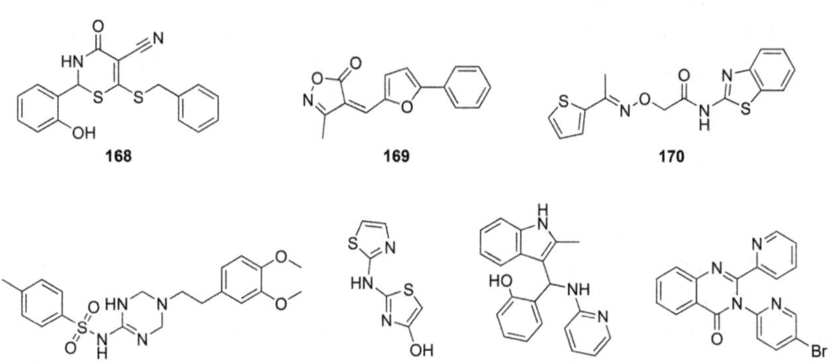

Figure 8.17 General kinase inhibitors repurposed for targeting IspE (Compounds **168–174**).

of **168** and the isoxazole of **169** did find active analogues, however, the results did not agree with the predicted simulated docking binding mode. In a second approach taken in the same study, a computational high throughput screen was performed with two million compounds focusing on the CDP-ME binding site. Of the 210 identified of hits, only ten were procured and tested, and the best activity was seen with **170** and **171**, which demonstrated 65% and 80% inhibition of *E. coli* IspE at 20 μM, respectively. A separate study took a very similar approach to try to find novel inhibitors of IspE, wherein both a virtual screen and a focused biochemical screen were performed.[151] In the virtual high throughput screen, a library of over four millions compounds was triaged based on physicochemical properties and filtered using a protein-based pharmacophore to yield approximately 43 000 structures that were docked into the *A. aeolicus* crystal structure at the cytidine binding site. There were 566 hits that satisfied particular criteria, fourteen were selected for biochemical follow-up assays, and only six had measurable IC_{50}s. Five of the six compounds had activity in the mM range on *E. coli* IspE, but the best compound, **172**, had an IC_{50} of 160 μM. In the biochemical screen, a kinase-specific library of approximately six thousand molecules was screened on *E. coli* IspC, yielding two compounds **173** and **174** that had IC_{50}s of 19 μM and 2.5 μM, respectively. Hit expansion around these three new scaffolds did not yield any derivative with increased potency. Despite the identification of hits that disrupt IspE function in a biochemical assay, the combination of virtual and biochemical screening produced conflicting data. SAR was inconsistent with predicted binding modes and biochemical hits identified from kinase-specific libraries were routinely docked *in silico* at the substrate cytosine pocket rather than the ATP binding pocket. Additional studies are required to determine enzyme-inhibitor co-crystal structures and understand the physiological consequence of treating cells with these inhibitors in order to validate these compounds as specific IspE inhibitors.

Attempts to definitively target the ATP-binding site in IspE led to the synthesis of various 8-substituted *syn*-configured adenosine derivatives.[152] Using the crystal structure of *E. coli* IspE bound to CDP-ME and the non-hydrolyzable ATP analogue adenosine 5′-(β,γ-imido)triphosphate (AMP-PNP), structure-based drug design was employed to replace the triphosphate of ATP with a more hydrophobic moiety capable of interacting with the glycine-rich loop, and substitution at the 8-position of the adenosine was introduced in order to force equilibrium towards the *syn* rather than the *anti* configuration and create an additional favorable halogen or hydrogen bonding interaction with the backbone carbonyl of Asp-64. Amino, bromo, and azido substituents were introduced at the 8-position and tested in combination with either a 2′,3′-*O*-isopropylidene protected ribosyl moiety with no 5′ substitution, or the deprotected ribosyl analogue containing amide, triazole, or sulfonamide substitutions at the 5′ position. Analogues with larger substituents at the 5′ position or an 8-bromo substituent on the adenosine ring did not have measurable activity on IspE, but several of the 2′,3′-*O*-isopropylidenated

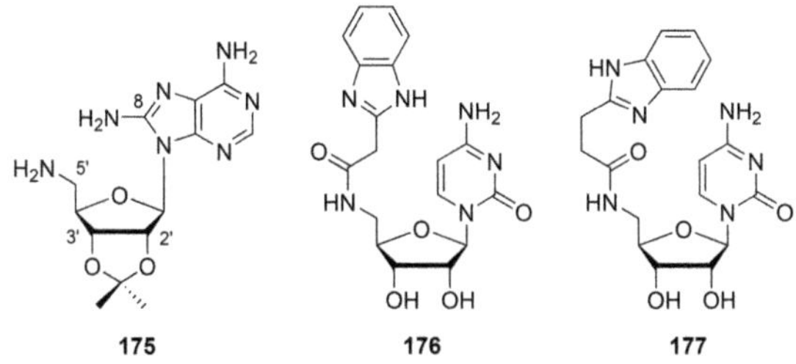

Figure 8.18 Nucleoside-based inhibitors of IspE (Compounds **175–177**).

derivatives containing either a C8 amine or azide had inhibitory activity with IC_{50}s of several hundred μM. The best compound **175** had an IC_{50} of 265 μM (Figure 8.18). These results are similar to those found for cytidine-based inhibitors **176** and **177** targeting the substrate binding pocket, which also possessed weak inhibitory activity on *E. coli* IspE with IC_{50}s of 2.0 mM and 1.7 mM, respectively.[153] Crystal structures *A. aeolicus* IspE bound to ligands **176** and **177** confirmed the binding mode as overlapping that of the cytidine moiety of CDP-ME but revealed that the benzimidazole is disordered and makes few interactions with the rest of the protein. These results were corroborated by mechanistic studies demonstrating that these compounds had a competitive mode of inhibition. Clearly, these nucleoside-based compounds targeting either the substrate- or ATP-binding pockets will need to be significantly optimized in order to enhance their affinity and improve inhibitory activity. Furthermore, achieving specificity for inhibitors based on natural ligands utilized by countless enzymes in all organisms may prove challenging.

8.3.6 Inhibiting IspF

IspF catalyzes the intramolecular nucleophilic attack of the 2-phosphate of CDP-ME2P onto the β-phosphate, yielding the cyclized product MEcPP with concomitant loss of CMP. This reaction requires a divalent cation to properly orient the substrate during catalysis and stabilize the pentavalent transition state.[154,155] Initial attempts to identify inhibitors of this reaction focused on cytosine nucleotide substrate analogues. The diphosphate of CDP was linked to an aromatic residue through an appropriately sized linker in order reach into a hydrophobic cleft responsible for binding the methyl-D-erythitol moiety of the substrate and product.[156] Anthranilate and dansyl moieties were chosen to facilitate the development of a fluorescence-based inhibition assay. Of the three compounds synthesized, the one with the best affinity for *E. coli* IspF, **178**, had an IC_{50} of 3.0 mM (Figure 8.19). In order to explore the diversity of chemical space capable of binding IspF, a fragment-based screen using metabolite-like molecules was performed on *Burkholderia pseudomallei* IspF.[157] After screening IspF with 1500 compounds by X-ray crystallography, six cytosine derivatives and three small heteroaromatic hits that coordinated to the catalytic zinc ion were found. Additional screening of 390 fragments

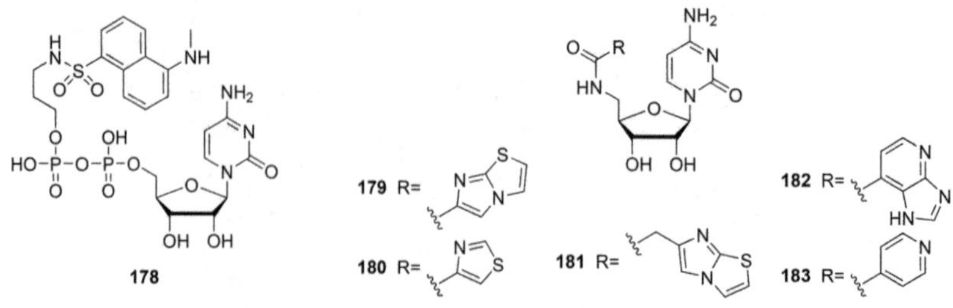

Figure 8.19 Substrate-based cytidine analogues inhibiting IspF (Compounds **178–183**).

by NMR followed by X-ray crystallography confirmation identified one more zinc-coordinator and two molecules, which bind outside of the active site in a hydrophobic region adjacent to the zinc. The discovery of alternative zinc-chelating fragments led to the synthesis of several fusion molecules in which the diphosphate of CDP was replaced with aromatic nitrogen-containing zinc-binders. The five fusion compounds **179–183** were demonstrated to have K_Ds from 70 μM from 200 μM, which was like the K_D of 75 μM observed for CDP, but higher than the K_D of 15 μM observed for **178** on *E. coli* IspF. None had antibacterial activity against *Burkholderia thailandensis* when tested up to concentrations of 1 mM. As with nucleoside-based inhibitors tested on IspE, compounds containing large portions of the natural ligand for IspF are likely poor scaffolds for the design of novel inhibitors and will potentially suffer from selectivity issues; however, these molecules may represent useful tools for validating important residues within the active-site that could facilitate binding of alternative ligands, thus guiding the design of future inhibitors with novel chemotypes.

Efforts to optimize the first-generation of cytidine-based inhibitors using crystallographic information obtained from **178** led to the synthesis of a series of molecules in which the cytosine moiety was replaced by 2,5-diaminopyridine or 5-amino-1H-imidazo[4,5-*b*]pyridine.[158] Employing structure-based drug design, hydrophobic substituents were appended to these alternate cores, as well as cytosine, through an exit vector that would allow interaction with the methyl-D-erythitol binding pocket while bypassing the ribosylpyrophosphate binding site. Although none of the imidazopyrimidines had IC_{50}s below 1 mM when tested on *E. coli* IspF, cytosines **184–186** and diaminopyrimine **187** did have inhibitory activity with IC_{50}s of approximately 0.5 mM (Figure 8.20). These molecules demonstrate that novel ligands can occupy the CDP pocket without engaging the ribose and diphosphate binding sites. A high throughput screen of *A. thaliana* IspF using a library of 40 000 molecules, identified a novel thiazolopyrimidine **188** that was demonstrated to inhibit IspF from multiple organisms including *M. tubercolusis* and *E. coli* with IC_{50}s of 6.1 μM and 32 μM, respectively.[159] Derivatives exploring different aryl substitutions on the pyrimidine and various halogen substitutions on the phenol mostly generated inhibitors with reduced activity, although

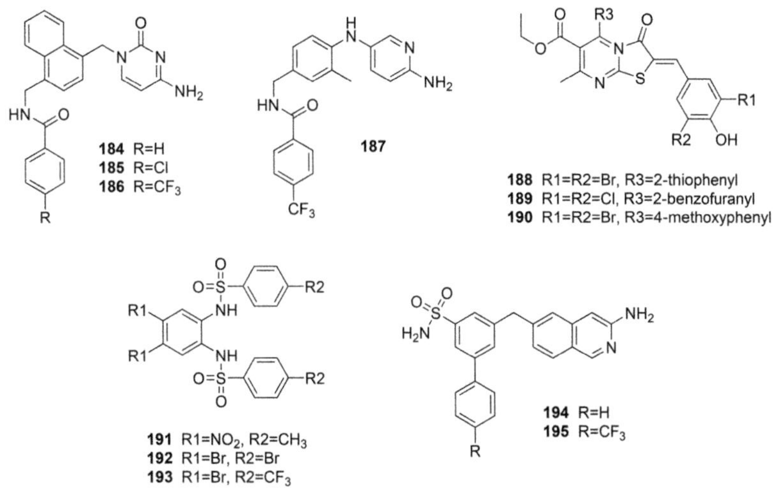

Figure 8.20 Inhibitors of IspF with novel scaffolds (Compounds **184–195**).

compound **189** was approximately equipotent to **188** and compound **190** was slightly more potent with IC_{50}s of 5.1 μM and 18 μM on *M. tubercolusis* and *E. coli* IspF, respectively. Although these molecules had antimalarial activity, they were not tested for antibacterial activity. The same screen also identified a bis-sulfonamide inhibitor of IspF, and although most follow-up work focused on enzyme from *A. thaliana* and *P. falciparum*, some data was presented for bacterial IspF from *B. pseudomallei* and *M. tuberculosis*.[160] A series of derivatives synthesized based on the initial hit **191** demonstrated that bromo substitution of the nitro groups was well tolerated, unsubstituted derivatives were less potent, and methyl substitution led to inactive compounds. Since these substitution patterns modulate the acidity of the sulfonamide nitrogen, with more active derivatives calculated to have a lower pK_a, the bis-sulfonamides are thought to bind in a deprotonated state. Various hydrophobic replacements of the tolyl moieties on the sulfonamides, including hexyl and substituted phenyl groups, were all found to retain activity, and the most potent derivatives, **192** and **193**, demonstrated an 8-fold improvement in IC_{50} compared to **191**, with IC_{50}s of 22 μM on *B. pseudomallei* IspF. Molecular modeling that docked the bis-sulfonamide compounds into crystal structures of IspF indicated that a single sulfonamide was coordinating to the active site zinc while the second sulfonamide was directing an aryl substituent into the cytosine binding pocket. Since asymmetric monosulfonamide derivatives were demonstrated to retain some inhibitory activity, rational design was used to create sterically unencumbered primary sulfonamides to facilitate improve chelation of the zinc while introducing an optimized *ortho* substituent that could more favorably interact with the cytosine binding pocket. Compounds **194** and **195** were successfully shown to inhibit *M. tuberculosis* IspF with IC_{50}s of 59 μM and 43 μM, respectively.

Thus, structure-based design was far more successful in improving activity of biochemical hits found from a high throughput screen than in creating inhibitors *de novo* starting from the endogenous substrate.

8.3.7 Inhibiting IspG and IspH

The last two steps of the MEP pathway are carried out by enzymes containing unique [4Fe–4S] cluster that are coordinated by three cysteine residues, leaving the fourth iron lacking contacts to any protein sidechains. Both IspG and IspH catalyze reductive dehydroxylation, leading to loss of the C-3 and C-1 hydroxyls of MEcPP, respectively, in formation of the final products of the MEP pathway, IPP and DMAPP. The first compounds designed to inhibit these enzymes were substrate analogues in which the diphosphate was replaced with substituted carbamates or *N*-acyl-*N*′-oxy sulfamates.[161] These compounds were extremely weak inhibitors, with the best compounds **196** and **197** displaying 36% inhibition of *Thermus thermophilus* IspG and 39% inhibition of *A. aeolicus* IspH at 1 mM, respectively (Figure 8.21). Subsequent attempts to develop inhibitors sought to explore unconventional ligands, wherein binding would be mediated by interaction with the [4Fe–4S] cluster rather than amino acids that recognize substrate.[162] Such an approach has the benefit of potentially targeting both IspG and IspH simultaneously, perhaps leading to antibacterial synergy and a higher barrier to resistance. Testing of small molecules and ions for direct binding to *A. aeolicus* IspH by EPR spectroscopy revealed that CN^-, but not CO, N_3^-, or MeCN, were able to bind end-on to the fourth iron, and that propargyl alcohol, propargylamine and propiolic acid were able to bind in a sideways-on mode. These fragments were linked to diphosphate, as well as linear and branched bisphosphonates, to improve binding to the active site. Although the bisphosphonates, which were supposed to act as isosteres of diphosphate, showed weak, if any, inhibition of IspH, all of the diphosphate inhibitors were active, with IC_{50}s ranging from 0.45 μM to 245 μM for **198–203**. Since the reaction mechanism of IspH is proposed to involve formation of an allylic cation, several pyridine diphosphate derivatives were also synthesized and tested for inhibition of IspH.[162] The *ortho*-pyridinium was the least active with an IC_{50} of 1.2 mM,

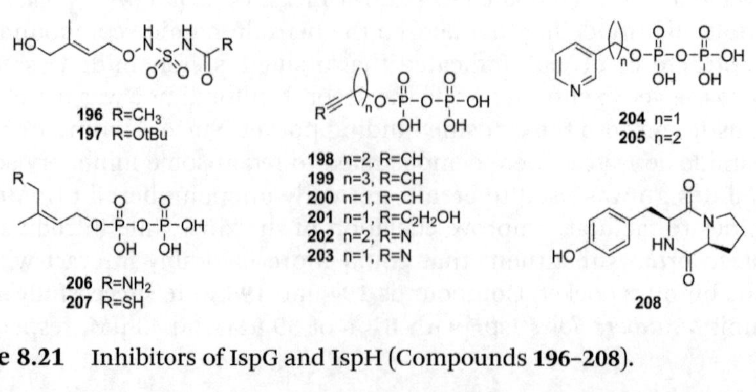

Figure 8.21 Inhibitors of IspG and IspH (Compounds **196–208**).

the *para*-pyridinium was modestly active with an IC_{50} of 149 μM, and the *meta*-pyridinium **204** was the most active with an IC_{50} of 38 μM. Analogues made to test the optimum distance to the diphosphate identified **205** with an ethyl linker as the most potent derivative with an IC_{50} of 9.1 μM. Differences between the regioisomers of the pyridine diphosphates along with molecular docking of **205** to IspH originally suggested that **205** binds in the vicinity of the fourth iron but mostly interacts with Glu-126; however subsequent hyperfine sublevel correlation (HYSCORE) spectroscopy demonstrated that the pyridine nitrogen actually binds in the neutral state to the fourth iron through a Lewis acid/base mechanism.[163] However, allylic diphosphates are proposed to undergo π complex formation with the fourth iron, potentially producing a metallocycle between acetylene and the [Fe4–S4] cluster. Intriguingly, as had been postulated based on the binding mode, many of these inhibitors had similar activity on *E. coli* IspG.[164,165] The bisphosphonate derivatives had IC_{50}s in the low single digit mM range, diphosphate derivatives with alkane or alkene linked to a carboxylate had IC_{50}s of approximately 400 μM, pyridines **204** and **205** had IC_{50}s of 140 μM and 268 μM, respectively, and the alkynes **198–201** had the best activity with IC_{50}s between 0.77 μM and 4.9 μM. Although binding of acetylene inhibitors to IspG was found to occur through a similar π complex formation found for IspH, there are some subtle differences in the binding modes of these molecules to these enzymes. In the active site of IspH, **198** reacts with water bound to the fourth iron leading to formation of an enolate that hydrolyzes to the corresponding aldehyde in solution.[166] IspG does not catalyze this reaction because the acetylene side chain is oriented in the opposite direction in the enzyme and can sterically clash with the [Fe4–S4] cluster, leading to loss of the apical iron in co-crystal structures.[167]

Studies using Mössbauer spectroscopy demonstrated that the first step in IspH catalysis involves binding of HMBPP to the fourth iron through its hydroxyl group.[168] This led to the synthesis of amino and thio substrate analogues **206** and **207** with the intention of coordinating to the iron with a poorer leaving group (Figure 8.21). Indeed, both were determined to be competitive reversible tight binding inhibitors of *E. coli* IspH with IC_{50}s of 150 nM and 210 nM, respectively.[169] Interestingly, mechanistic studies demonstrated that **206** was a slower binder than **207**, possibly as a result of requiring deprotonation prior to interacting with the [Fe4–S4] cluster. Co-crystal structures proved that these molecules bound to the fourth iron and adopted a similar binding mode to the natural HMBPP ligand, though a second conformation for **206** was observed in which the amino group instead ion-paired with Glu-126.[170] As was discovered for the acetylene phosphate inhibitors, **204** and **205** also inhibit *E. coli* IspG with IC_{50}s of 2.5 μM and 1.4 μM, respectively.[171] Interestingly, these compounds demonstrate a diverse array of binding modes towards IspG and IspH. When binding oxidized IspH, both **206** and **207** form direct Fe–N and Fe–S bonds, respectively.[168] However, when the [Fe4–S4] cluster is reduced, **206** forms π interactions with the fourth iron and strong Coulombic interactions with Glu-126, whereas **207** precludes reduction and remains directly bound to the oxidized iron.[171] In binding IspG, **206** neither directly bonds to nor forms a π complex with the fourth iron. Molecular modeling suggests that the inhibitor instead ion pairs with Asp-87, much as it does with Glu-126 in reduced IspH. However, **207** forms a π-bond with the fourth iron in reduced IspG in a similar fashion as **198**. The mechanistic insights derived from these substrate analogues will be useful in designing future dual-targeting inhibitors of IspG and IspH.

Cell-based phenotypic screening targeting the MEP pathway has identified a natural product **208** named maculosin, which putatively targets IspG (Figure 8.21).[172] Screening of *Actinomycetes* strains for production of metabolites capable of inhibiting growth of *E. coli*, which utilizes the MEP pathway, but not *S. aureus*, which utilizes the MVA pathway, resulted in the identification of a strain that produces **208**. Notwithstanding weak antibacterial activity, addition of IPP to the culture media was able to rescue growth deficiencies in a similar manner as rescue after treatment with **21**, indicating that **208** was targeting an early step in isoprenoid biosynthesis. Affinity pulldown with a biotinylated-**208** using lysates from seven *E. coli* strains individually overexpressing one of the MEP pathway enzymes identified IspG as the only enzyme in the pathway that interacted with **208**. Although direct biochemical inhibition was not measured, quartz crystal microbalance experiments gave an estimated K_a of 0.22 μM. The interaction with **208** was somewhat specific, as binding was abrogated when testing a cyclo(L-Phe-L-Pro) analogue lacking the hydroxyl. Given that structure-based drug design for IspG and IspH has succeeded in generating inhibitors of limited chemical diversity, most of which are diphosphate derivatives that closely resemble the natural substrate, such alternative screening approaches may be necessary to more fully probe chemical space. The use of phenotypic assays is advantageous in that hits will start with cellular activity and a single screen can simultaneously target the MEP pathway as a whole.

8.4 Alternate Targets Utilizing IPP Precursors

8.4.1 Inhibiting IDI

The final enzyme in the MEP pathway, IspH, catalyzes the formation of IPP and DMAPP in a 5 : 1 ratio, however, in the final step of the MVA pathway, DPMD only catalyzes the formation of IPP.[14] Since DMAPP serves as the initial electrophilic substrate for chain elongation reactions that ultimately produce more complex isoprenoids, isomerization is an essential step following IPP formation by the MVA pathway. There are two structurally distinct forms of IDI. The first, termed IDI-1, was originally discovered in the 1950s[173] and is a zinc metalloprotein that also requires Mg^{2+} for catalysis. The second, termed IDI-2, was discovered much later in 2001[16] and is a flavoenzyme that requires FMN and Mg^{2+} for catalysis. Although there is no strict association between isoprenoid pathway and IDI isoform, IDI-2 is essential in several serious Gram-positive pathogens that use the MVA pathway, including

Staphylococci, *Streptococci*, and *Enterococci*. Conversely, the IDI-1 isoform is found in many bacteria that utilize the MEP pathway, such as *E. coli*, *Vibrio cholera*, and *M. tuberculosis*, where function is not essential but presumably helps balance the pools of DMAPP and IPP.[174] Given the essential nature of IPP isomerase in several key ESKAPE pathogens and the fact that Eukaryota encode the IDI-1 isoform of the enzyme, IDI-2 serves as a potential novel selective therapeutic target for the development of new antibacterials. Early mechanistic studies designed to assess whether IDI-2-catalyzed isomerization proceeds *via* a carbocation-type intermediate demonstrated that 3,4-epoxy-3-methylbutyl diphosphate **209** (Figure 8.22), a mechanism based inhibitor of *E. coli* IDI-1 was a slow inactivating inhibitor of *Methanocaldococcus jannaschii* IDI-2 with a K_I of 57 mM.[175] Other mechanistic experiments exploring the activity of epoxy, fluoro, diene, alkyne, and allene substrate analogues on *T. thermophilus* IDI-2 found additional irreversible inhibitors. Certain ligands such as **209** and **210**, which had K_Is of 54 μM and 1.4 μM, respectively, were not substrates for IDI-2 catalysis, whereas other inhibitors such as **211** and **212**, which had K_Is of 8 μM and 7.4 μM, respectively, could be isomerized by IDI-2. There was a similar difference in catalysis observed for reversible inhibitors of IDI-2. Compounds **213** and **214**, which had K_Is of 48 μM and 36 μM, respectively, were not substrates for IDI-2, whereas compounds **215–217**, which had K_Is of 12 μM, 48.6 μM, and 383 μM, respectively, underwent isomerization by IDI-2.[176–178] The difference between inhibitors which could or could not act as substrates delineated that isomerization by IDI-2 proceeded *via* a protonation/deprotonation mechanism, consistent with a carbocation intermediate, and irreversible inhibitors that formed covalent adducts to the N5 isoalloxazine ring of FMN demonstrated that this cofactor was acting as the general acid/base catalyst rather than an amino acid side chain in the protein.[179] As with diphosphate substrate analogue inhibitors of various steps in the MEP pathway, these molecules are a first step towards understanding the catalytic mechanism of IDI-2 and identifying key residues in the active site. Further optimization will be required to alter the diphosphate motif in order to create more drug-like molecules which are capable of bacterial cell entry.

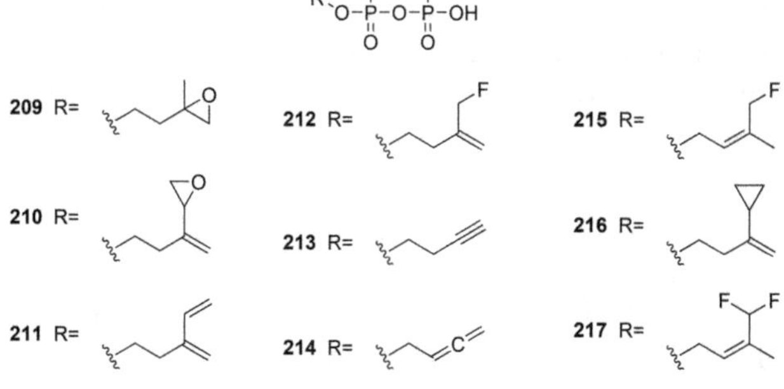

Figure 8.22 Inhibitors of IDI (Compounds **209–217**).

8.4.2 Inhibiting UppS

One of the most critical isoprenoids synthesized by bacteria is bactoprenol. This lipid carrier plays a central role in the biosynthesis of many essential cell envelope structures, in particularly peptidoglycan. Bactoprenol is synthesized from the sequential condensation of 8 IPP monomers with farnesyl pyrophosphate by the enzyme undecaprenyl pyrophosphate synthase (UppS). UppS is unique because it is a *Z*-type prenyltransferase and catalyzes *cis*-double bond formation between IPP monomers, whereas most other prenyltransferases, such as farnesyl pyrophosphate (FPP) synthase and geranylgeranyl pyrophosphate (GGPP) synthease, are *E*-type and condense new IPP units in a *trans* configuration. Since *Z*-type prenyltransferases share no sequence or structural homology to *E*-type prenyltransferases[180] and UppS is not produced by humans,[181] there has been significant interest in UppS as a novel selective antibacterial target. Furthermore, because cell wall biosynthesis relies on availability of bactoprenol, inhibitors of UppS could potentially synergize with β-lactams and vancomycin, some of the most successful antibiotics introduced into clinical practice.[182] As with other isoprenoid biosynthetic enzymes, the first compounds made to target UppS were substrate analogues. These included alkenyl dicarboxylic acids modeled after maleate that were mostly inactive except for **218** which had an IC50 of 135 μM against *E. coli* UppS,[183] farnesyl thiopyrophosphate,[184] 3-desmethyl FPP and *Z*-GPP,[185] and polyaryl bisphosphonates that had a range of potencies, the best being **219** and **220** with IC_{50}s of ~0.6 μM against *E. coli* UppS (Figure 8.23).[186] Despite demonstrating some potency in biochemical assays, these substrate analogues are unlikely to progress as viable leads for the development of antibacterials because they contain cell impermeable diphosph(on)ate moieties and were demonstrated to be nonspecific, inhibiting even *E*-type prenyltransferases. Additionally, co-crystal structures of various bisphosphonates bound to UppS demonstrate that these molecules can occupy as many as four distinct binding sites, presumably reflecting the similar structure of multiple ligands that the protein binds during iterative catalysis. Utilizing structure-based design to optimize inhibitors that can simultaneously occupy multiple sites presents a difficult challenge.

Virtual screening has proven rather successful in identifying inhibitors of UppS lacking bisphosphonate or diphosphate moieties. Using the Maybridge Chemical Company compound database, over 58 000 molecules were virtually screened against the crystal structures of *E. coli* and *Helicobacter pylori* UppS, and of twenty-six potential hits, two were validated as biochemical inhibitors

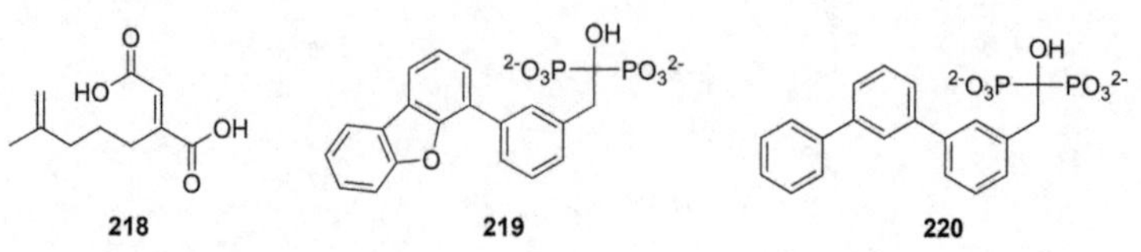

Figure 8.23 Substrate-based inhibitors of UppS (Compounds **218–220**).

of both enzymes.[187] Although **221** and **222** were equipotent against UppS from *H. pylori* with IC_{50}s of ~360 μM, **221** was more potent on *E. coli* UppS with an IC_{50} of 71 μM and **222** was almost inactive (Figure 8.24). The published crystal structure of *Trypanosoma brucei* FPP synthase along with the top three conformations derived from molecular dynamics simulation were used to dock 2000 compounds from the National Cancer Institute Diversity

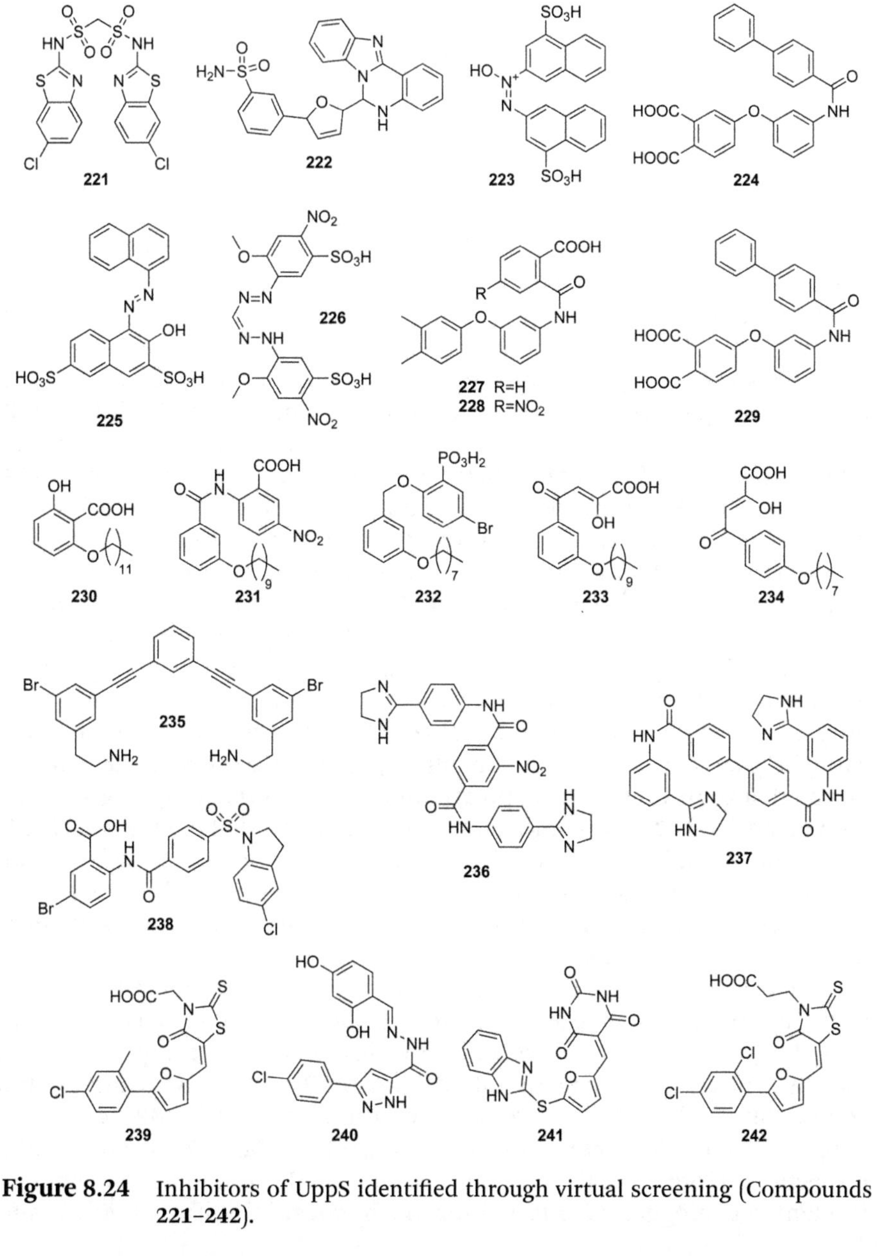

Figure 8.24 Inhibitors of UppS identified through virtual screening (Compounds **221–242**).

Set I, yielding a list of 18 top hits.[188] When counterscreened against UppS, compounds **223–226** displayed inhibitory activity. **223** and **224** were fairly potent inhibitors with IC_{50}s between 3 μM and 7 μM on both *S. aureus* and *E. coli* UppS, whereas **225** and **226** were less potent with IC_{50}s of 14 μM and 40 μM against *S. aureus* and *E. coli* UppS, respectively. Interestingly, despite being identified from a screen on FPP synthase, **224–226** did not show measurable inhibition of FPP synthase from three different organisms, though **223** did demonstrate activity with IC_{50}s of 21 μM, 46 μM, and 237 μM against enzymes from *T. brucei*, *S. aureus*, and *Homo sapiens*, respectively. Further hit development led to the synthesis of various benzoates, phosphonates, diketoacids, and cationic compounds that inhibited *E. coli* and *S. aureus* UppS.[189] The smaller benzoates **227** and **228** were fairly weak inhibitors with IC_{50}s between 35 μM and 170 μM, but benzoates **229–231**, phosphonate **232**, and diketoacids **233** and **234** containing a large linear hydrophobic sidechain were all potent inhibitors with IC_{50}s in the range of 0.5–7 μM. Bisamine **235** and bisamidine **236** had similar activity, and bisamidine **237** was the most potent compound tested, with IC_{50}s of 0.1 μM on both *E. coli* and *S. aureus* UppS. It was noted that these structures were similar to a previously reported anthranilate **238** with cellular activity against *S. aureus*.[190] The mechanism of action was not fully elucidated but there was evidence of cell wall inhibition.[191] Testing of **238** in these assays revealed that it could inhibit UppS with an IC_{50} of 1.5 μM.[189] Compounds **229, 235, 236,** and **237** had antibacterial activity against *E. coli* with MICs between 4–16 μg mL^{-1} and compounds **233, 234, 237,** and **238** had antibacterial activity against *S. aureus* with MICs between 0.25–1 μg mL^{-1}. Co-crystal structures of *E. coli* UppS bound to all of these inhibitors, except for **237**, were solved, and it was found that whereas weak benzoic acids **227** and **228** bound to site 3 near the catalytic center, all potent compounds bound to allosteric site 4, and potentially additional sites, in the structure. This suggests that site 4 occupancy may be required for potent enzyme inhibition and may offer the best opportunity for structure-based drug design. For the most potent analogue **237**, *in vitro* synergy was observed with the β-lactam antibiotic methicillin, providing some validation to the notion that UppS inhibitors might synergize with cell wall-targeting antibiotics. Furthermore, **237** was efficacious at 10 mpk in a lethal mouse model of *S. aureus* infection, leading to survival of all mice in the treatment group compared to complete mortality in the vehicle control group after seven days. Testing additional analogues of **237** failed to identify bisamidines with more potent inhibition of UppS.[192]

Retrospective analysis of twelve crystal structures and one-hundred and twelve inhibitors from the aforementioned studies led to the selection of two crystal structures on which virtual screening of 450 000 compounds from the ChemBridge Experimental Library was performed.[193] Assaying the top 100 hits in biochemical assays identified three molecules, **239–241**, which were found to inhibit *S. aureus* UppS with IC_{50}s of 2.7 μM, 13.7 μM, and 6.7 μM, respectively (Figure 8.24). In order to explore SAR around **239**, a series of rhodanine analogues were tested for biochemical inhibition of *E. coli* and

S. aureus UppS, antibacterial activity on *B. subtilis* and *S. aureus*, and cytotoxicity on *S. cerevisiae*. Most analogues completely lost activity, a few were less potent, but **242** demonstrated equipotent activity on *S. aureus* UppS and a 100-fold increase in potency on *E. coli* UppS with an IC_{50} of 2.1 μM. Both **239** and **242** had antibacterial activity on a variety of Gram-positive bacteria with MICs between 1–8 μg mL^{-1} and 0.125–4 μg mL^{-1}, respectively, on *B. subtilis*, *B. anthracis*, *S. aureus*, *E. faecalis*, and *Listeria monocytogenes*. Interestingly, **239** demonstrated dramatic antibacterial synergy with methicillin against *S. aureus*, but not with vancomycin against *E. faecalis* or with ampicillin against *B. anthracis*. Just as **239** was similar in structure to the drug epalrestat, the HIV integrase inhibitor elvitegravir inspired the synthesis of diketoacids for potential inhibition of UppS.[194] HIV-1 integrase contains an active site Asp/Mg^{2+} domain to which several integrase inhibitors bind; a motif which is similar to the active site Asp/Mg^{2+} motif of prenyltransferases responsible for coordination of the substrate diphosphate. Thirty eight compounds containing amide-diketo acid, dihydropyridone-3-carboxylic acid, or aryl-diketo acid head groups linked to a hydrophobic tail were synthesized, of which twenty-eight had IC_{50}s less than 10 μM against *S. aureus* and *E. coli* UppS. Two of the most active amide-diketo acids **243** and **244** had IC_{50}s of approximately 0.5 μM and the two aryl-diketo acids **245** and **246** had IC_{50}s between 0.5–2.0 μM against both enzymes (Figure 8.25). The co-crystal structure of **243** bound to *E. coil* UppS revealed that this inhibitor occupies site 1, the FPP binding site. The hydrophobic tail of **243** maps closely to the backbone of FPP, with the diketo-acid in close proximity to the diphosphate moiety, confirming the predicted binding mode. Although the amide-diketo acids did not have antibacterial activity, **245** and **246** had promising Gram-positive activity, with MICs of 0.5 μg mL^{-1} against *S. aureus* and *B. anthracis*, 4 μg mL^{-1} against *L. monocytogenes* and *E. faecium*, and 1 μg mL^{-1} against *S. pyogenes*.

High throughput screening has also been successfully used to identify inhibitors of UppS. Assaying enzymatic inhibition of *S. pneumoniae* UppS, tetramic acid **247** was identified with an IC_{50} of 19 μM (Figure 8.26).[195] This class of compound binds to an allosteric site in the open conformation of the substrate-free protein and permits binding of either IPP or FPP but precludes binding of both, thus preventing catalysis.[196] SAR analysis demonstrated that alteration of the tetramic acid core was tolerated, as *N*-alkylated tetramic acids, tetronic acids, and dihydropyridin-2-ones all yielded potent inhibition of UppS.[195] Derivatization of the acyclic amide favored aryl-phenyl or cycloalky-phenyl groups, with smaller aromatic and aliphatic groups losing

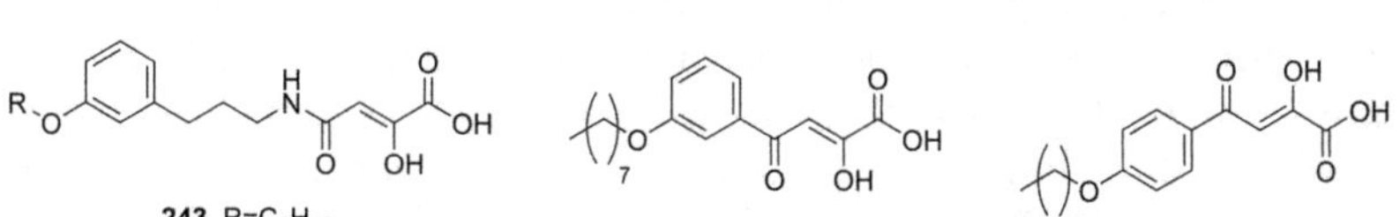

243 R=C_6H_{13}
244 R=p-tosyl

245

246

Figure 8.25 Inhibitors of UppS synthesized based on the structure of the HIV-1 integrase inhibitor elvitegravir (Compounds **243–246**).

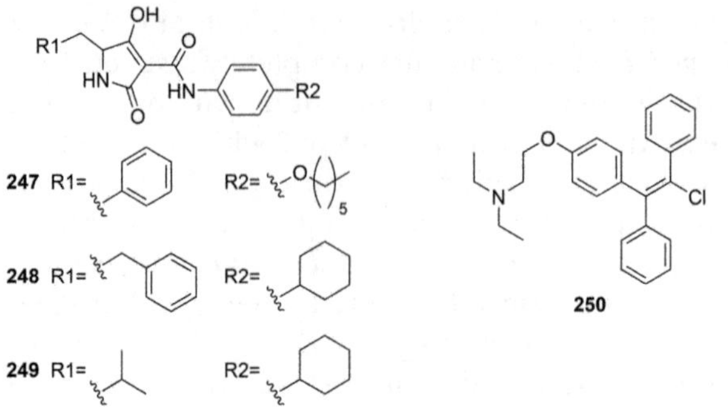

Figure 8.26 Inhibitors of UppS identified through high throughput screening (Compounds **247–250**).

activity. Substitution of the tetramate benzyl moiety with other hydrophobic substituents such as isobutyl and phenyl yielded similarly potent compounds. Most of the analogues generated were more potent than **247** with IC_{50}s on UppS between 0.1 μM and 2 μM and many had antibacterial activity on several Gram-positive pathogens with MICs in the range of 0.5–64 μg mL^{-1}. The best overall tetramic acid inhibitors **248** and **249** had IC_{50}s of 0.16 μM and 1.3 μM against UppS, and MICs of 4 μg mL^{-1} and 2 μg mL^{-1} against *E. faecalis*, 4 μg mL^{-1} and 0.5 μg mL^{-1} against *S. aureus*, and 1 μg mL^{-1} against *S. pneumoniae*, respectively. Phenotypic screening of 1600 off-patent US FDA-approved molecules on *S. aureus* for intrinsically bioactive molecules that could suppress the lethal activity of targocil, an inhibitor of the wall teichoic acid (WTA) transporter TarG, identified clomiphene **250**.[197] Since late stage proteins involved in WTA biogenesis are conditionally essential and become dispensable when early nonessential biosynthetic steps are blocked, antagonism with the late stage inhibitor targocil suggested that **250** was inhibiting initiation of WTA formation by targeting an earlier more central metabolic process. Unlike **21**, for which *B. subtilis* antibacterial activity could be suppressed by addition of FPP, IPP, or bactoprenol, only bactoprenol could suppress the activity of **250** when exogenously added to cells. This indicated that **250** was inhibiting an essential process downstream of IPP and FPP biosynthesis, as opposed to **21** which inhibits the upstream MEP pathway enzyme IspC. UppS lies between these two metabolic processes, utilizing upstream metabolites IPP and FPP as substrates to synthesize bactoprenol, the carrier lipid upon which downstream WTA is synthesized. Overexpression of UppS in *E. coli* rendered cells two-fold more resistant to inhibition by **250**. Furthermore, treatment of *S. aureus* with **250** resulted in accumulation of staphyloxanthin, a carotenoid that is synthesized from FPP, the pools of which could be potentially elevated, or at least diverted, upon inhibition of UppS. Ultimately, **250** was shown to inhibit purified UppS in biochemical assays, with IC_{50}s of 7.5 μM and 15 μM on the S. *aureus* and *E. coli* enzymes, respectively. A co-crystal structure determined that **250** binds to *E. coli* UppS in a hydrophobic pocket that spans a region that is occupied by the hydrophobic tail of FPP in site 1 and the aromatic sidechain of bisphosphonate **219** in site 4. Similar to **239, 250** was shown to be synergistic with a range of cell wall-targeting antibiotics, including β-lactams, vancomycin, bacitracin, cycloserine, and fosfomycin, further supporting UppS as the target and validating the interdependence between isoprenoid and peptidoglycan biosynthesis.

8.4.3 Inhibiting Staphyloxanthin Biosynthesis

In addition to playing an essential role in primary metabolism, many organisms divert isoprenoid precursors to produce secondary metabolites. One of these is the carotenoid staphyloxanthin, a polyene golden pigment for which the producing organism, *S. aureus*, is named.[198] Although staphyloxanthin is not essential, it is an important virulence factor that protects *S. aureus* from the host innate immune system, acting as an antioxidant against reactive oxygen species produced by neutrophils during infection.[199,200] The first committed step in staphyloxanthin biosynthesis is the head-to-head condensation of two molecules of FPP to form presqualene diphosphate (PSPP), the same intermediate as that of the first step in cholesterol and ergosterol biosynthesis. This reaction catalyzed by CrtM ultimately yields dehydrosqualene, whereas in cholesterol biosynthesis PSPP is converted to squalene. Indeed, CrtM and human squalene synthase are very similar with only a 5.5 Å Cα RMS deviation between the two structures.[13] Given the structural homology between the proteins, eight squalene synthase inhibitors originally synthesized as cholesterol lowering agents were screened against CrtM, and three phosphonosulfonates were identified as potent inhibitors of CrtM. The most active compound, **251**, had a K_I of 1.5 nM on *S. aureus* CrtM, could inhibit staphyloxanthin production in cell culture with an IC_{50} of 110 nM, and, as expected based on targeting a nonessential virulence factor, had no antibacterial activity up to a concentration of 2 mM (Figure 8.27). Despite the lack of antibacterial activity, there were dramatic biological consequences for inhibiting staphyloxanthin production. *S. aureus* treated with 100 μM **251** were 15-times more susceptible to killing when treated with 1.5% hydrogen

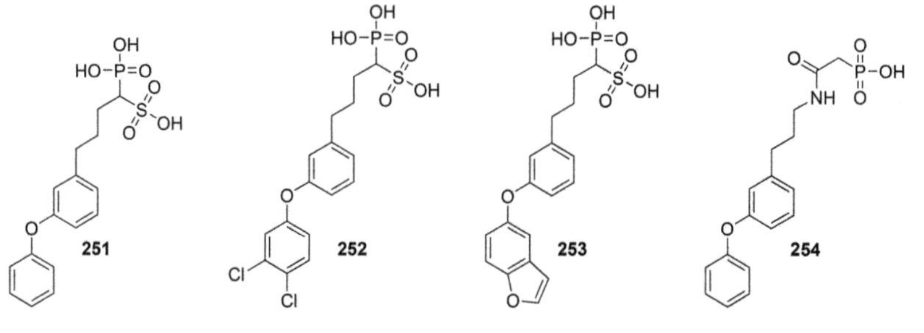

Figure 8.27 Inhibitors of staphyloxanthin production targeting CrtM (Compounds **251–254**).

peroxide and 4-times more susceptible to killing by fresh human whole blood. Furthermore, dosing 0.5 mg of **251** twice a day in a 72 h systemic *S. aureus* murine model of infection led to a 98% reduction in bacterial burden in the treatment group *versus* control. The activity of **251** led to the synthesis of various analogues in order to elucidate SAR.[201] Although bisphosphonates were an order of magnitude more potent than the phosphonosulfonates, they had no cellular activity. The *S*-enantiomer of **251** was found to be 30-fold more potent than the *R*-enantiomer, in line with previous results obtained for inhibition of squalene synthase.[202] Decreasing the linker between the phosphonosulfonate and aryl side chain negatively impacted activity, with one- and two-carbon linkers yielding a 1200 and 360-fold increase in IC_{50}, respectively. Substitution of the diphenyl ether oxygen with a nitrogen or carbon, or completely replacing the tail with a bisphenyl moiety led to a fourfold drop in inhibition on average. The only modifications, which increased both enzyme and cellular activity, were substitutions of the terminal phenyl group with *para* and/or *meta* hydrophobic alkyl groups or halogens. The most potent compounds were **252** and **253**, which had IC_{50}s of 2.2 μM and 2.1 μM on *S. aureus* CrtM, and IC_{50}s of 11 nM and 15 nM on cellular staphyloxanthin production, respectively. Additional studies investigating replacement of the sulfonate in order to reduce the net negative charge on the molecule identified phosphonoacetamides as potent inhibitors.[203] The best compound **254** was slightly less potent than **251** with a K_I of 40 nM against *S. aureus* CrtM and an IC_{50} of 8 nM on cellular staphyloxanthin production; however it was similarly efficacious in a systemic *S. aureus* murine model of infection, leading to a 96% reduction in bacterial burden *versus* control after 3 d.

Inspired by the results with **251**, phenotypic screening for additional inhibitors of staphyloxanthin production identified the antifungal naftifine **255**, which had a cellular IC_{50} of 296 nM (Figure 8.28).[204] Like **251**, **255** treatment made *S. aureus* 130-fold more susceptible to killing by hydrogen peroxide and 20-fold more susceptible to killing by whole blood. Furthermore, **255** reduced bacterial organ burden by 1–2 $\log_{10}$s in a systemic murine model of infection and protected mice from a lethal challenge of various strains of *S. aureus*, leading to 70% survival after 12 d *vs.* 100% mortality after 1 d in the vehicle-only control. Through a series of metabolite identification experiments, **255** was demonstrated to cause accumulation of the same staphyloxanthin intermediates as a *crtN* knockout *S. aureus* strain or an *E. coli* strain overexpressing CrtM, which synthesizes the dehydrosqualene substrate of CrtN. Additionally, overexpression of CrtN resulted in a 71-fold increase in

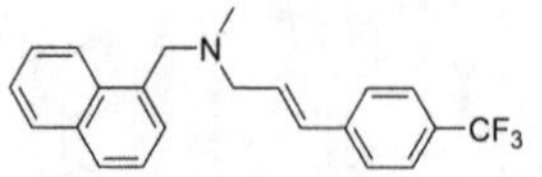

255

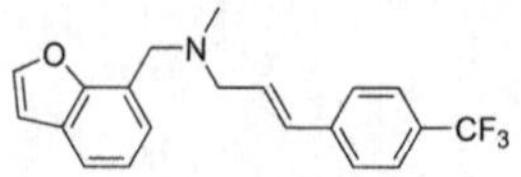

256

Figure 8.28 Inhibitors of staphyloxanthin production targeting CrtN (Compounds 255–256).

IC_{50} of cellular staphyloxanthin production for **255**, but not the CrtM inhibitor **251**. CrtN is a 4,4′-diapophytoene desaturase, responsible for catalyzing the conversion of dehydrosqualene to 4,4′-diaponeurosporene in the step after CrtM on route to staphyloxanthin biosynthesis. Ultimately, **255** was shown to bind CrtN through thermal shift assays and be a reversible competitive inhibitor of CrtN with an IC_{50} of 8.8 μM. Efforts to chemically modify the scaffold of **255** to increase potency and improve bioavailability revealed that the naphthalenyl moiety was important but could be replaced with a similar benzofuran, the *N*-methyl substituent was critical and could not tolerate removal or extension, the allyl linker could be lengthened or further oxidized to a propargyl but did not tolerate branching or saturation, and substitution of the phenyl ring at the *para* position or exchange with a naphthalene-2-yl increased activity but replacement with cycloalkyl or heteroaromatic groups led to a loss of activity.[205] The most potent analogue identified, **256**, had an IC_{50} of 338 nM against *S. aureus* CrtN and an IC_{50} ranging from 0.38 nM to 5.4 nM against cellular staphyloxanthin production, depending on the strain of *S. aureus* tested. Like **255**, **256** had dramatic effects on sensitizing *S. aureus* to hydrogen peroxide and whole blood killing and attenuated virulence in a systemic murine model of infection, leading to a 96.6–99.97% reduction in bacterial organ burden for multiple strains across multiple sites.

8.5 Conclusions

Despite the rich target space isoprenoid biosynthesis represents, not a single antibiotic inhibiting this essential pathway has reached clinical utility. However, isoprenoid biosynthesis remains a chemically validated target with several natural product inhibitors of both the MEP and MVA pathways identified. Furthermore, in human biology, some of the most successful drugs used in treating cardiovascular and bone metabolism diseases disrupt isoprenoid production, inspiring the notion that similar success can be achieved in the antibacterial space. Indeed, there are ample resources now available to facilitate the discovery and development of novel isoprenoid-targeting compounds. The biosynthetic pathways for the common IPP and DMAPP precursors have been delineated, crystal structures exist for all involved proteins, often with bound ligands, and biochemical assays have been developed to evaluate enzyme inhibition. Much of the lack of antibacterial activity observed for previously designed inhibitors can be attributed to an emphasis on designing substrate analogues that often contain negatively charged diphosphate or nucleotide moieties and lack the ability to penetrate bacterial membranes. Focusing on more drug-like chemical space should present an opportunity to identify more promising inhibitors to these targets which have already been interdicted.

Perhaps the most powerful resource available in targeting isoprenoid biosynthesis is phenotypic screening. The existence of two distinct pathways yielding a common metabolite allows for the creation of synthetic organisms in which endogenous isoprenoid biosynthesis is bypassed upon expression of the alternate pathway, thereby suppressing the activity of any screening hit targeting circumvented enzymes. Likewise, several isoprenoid intermediates and precursors can be exogenously supplemented to cells and incorporated into downstream metabolite production, suppressing inhibition of any upstream biosynthetic enzymes. Also, the central role isoprenoids play in bacterial physiology results in several synergistic and antagonistic interactions with other metabolic pathways, many of which have cognate inhibitors with which chemical genetic screens can be performed. These screening approaches, coupled with previously utilized targeted gene knockdown or overexpression strains, can facilitate identification of antibacterial hits directed towards isoprenoid biosynthesis. Ultimately, such compounds could be developed into the next antibiotic and would represent a much needed new mechanism of action.

References

1. H. Katsuki and K. Bloch, *J. Biol. Chem.*, 1967, **242**, 222–227.
2. F. Lynen, *Pure. Appl. Chem.*, 1967, **14**, 137–167.
3. A. Endo, M. Kuroda and Y. Tsujita, *J. Antibiot.*, 1976, **29**, 1346–1348.
4. A. Endo, *J. Lipid Res.*, 1992, **33**, 1569–1582.
5. R. G. Russell, *Ann. N. Y. Acad. Sci.*, 2006, **1068**, 367–401.
6. N. S. Moorthy, S. F. Sousa, M. J. Ramos and P. A. Fernandes, *Curr. Med. Chem.*, 2013, **20**, 4888–4923.
7. A. Bouhss, A. E. Trunkfield, T. D. Bugg and D. Mengin-Lecreulx, *FEMS Microbiol. Rev.*, 2008, **32**, 208–233.
8. R. Meganathan, *FEMS Microbiol. Lett.*, 2001, **203**, 131–139.
9. T. Mogi, K. Saiki and Y. Anraku, *Mol. Microbiol.*, 1994, **14**, 391–398.
10. K. M. Thompson and S. Gottesman, *J. Bacteriol.*, 2014, **196**, 754–761.
11. B. C. Persson, B. Esberg, O. Olafsson and G. R. Bjork, *Biochimie*, 1994, **76**, 1152–1160.
12. J. P. Saenz, D. Grosser, A. S. Bradley, T. J. Lagny, O. Lavrynenko, M. Broda and K. Simons, *Proc. Natl. Acad. Sci. U. S. A.*, 2015, **112**, 11971–11976.
13. C. I. Liu, G. Y. Liu, Y. Song, F. Yin, M. E. Hensler, W. Y. Jeng, V. Nizet, A. H. Wang and E. Oldfield, *Science*, 2008, **319**, 1391–1394.
14. J. Perez-Gil and M. Rodriguez-Concepcion, *Biochem. J.*, 2013, **452**, 19–25.
15. F. M. Hahn, A. P. Hurlburt and C. D. Poulter, *J. Bacteriol.*, 1999, **181**, 4499–4504.
16. K. Kaneda, T. Kuzuyama, M. Takagi, Y. Hayakawa and H. Seto, *Proc. Natl. Acad. Sci. U. S. A.*, 2001, **98**, 932–937.
17. M. Rohmer, *Lipids*, 2008, **43**, 1095–1107.
18. J. N. Pendleton, S. P. Gorman and B. F. Gilmore, *Expert Rev. Anti-Infect. Ther.*, 2013, **11**, 297–308.
19. H. W. Boucher, G. H. Talbot, J. S. Bradley, J. E. Edwards, D. Gilbert, L. B. Rice, M. Scheld, B. Spellberg and J. Bartlett, *Clin. Infect. Dis.*, 2009, **48**, 1–12.

20. S. R. Putra, A. Disch, J. M. Bravo and M. Rohmer, *FEMS Microbiol. Lett.*, 1998, **164**, 169–175.
21. S. Heuston, M. Begley, C. G. Gahan and C. Hill, *Microbiology*, 2012, **158**, 1389–1401.
22. E. I. Wilding, J. R. Brown, A. P. Bryant, A. F. Chalker, D. J. Holmes, K. A. Ingraham, S. Iordanescu, C. Y. So, M. Rosenberg and M. N. Gwynn, *J. Bacteriol.*, 2000, **182**, 4319–4327.
23. L. Bjorkhem-Bergman, P. Bergman, J. Andersson and J. D. Lindh, *PLoS One*, 2010, **5**, e10702.
24. I. M. Tleyjeh, T. Kashour, F. A. Hakim, V. A. Zimmerman, P. J. Erwin, A. J. Sutton and T. Ibrahim, *Arch. Intern. Med.*, 2009, **169**, 1658–1667.
25. S. Jerwood and J. Cohen, *J. Antimicrob. Chemother.*, 2008, **61**, 362–364.
26. P. Bergman, C. Linde, K. Putsep, A. Pohanka, S. Normark, B. Henriques-Normark, J. Andersson and L. Bjorkhem-Bergman, *PLoS One*, 2011, **6**, e24394.
27. M. K. Jain and P. M. Ridker, *Nat. Rev. Drug Discovery*, 2005, **4**, 977–987.
28. D. C. Aldridge, D. Giles and W. B. Turner, *J. Chem. Soc. Perkin 1*, 1971, **23**, 3888–3891.
29. S. Omura, H. Tomoda, H. Kumagai, M. D. Greenspan, J. B. Yodkovitz, J. S. Chen, A. W. Alberts, I. Martin, S. Mochales and R. L. Monaghan, *et al.*, *J. Antibiot.*, 1987, **40**, 1356–1357.
30. H. Tomoda, H. Kumagai, Y. Takahashi, Y. Tanaka, Y. Iwai and S. Omura, *J. Antibiot.*, 1988, **41**, 247–249.
31. D. A. Skaff, K. X. Ramyar, W. J. McWhorter, M. L. Barta, B. V. Geisbrecht and H. M. Miziorko, *Biochemistry*, 2012, **51**, 4713–4722.
32. C. J. Balibar, X. Shen and J. Tao, *J. Bacteriol.*, 2009, **191**, 851–861.
33. S. Ferrand, J. Tao, X. Shen, D. McGuire, A. Schmid, J. F. Glickman and U. Schopfer, *J. Biomol. Screening*, 2011, **16**, 637–646.
34. J. B. Bonanno, C. Edo, N. Eswar, U. Pieper, M. J. Romanowski, V. Ilyin, S. E. Gerchman, H. Kycia, F. W. Studier, A. Sali and S. K. Burley, *Proc. Natl. Acad. Sci. U. S. A.*, 2001, **98**, 12896–12901.
35. J. L. Andreassi 2nd, K. Dabovic and T. S. Leyh, *Biochemistry*, 2004, **43**, 16461–16466.
36. T. Kudoh, C. S. Park, S. T. Lefurgy, M. Sun, T. Michels, T. S. Leyh and R. B. Silverman, *Bioorg. Med. Chem.*, 2010, **18**, 1124–1134.
37. S. T. Lefurgy, S. B. Rodriguez, C. S. Park, S. Cahill, R. B. Silverman and T. S. Leyh, *J. Biol. Chem.*, 2010, **285**, 20654–20663.
38. J. F. Nave, H. d'Orchymont, J. B. Ducep, F. Piriou and M. J. Jung, *Biochem. J.*, 1985, **227**, 247–254.
39. J. E. Reardon and R. H. Abeles, *Biochemistry*, 1987, **26**, 4717–4722.
40. G. B. Quistad, D. C. Cerf, D. A. Schooley and G. B. Staal, *Nature*, 1981, **289**, 176–177.
41. F. M. Singer, J. P. Januszka and A. Borman, *Proc. Soc. Exp. Biol. Med.*, 1959, **102**, 370–373.
42. S. Dhe-Paganon, J. Magrath and R. H. Abeles, *Biochemistry*, 1994, **33**, 13355–13362.

43. I. Vlattas, J. Dellureficio, E. Ku, R. Bohacek and X. L. Zhang, *Bioorg. Med. Chem. Lett.*, 1996, **6**, 2091–2096.
44. J. K. Addo, D. A. Skaff and H. M. Miziorko, *J. Mol. Model.*, 2016, **22**, 13.
45. M. L. Barta, W. J. McWhorter, H. M. Miziorko and B. V. Geisbrecht, *Biochemistry*, 2012, **51**, 5611–5621.
46. M. L. Barta, D. A. Skaff, W. J. McWhorter, T. J. Herdendorf, H. M. Miziorko and B. V. Geisbrecht, *J. Biol. Chem.*, 2011, **286**, 23900–23910.
47. S. Kang, M. Watanabe, J. C. Jacobs, M. Yamaguchi, S. Dahesh, V. Nizet, T. S. Leyh and R. B. Silverman, *Eur. J. Med. Chem.*, 2015, **90**, 448–461.
48. D. A. Skaff, W. J. McWhorter, B. V. Geisbrecht, G. J. Wyckoff and H. M. Miziorko, *Arch. Biochem. Biophys.*, 2015, **566**, 1–6.
49. Y. Kuroda, M. Okuhara, T. Goto, M. Okamoto, H. Terano, M. Kohsaka, H. Aoki and H. Imanaka, *J. Antibiot.*, 1980, **33**, 29–35.
50. M. Okuhara, Y. Kuroda, T. Goto, M. Okamoto, H. Terano, M. Kohsaka, H. Aoki and H. Imanaka, *J. Antibiot.*, 1980, **33**, 24–28.
51. Y. Shigi, *J. Antimicrob. Chemother.*, 1989, **24**, 131–145.
52. T. Kuzuyama, T. Shimizu, S. Takahashi and H. Seto, *Tetrahedron Lett.*, 1998, **39**, 7913–7916.
53. J. Zeidler, J. Schwender, C. Muller, J. Wiesner, C. Weidemeyer, E. Beck, H. Jomaa and H. K. Lichtenthaler, *Z. Naturforsch. C*, 1998, **53**, 980–986.
54. G. Sandmann and P. Boger, *Z. Naturforsch. C*, 1986, **41**, 729–732.
55. C. Mueller, J. Schwender, J. Zeidler and H. K. Lichtenthaler, *Biochem. Soc. Trans.*, 2000, **28**, 792–793.
56. Y. Ferhatoglu and M. Barrett, *Pestic. Biochem. Physiol.*, 2006, **85**, 7–14.
57. Y. Matsue, H. Mizuno, T. Tomita, T. Asami, M. Nishiyama and T. Kuzuyama, *J. Antibiot.*, 2010, **63**, 583–588.
58. Q. Du, H. Wang and J. Xie, *Int. J. Biol. Sci.*, 2011, **7**, 41–52.
59. R. E. Hill, K. Himmeldirk, I. A. Kennedy, R. M. Pauloski, B. G. Sayer, E. Wolf and I. D. Spenser, *J. Biol. Chem.*, 1996, **271**, 30426–30435.
60. S. Xiang, G. Usunow, G. Lange, M. Busch and L. Tong, *J. Biol. Chem.*, 2007, **282**, 2676–2682.
61. J. L. Mao, H. J. Eoh, R. He, Y. H. Wang, B. J. Wan, S. G. Franzblau, D. C. Crick and A. P. Kozikowski, *Bioorg. Med. Chem. Lett.*, 2008, **18**, 5320–5323.
62. H. Zhao, L. P. de Carvalho, C. Nathan and O. Ouerfelli, *Bioorg. Med. Chem. Lett.*, 2010, **20**, 6472–6474.
63. T. Masini, B. Lacy, L. Monjas, D. Hawksley, A. R. de Voogd, B. Illarionov, A. Iqbal, F. J. Leeper, M. Fischer, M. Kontoyianni and A. K. Hirsch, *Org. Biomol. Chem.*, 2015, **13**, 11263–11277.
64. L. M. Eubanks and C. D. Poulter, *Biochemistry*, 2003, **42**, 1140–1149.
65. R. Kluger and K. Tittmann, *Chem. Rev.*, 2008, **108**, 1797–1833.
66. L. A. Brammer, J. M. Smith, H. Wade and C. F. Meyers, *J. Biol. Chem.*, 2011, **286**, 36522–36531.
67. L. A. Brammer and C. F. Meyers, *Org. Lett.*, 2009, **11**, 4748–4751.
68. M. Schurmann, M. Schurmann and G. A. Sprenger, *J. Mol. Catal. B: Enzym.*, 2002, **19**, 247–252.
69. J. M. Smith, R. J. Vierling and C. F. Meyers, *MedChemComm*, 2012, **3**, 65–67.

70. F. Morris, R. Vierling, L. Boucher, J. Bosch and C. L. F. Meyers, *ChemBioChem*, 2013, **14**, 1309–1315.
71. J. M. Smith, N. V. Warrington, R. J. Vierling, M. L. Kuhn, W. F. Anderson, A. T. Koppisch and C. L. F. Meyers, *J. Antibiot.*, 2014, **67**, 77–83.
72. D. Bartee, F. Morris, A. Al-khouja and C. L. F. Meyers, *ChemBioChem*, 2015, **16**, 1771–1781.
73. D. Hayashi, N. Kato, T. Kuzuyama, Y. Sato and J. Ohkanda, *Chem. Commun.*, 2013, **49**, 5535–5537.
74. J. Wungsintaweekul, S. Herz, S. Hecht, W. Eisenreich, R. Feicht, F. Rohdich, A. Bacher and M. H. Zenk, *Eur. J. Biochem.*, 2001, **268**, 310–316.
75. J. Perez-Gil, E. M. Uros, S. Sauret-Gueto, L. M. Lois, J. Kirby, M. Nishimoto, E. E. Baidoo, J. D. Keasling, A. Boronat and M. Rodriguez-Concepcion, *PLoS One*, 2012, **7**, e43775.
76. S. Sauret-Gueto, E. M. Uros, E. Ibanez, A. Boronat and M. Rodriguez-Concepcion, *FEBS Lett.*, 2006, **580**, 736–740.
77. N. Campos, M. Rodriguez-Concepcion, S. Sauret-Gueto, F. Gallego, L. M. Lois and A. Boronat, *Biochem. J.*, 2001, **353**, 59–67.
78. K. Kakinuma, Y. Dekishima, Y. Matsushima, T. Eguchi, N. Misawa, M. Takagi, T. Kuzuyama and H. Seto, *J. Am. Chem. Soc.*, 2001, **123**, 1238–1239.
79. T. Masini, B. S. Kroezen and A. K. H. Hirsch, *Drug Discovery Today*, 2013, **18**, 1256–1262.
80. S. Yajima, T. Nonaka, T. Kuzuyama, H. Seto and K. Ohsawa, *J. Biochem.*, 2002, **131**, 313–317.
81. M. Okuhara, Y. Kuroda, T. Goto, M. Okamoto, H. Terano, M. Kohsaka, H. Aoki and H. Imanaka, *J. Antibiot.*, 1980, **33**, 13–17.
82. H. Jomaa, J. Wiesner, S. Sanderbrand, B. Altincicek, C. Weidemeyer, M. Hintz, I. Turbachova, M. Eberl, J. Zeidler, H. K. Lichtenthaler, D. Soldati and E. Beck, *Science*, 1999, **285**, 1573–1576.
83. M. Lanaspa, C. Moraleda, S. Machevo, R. Gonzalez, B. Serrano, E. Macete, P. Cistero, A. Mayor, D. Hutchinson, P. G. Kremsner, P. Alonso, C. Menendez and Q. Bassat, *Antimicrob. Agents Chemother.*, 2012, **56**, 2923–2928.
84. S. Borrmann, S. Issifou, G. Esser, A. A. Adegnika, M. Ramharter, P. B. Matsiegui, S. Oyakhirome, D. P. Mawili-Mboumba, M. A. Missinou, J. F. J. Kun, H. Jomaa and P. G. Kremsner, *J. Infect. Dis.*, 2004, **190**, 1534–1540.
85. S. Borrmann, A. A. Adegnika, P. B. Matsiegui, S. Issifou, A. Schindler, D. P. Mawili-Mboumba, T. Baranek, J. Wiesner, H. Jomaa and P. G. Kremsner, *J. Infect. Dis.*, 2004, **189**, 901–908.
86. S. Borrmann, I. Lundgren, S. Oyakhirome, B. Impouma, P. B. Matsiegui, A. A. Adegnika, S. Issifou, J. F. J. Kun, D. Hutchinson, J. Wiesner, H. Jomaa and P. G. Kremsner, *Antimicrob. Agents Chemother.*, 2006, **50**, 2713–2718.
87. S. Oyakhirome, S. Issifou, P. Pongratz, F. Barondi, M. Rarnharter, J. F. Kun, M. A. Missinou, B. Lell and P. G. Kremsner, *Antimicrob. Agents Chemother.*, 2007, **51**, 1869–1871.

88. S. Borrmann, A. A. Adegnika, F. Moussavou, S. Oyakhirome, G. Esser, P. B. Matsiegui, M. Ramharter, I. Lundgren, M. Kombila, S. Issifou, D. Hutchinson, J. Wiesner, H. Jomaa and P. G. Kremsner, *Antimicrob. Agents Chemother.*, 2005, **49**, 3749–3754.
89. M. Ramharter, S. Oyakhirome, P. K. Klouwenberg, A. A. Adegnika, S. T. Agnandji, M. A. Missinou, P. B. Matsiegui, B. Mordmuller, S. Borrmann, J. F. Kun, B. Lell, S. Krishna, W. Graninger, S. Issifou and P. G. Kremsner, *Clin. Infect. Dis.*, 2005, **40**, 1777–1784.
90. T. Masini and A. K. Hirsch, *J. Med. Chem.*, 2014, **57**, 9740–9763.
91. Y. Sakamoto, S. Furukawa, H. Ogihara and M. Yamasaki, *Biosci., Biotechnol., Biochem.*, 2003, **67**, 2030–2033.
92. R. S. Mackie, E. S. McKenney and M. L. van Hoek, *Front. Microbiol.*, 2012, **3**, 226.
93. A. Hemmerlin, D. Tritsch, P. Hammann, M. Rohmer and T. J. Bach, *Biochimie*, 2014, **99**, 54–62.
94. M. Andaloussi, L. M. Henriksson, A. Wieckowska, M. Lindh, C. Bjorkelid, A. M. Larsson, S. Suresh, H. Iyer, B. R. Srinivasa, T. Bergfors, T. Unge, S. L. Mowbray, M. Larhed, T. A. Jones and A. Karlen, *J. Med. Chem.*, 2011, **54**, 4964–4976.
95. R. K. Dhiman, M. L. Schaeffer, A. M. Bailey, C. A. Testa, H. Scherman and D. C. Crick, *J. Bacteriol.*, 2005, **187**, 8395–8402.
96. A. C. Brown and T. Parish, *BMC Microbiol.*, 2008, **8**, 78.
97. E. Uh, E. R. Jackson, G. San Jose, M. Maddox, R. E. Lee, R. E. Lee, H. I. Boshoff and C. S. Dowd, *Bioorg. Med. Chem. Lett.*, 2011, **21**, 6973–6976.
98. E. S. McKenney, M. Sargent, H. Khan, E. Uh, E. R. Jackson, G. San Jose, R. D. Couch, C. S. Dowd and M. L. van Hoek, *PLoS One*, 2012, **7**, e38167.
99. J. Perruchon, R. Ortmann, M. Altenkamper, K. Silber, J. Wiesner, H. Jomaa, G. Klebe and M. Schlitzer, *ChemMedChem*, 2008, **3**, 1232–1241.
100. M. Andaloussi, M. Lindh, C. Bjorkelid, S. Suresh, A. Wieckowska, H. Iyer, A. Karlen and M. Larhed, *Bioorg. Med. Chem. Lett.*, 2011, **21**, 5403–5407.
101. B. Lou and K. Yang, *Mini-Rev. Med. Chem.*, 2003, **3**, 609–620.
102. J. Zhang, L. Zhang, X. Li and W. Xu, *Curr. Med. Chem.*, 2012, **19**, 2038–2050.
103. J. N. Sangshetti, F. A. Khan and D. B. Shinde, *Curr. Med. Chem.*, 2015, **22**, 214–236.
104. T. Bodin, A. C. Conibear, G. L. Blatch, K. A. Lobb and P. T. Kaye, *Bioorg. Med. Chem.*, 2011, **19**, 1321–1327.
105. Y. H. Woo, R. P. M. Fernandes and P. J. Proteau, *Bioorg. Med. Chem.*, 2006, **14**, 2375–2385.
106. R. Chofor, M. D. P. Risseeuw, J. Pouyez, C. Johny, J. Wouters, C. S. Dowd, R. D. Couch and S. Van Calenbergh, *Molecules*, 2014, **19**, 2571–2587.
107. L. Kuntz, D. Tritsch, C. Grosdemange-Billiard, A. Hemmerlin, A. Willem, T. J. Bacht and M. Rohmer, *Biochem. J.*, 2005, **386**, 127–135.
108. D. Giessmann, P. Heidler, T. Haemers, S. Van Calenbergh, A. Reichenberg, H. Jomaa, C. Weidemeyerd, S. Sanderbrand, J. Wiesner and A. Link, *Chem. Biodiversity*, 2008, **5**, 643–656.

109. R. Ortmann, J. Wiesner, K. Silber, G. Klebe, H. Jomaa and M. Schlitzer, *Arch. Pharm.*, 2007, **340**, 483–490.
110. G. S. Jose, E. R. Jackson, E. Uh, C. Johny, A. Haymond, L. Lundberg, C. Pinkham, K. Kehn-Hall, H. I. Boshoff, R. D. Couch and C. S. Dowd, *MedChemComm*, 2013, **4**, 1099–1104.
111. A. M. Jansson, A. Wieckowska, C. Bjorkelid, S. Yahiaoui, S. Sooriyaarachchi, M. Lindh, T. Bergfors, S. Dharavath, M. Desroses, S. Suresh, M. Andaloussi, R. Nikhil, S. Sreevalli, B. R. Srinivasa, M. Larhed, T. A. Jones, A. Karlen and S. L. Mowbray, *J. Med. Chem.*, 2013, **56**, 6190–6199.
112. K. Hemmi, H. Takeno, M. Hashimoto and T. Kamiya, *Chem. Pharm. Bull.*, 1982, **30**, 111–118.
113. E. R. Jackson, G. San Jose, R. C. Brothers, E. K. Edelstein, Z. Sheldon, A. Haymond, C. Johny, H. I. Boshoff, R. D. Couch and C. S. Dowd, *Bioorg. Med. Chem. Lett.*, 2014, **24**, 649–653.
114. T. Haemers, J. Wiesner, R. Busson, H. Jomaa and S. Van Calenbergh, *Eur. J. Org. Chem.*, 2006, 3856–3863.
115. V. Devreux, J. Wiesner, J. L. Goeman, J. Van der Eycken, H. Jomaa and S. Van Calenbergh, *J. Med. Chem.*, 2006, **49**, 2656–2660.
116. T. Haemers, J. Wiesner, D. Giessmann, T. Verbrugghen, U. Hillaert, R. Ortmann, H. Jomaa, A. Link, M. Schlitzer and S. Van Calenbergh, *Bioorg. Med. Chem.*, 2008, **16**, 3361–3371.
117. N. Katayama, S. Tsubotani, Y. Nozaki, S. Harada and H. Ono, *J. Antibiot.*, 1990, **43**, 238–246.
118. T. Haemers, J. Wiesner, S. Van Poecke, J. Goeman, D. Henschker, E. Beck, H. Jomaa and S. Van Calenbergh, *Bioorg. Med. Chem. Lett.*, 2006, **16**, 1888–1891.
119. A. Nordqvist, C. Bjorkelid, M. Andaloussi, A. M. Jansson, S. L. Mowbray, A. Karlen and M. Larhed, *J. Org. Chem.*, 2011, **76**, 8986–8998.
120. T. Verbrugghen, P. Cos, L. Maes and S. Van Calenbergh, *J. Med. Chem.*, 2010, **53**, 5342–5346.
121. T. Verbrugghen, P. Vandurm, J. Pouyez, L. Maes, J. Wouters and S. Van Calenbergh, *J. Med. Chem.*, 2013, **56**, 376–380.
122. J. Xue, J. S. Diao, G. B. Cai, L. S. Deng, B. S. Zheng, Y. Yao and Y. C. Song, *ACS Med. Chem. Lett.*, 2013, **4**, 278–282.
123. A. Kunfermann, C. Lienau, B. Illarionov, J. Held, T. Grawert, C. T. Behrendt, P. Werner, S. Hahn, W. Eisenreich, U. Riederer, B. Mordmuller, A. Bacher, M. Fischer, M. Groll and T. Kurz, *J. Med. Chem.*, 2013, **56**, 8151–8162.
124. J. F. Hoeffler, D. Tritsch, C. Grosdemange-Billiard and M. Rohmer, *Eur. J. Biochem.*, 2002, **269**, 4446–4457.
125. A. Wong, J. W. Munos, V. Devasthali, K. A. Johnson and H. W. Liu, *Org. Lett.*, 2004, **6**, 3625–3628.
126. C. Phaosiri and P. J. Proteau, *Bioorg. Med. Chem. Lett.*, 2004, **14**, 5309–5312.
127. J. R. Walker and C. D. Poulter, *J. Org. Chem.*, 2005, **70**, 9955–9959.
128. S. Yajima, K. Hara, J. M. Sanders, F. L. Yin, K. Ohsawa, J. Wiesner, H. Jomaa and E. Oldfield, *J. Am. Chem. Soc.*, 2004, **126**, 10824–10825.

129. L. S. Deng, S. Sundriyal, V. Rubio, Z. Z. Shi and Y. C. Song, *J. Med. Chem.*, 2009, **52**, 6539–6542.
130. L. S. Deng, K. Endo, M. Kato, G. Cheng, S. Yajima and Y. C. Song, *ACS Med. Chem. Lett.*, 2011, **2**, 165–170.
131. L. S. Deng, J. S. Diao, P. H. Chen, V. Pujari, Y. Yao, G. Cheng, D. C. Crick, B. V. V. Prasad and Y. C. Song, *J. Med. Chem.*, 2011, **54**, 4721–4734.
132. S. B. Richard, A. M. Lillo, C. N. Tetzlaff, M. E. Bowman, J. P. Noel and D. E. Cane, *Biochemistry*, 2004, **43**, 12189–12197.
133. A. Kunfermann, M. Witschel, B. Illarionov, R. Martin, M. Rottmann, H. W. Hoffken, M. Seet, W. Eisenreich, H. J. Knolker, M. Fischer, A. Bacher, M. Groll and F. Diederich, *Angew. Chem., Int. Ed.*, 2014, **53**, 2235–2239.
134. M. C. Witschel, H. W. Hoffken, M. Seet, L. Parra, T. Mietzner, F. Thater, R. Niggeweg, F. Rohl, B. Illarionov, F. Rohdich, J. Kaiser, M. Fischer, A. Bacher and F. Diederich, *Angew. Chem., Int. Ed.*, 2011, **50**, 7931–7935.
135. M. Witschel, F. Rohl, R. Niggeweg and T. Newton, *Pest Manage. Sci.*, 2013, **69**, 559–563.
136. W. Wu, Z. Herrera, D. Ebert, K. Baska, S. H. Cho, J. L. DeRisi and E. Yeh, *Antimicrob. Agents Chemother.*, 2015, **59**, 356–364.
137. L. S. Imlay, C. M. Armstrong, M. C. Masters, T. Li, K. E. Price, R. L. Edwards, K. M. Mann, L. X. Li, C. L. Stallings, N. G. Berry, P. M. O'Neill and A. R. Odom, *ACS Infect. Dis.*, 2015, **1**, 157–167.
138. C. Bjorkelid, T. Bergfors, L. M. Henriksson, A. L. Stern, T. Unge, S. L. Mowbray and T. A. Jones, *Acta Crystallogr., Sect. D: Biol. Crystallogr.*, 2011, **67**, 403–414.
139. M. Gabrielsen, J. Kaiser, F. Rohdich, W. Eisenreich, R. Laupitz, A. Bacher, C. S. Bond and W. N. Hunter, *FEBS J.*, 2006, **273**, 1065–1073.
140. R. A. Varikoti, R. P. Gangwal, G. V. Dhoke, R. V. Krishnan and A. T. Sangamwar, *Nature Precedings*, 2012.
141. A. M. Lillo, C. N. Tetzlaff, F. J. Sangari and D. E. Cane, *Bioorg. Med. Chem. Lett.*, 2003, **13**, 737–739.
142. P. Gao, Y. H. Yang, C. L. Xiao, Y. S. Liu, M. L. Gan, Y. Guan, X. Q. Hao, J. Z. Meng, S. Zhou, X. J. Chen and J. F. Cui, *Eur. J. Pharmacol.*, 2012, **694**, 45–52.
143. A. K. Hirsch, S. Lauw, P. Gersbach, W. B. Schweizer, F. Rohdich, W. Eisenreich, A. Bacher and F. Diederich, *ChemMedChem*, 2007, **2**, 806–810.
144. A. K. Hirsch, M. S. Alphey, S. Lauw, M. Seet, L. Barandun, W. Eisenreich, F. Rohdich, W. N. Hunter, A. Bacher and F. Diederich, *Org. Biomol. Chem.*, 2008, **6**, 2719–2730.
145. A. P. Schutz, S. Osawa, J. Mathis, A. K. H. Hirsch, B. Bernet, B. Illarionov, M. Fischer, A. Bacher and F. Diederich, *Eur. J. Org. Chem.*, 2012, 3278–3287.
146. A. P. Schutz, S. Locher, B. Bernet, B. Illarionov, M. Fischer, A. Bacher and F. Diederich, *Eur. J. Org. Chem.*, 2013, 880–887.
147. P. Mombelli, C. Le Chapelain, N. Munzinger, E. Joliat, B. Illarionov, W. B. Schweizer, A. K. H. Hirsch, M. Fischer, A. Bacher and F. Diederich, *Eur. J. Org. Chem.*, 2013, 1068–1079.
148. A. K. H. Hirsch, F. Diederich, M. Antonietti and H. G. Borner, *Soft Matter*, 2010, **6**, 88–91.

149. K. J. Wierenga, K. Lai, P. Buchwald and M. S. Tang, *J. Biomol. Screening*, 2008, **13**, 415–423.
150. M. Tang, S. I. Odejinmi, Y. M. Allette, H. Vankayalapati and K. Lai, *Bioorg. Med. Chem.*, 2011, **19**, 5886–5895.
151. N. Tidten-Luksch, R. Grimaldi, L. S. Torrie, J. A. Frearson, W. N. Hunter and R. Brenk, *PLoS One*, 2012, 7(4), e35792.
152. M. Harder, E. Schafer, T. Kumin, B. Illarionov, A. Bacher, M. Fischer, F. Diederich and B. Bernet, *Eur. J. Org. Chem.*, 2016, 402.
153. C. M. Crane, A. K. H. Hirsch, M. S. Alphey, T. Sgraja, S. Lauw, V. Illarionova, F. Rohdich, W. Eisenreich, W. N. Hunter, A. Bacher and F. Diederich, *ChemMedChem*, 2008, **3**, 91–101.
154. M. Takagi, T. Kuzuyama, K. Kaneda, H. Watanabe, T. Dairi and H. Seto, *Tetrahedron Lett.*, 2000, **41**, 3395–3398.
155. S. Herz, J. Wungsintaweekul, C. A. Schuhr, S. Hecht, H. Luttgen, S. Sagner, M. Fellermeier, W. Eisenreich, M. H. Zenk, A. Bacher and F. Rohdich, *Proc. Natl. Acad. Sci. U. S. A.*, 2000, **97**, 2486–2490.
156. C. M. Crane, J. Kaiser, N. L. Ramsden, S. Lauw, F. Rohdich, W. Eisenreich, W. N. Hunter, A. Bacher and F. Diederich, *Angew. Chem., Int. Ed.*, 2006, **45**, 1069–1074.
157. D. W. Begley, R. C. Hartley, D. R. Davies, T. E. Edwards, J. T. Leonard, J. Abendroth, C. A. Burris, J. Bhandari, P. J. Myler, B. L. Staker and L. J. Stewart, *J. Struct. Funct. Genomics*, 2011, **12**, 63–76.
158. C. Baumgartner, C. Eberle, F. Diederich, S. Lauw, F. Rohdich, W. Eisenreich and A. Barber, *Helv. Chim. Acta*, 2007, **90**, 1043–1068.
159. J. G. Geist, S. Lauw, V. Illarionova, B. Illarionov, M. Fischer, T. Grawert, F. Rohdich, W. Eisenreich, J. Kaiser, M. Groll, C. Scheurer, S. Wittlin, J. L. Alonso-Gomez, W. B. Schweizer, A. Bacher and F. Diederich, *ChemMedChem*, 2010, **5**, 1092–1101.
160. J. Thelemann, B. Illarionov, K. Barylyuk, J. Geist, J. Kirchmair, P. Schneider, L. Anthore, K. Root, N. Trapp, A. Bacher, M. Witschel, R. Zenobi, M. Fischer, G. Schneider and F. Diederich, *ChemMedChem*, 2015, **10**, 2090–2098.
161. S. Van Hoof, C. J. Lacey, R. C. Rohrich, J. Wiesner, H. Jomaa and S. Van Calenbergh, *J. Org. Chem.*, 2008, **73**, 1365–1370.
162. K. Wang, W. Wang, J. H. No, Y. Zhang, Y. Zhang and E. Oldfield, *J. Am. Chem. Soc.*, 2010, **132**, 6719–6727.
163. W. X. Wang, J. K. Li, K. Wang, T. I. Smirnova and E. Oldfield, *J. Am. Chem. Soc.*, 2011, **133**, 6525–6528.
164. W. X. Wang, J. K. Li, K. Wang, C. C. Huang, Y. Zhang and E. Oldfield, *Proc. Natl. Acad. Sci. U. S. A.*, 2010, **107**, 11189–11193.
165. Y. L. Liu, F. Guerra, K. Wang, W. X. Wang, J. K. Li, C. C. Huang, W. Zhu, K. Houlihan, Z. Li, Y. Zhang, S. K. Nair and E. Oldfield, *Proc. Natl. Acad. Sci. U. S. A.*, 2012, **109**, 8558–8563.
166. I. Span, K. Wang, W. X. Wang, Y. H. Zhang, A. Bacher, W. Eisenreich, K. Li, C. Schulz, E. Oldfield and M. Groll, *Nat. Commun.*, 2012, **3**, 1042.
167. F. Quitterer, A. Frank, K. Wang, G. D. Rao, B. O'Dowd, J. K. Li, F. Guerra, S. Abdel-Azeim, A. Bacher, J. Eppinger, E. Oldfield and M. Groll, *J. Mol. Biol.*, 2015, **427**, 2220–2228.

168. A. Ahrens-Botzong, K. Janthawornpong, J. A. Wolny, E. N. Tambou, M. Rohmer, S. Krasutsky, C. D. Poulter, V. Schunemann and M. Seemann, *Angew. Chem., Int. Ed.*, 2011, **50**, 11976–11979.
169. K. Janthawornpong, S. Krasutsky, P. Chaignon, M. Rohmer, C. D. Poulter and M. Seemann, *J. Am. Chem. Soc.*, 2013, **135**, 1816–1822.
170. I. Span, K. Wang, W. X. Wang, J. Jauch, W. Eisenreich, A. Bacher, E. Oldfield and M. Groll, *Angew. Chem., Int. Ed.*, 2013, **52**, 2118–2121.
171. F. Guerra, K. Wang, J. K. Li, W. X. Wang, Y. L. Liu, S. Amin and E. Oldfield, *Chem. Sci.*, 2014, **5**, 1642–1649.
172. K. Nakagawa, K. Takada and N. Imamura, *Biosci., Biotechnol., Biochem.*, 2013, **77**, 1449–1454.
173. B. W. Agranoff, H. Eggerer, U. Henning and F. Lynen, *J. Biol. Chem.*, 1960, **235**, 326–332.
174. J. de Ruyck, J. Wouters and C. D. Poulter, *Curr. Enzyme Inhib.*, 2011, **7**(2), 79–95.
175. T. Hoshino, H. Tamegai, K. Kakinuma and T. Eguchi, *Bioorg. Med. Chem.*, 2006, **14**, 6555–6559.
176. J. R. Walker, S. C. Rothman and C. D. Poulter, *J. Org. Chem.*, 2008, **73**, 726–729.
177. N. K. Sharma, J. J. Pan and C. D. Poulter, *Biochemistry*, 2010, **49**, 6228–6233.
178. S. C. Rothman, J. B. Johnston, S. Lee, J. R. Walker and C. D. Poulter, *J. Am. Chem. Soc.*, 2008, **130**, 4906–4913.
179. T. Nagai, H. Unno, M. W. Janczak, T. Yoshimura, C. D. Poulter and H. Hemmi, *Proc. Natl. Acad. Sci. U. S. A.*, 2011, **108**, 20461–20466.
180. T. P. Ko, Y. K. Chen, H. Robinson, P. C. Tsai, Y. G. Gao, A. P. C. Chen, A. H. J. Wang and P. H. Liang, *J. Biol. Chem.*, 2001, **276**, 47474–47482.
181. C. M. Apfel, S. Takacs, M. Fountoulakis, M. Stieger and W. Keck, *J. Bacteriol.*, 1999, **181**, 483–492.
182. L. L. Silver, *Ann. N. Y. Acad. Sci.*, 2013, **1277**, 29–53.
183. A. A. Scholte, L. M. Eubanks, C. D. Poulter and J. C. Vederas, *Bioorg. Med. Chem.*, 2004, **12**, 763–770.
184. R. T. Guo, T. P. Ko, A. P. C. Chen, C. J. Kuo, A. H. J. Wang and P. H. Liang, *J. Biol. Chem.*, 2005, **280**, 20762–20774.
185. K. Fujikura, Y. Maki, N. Ohya, M. Satoh and T. Koyama, *Biosci., Biotechnol., Biochem.*, 2008, **72**, 851–855.
186. R. T. Guo, R. Cao, P. H. Liang, T. P. Ko, T. H. Chang, M. P. Hudock, W. Y. Jeng, C. K. M. Chen, Y. H. Zhang, Y. C. Song, C. J. Kuo, F. L. Yin, E. Oldfield and A. H. J. Wang, *Proc. Natl. Acad. Sci. U. S. A.*, 2007, **104**, 10022–10027.
187. C. J. Kuo, R. T. Guo, I. L. Lu, H. G. Liu, S. Y. Wu, T. P. Ko, A. H. J. Wang and P. H. Liang, *J. Biomed. Biotechnol.*, 2008, 841312.
188. J. D. Durrant, R. Cao, A. A. Gorfe, W. Zhu, J. K. Li, A. Sankovsky, E. Oldfield and J. A. McCammon, *Chem. Biol. Drug Des.*, 2011, **78**, 323–332.
189. W. Zhu, Y. H. Zhang, W. Sinko, M. E. Hensler, J. Olson, K. J. Molohon, S. Lindert, R. Cao, K. Li, K. Wang, Y. Wang, Y. L. Liu, A. Sankovsky, C. A.

F. de Oliveira, D. A. Mitchell, V. Nizet, J. A. McCammon and E. Oldfield, *Proc. Natl. Acad. Sci. U. S. A.*, 2013, **110**, 123–128.
190. S. D. Larsen, M. R. Hester, J. C. Ruble, G. M. Kamilar, D. L. Romero, B. Wakefield, E. P. Melchior, M. T. Sweeney and K. R. Marotti, *Bioorg. Med. Chem. Lett.*, 2006, **16**, 6173–6177.
191. J. E. Mott, B. A. Shaw, J. F. Smith, P. D. Bonin, D. L. Romero, K. R. Marotti and A. A. Miller, *J. Antimicrob. Chemother.*, 2008, **62**, 720–729.
192. W. Zhu, Y. Wang, K. Li, J. Gao, C. H. Huang, C. C. Chen, T. P. Ko, Y. H. Zhang, R. T. Guo and E. Oldfield, *J. Med. Chem.*, 2015, **58**, 1215–1227.
193. W. Sinko, Y. Wang, W. Zhu, Y. Zhang, F. Feixas, C. L. Cox, D. A. Mitchell, E. Oldfield and J. A. McCammon, *J. Med. Chem.*, 2014, **57**, 5693–5701.
194. Y. H. Zhang, F. Y. Lin, K. Li, W. Zhu, Y. L. Liu, R. Cao, R. Pang, E. H. Lee, J. Axelson, M. Hensler, K. Wang, K. J. Molohon, Y. Wang, D. A. Mitchell, V. Nizet and E. Odfied, *ACS Med. Chem. Lett.*, 2012, **3**, 402–406.
195. S. Peukert, Y. C. Sun, R. Zhang, B. Hurley, M. Sabio, X. Y. Shen, C. Gray, J. Dzink-Fox, J. S. Tao, R. Cebula and S. Wattanasin, *Bioorg. Med. Chem. Lett.*, 2008, **18**, 1840–1844.
196. L. V. Lee, B. Granda, K. Dean, J. S. Tao, E. Liu, R. Zhang, S. Peukert, S. Wattanasin, X. L. Xie, N. S. Ryder, R. Tommasi and G. J. Deng, *Biochemistry*, 2010, **49**, 5366–5376.
197. M. A. Farha, T. L. Czarny, C. L. Myers, L. J. Worrall, S. French, D. G. Conrady, Y. Wang, E. Oldfield, N. C. Strynadka and E. D. Brown, *Proc. Natl. Acad. Sci. U. S. A.*, 2015, **112**, 11048–11053.
198. A. Pelz, K. P. Wieland, K. Putzbach, P. Hentschel, K. Albert and F. Gotz, *J. Biol. Chem.*, 2005, **280**, 32493–32498.
199. G. Y. Liu, A. Essex, J. T. Buchanan, V. Datta, H. M. Hoffman, J. F. Bastian, J. Fierer and V. Nizet, *J. Exp. Med.*, 2005, **202**, 209–215.
200. A. Clauditz, A. Resch, K. P. Wieland, A. Peschel and F. Gotz, *Infect. Immun.*, 2006, **74**, 4950–4953.
201. Y. C. Song, F. Y. Lin, F. L. Yin, M. Hensler, C. A. R. Poveda, D. Mukkamala, R. Cao, H. Wang, C. T. Morita, D. G. Pacanowska, V. Nizet and E. Oldfield, *J. Med. Chem.*, 2009, **52**, 976–988.
202. R. M. Lawrence, S. A. Biller, J. K. Dickson, J. V. H. Logan, D. R. Magnin, R. B. Sulsky, J. D. DiMarco, J. Z. Gougoutas, B. D. Beyer, S. C. Taylor, S. J. Lan, C. P. Ciosek, T. W. Harrity, K. G. Jolibois, L. K. Kunselman and D. A. Slusarchyk, *J. Am. Chem. Soc.*, 1996, **118**, 11668–11669.
203. Y. C. Song, C. I. Liu, F. Y. Lin, J. H. No, M. Hensler, Y. L. Liu, W. Y. Jeng, J. Low, G. Y. Liu, V. Nizet, A. H. J. Wang and E. Oldfield, *J. Med. Chem.*, 2009, **52**, 3869–3880.
204. F. F. Chen, H. X. Di, Y. X. Wang, Q. Cao, B. Xu, X. Zhang, N. Yang, G. J. Liu, C. G. Yang, Y. Xu, H. L. Jiang, F. L. Lian, N. X. Zhang, J. Li and L. F. Lan, *Nat. Chem. Biol.*, 2016, **12**, 174–179.
205. Y. X. Wang, F. F. Chen, H. X. Di, Y. Xu, Q. Xiao, X. H. Wang, H. W. Wei, Y. L. Lu, L. L. Zhang, J. Zhu, C. Q. Sheng, L. F. Lan and J. Li, *J. Med. Chem.*, 2016, **59**, 3215–3230.

Subject Index

Locators in *italic* refer to figures; those in **bold** to tables